CARL SAUER
ON CULTURE AND LANDSCAPE

Readings and Commentaries

Edited by WILLIAM M. DENEVAN
and KENT MATHEWSON

LOUISIANA STATE UNIVERSITY PRESS

BATON ROUGE

Published by Louisiana State University Press
Copyright © 2009 by Louisiana State University Press
All rights reserved
Manufactured in the United States of America

An LSU Press Paperback Original
First printing

DESIGNER: Michelle A. Neustrom
TYPEFACE: Warnock Pro, Bodoni SvtyTwo
TYPESETTER: J. Jarrett Engineering, Inc.
PRINTER AND BINDER: Thomson-Shore, Inc.

LIBRARY OF CONGRESS CATALOGING-IN-PUBLICATION DATA

Carl Sauer on culture and landscape : readings and commentaries / edited by William M. Denevan and Kent Mathewson.
   p. cm.
"An LSU Press paperback original"—T.p. verso.
Includes bibliographical references and index.
ISBN 978-0-8071-3394-1 (pbk. : alk. paper) 1. Sauer, Carl Ortwin, 1889–1975. 2. Geographers—United States—Biography. 3. Geography. 4. Cultural landscapes. 5. Human geography. I. Denevan, William M. II. Mathewson, Kent, 1946–
   G69.S29C38 2009
   910.92—c22

2008031230

The paper in this book meets the guidelines for permanence and durability of the Committee on Production Guidelines for Book Longevity of the Council on Library Resources. ∞

We are concerned with the . . . interrelation of . . . cultures and site, as expressed in the various landscapes of the world.

                Carl O. Sauer, "The Morphology of Landscape," 1925

We dedicate this book to our immediate family members
(Susie, Curt and Tori, and Kathy and Mimi)
and to the many members of Carl Sauer's
extended academic family.

Contents

Foreword, by Michael Williams   xi
Preface, by William M. Denevan   xvii

I. ABOUT CARL SAUER

1. Carl Ortwin Sauer, 1889–1975 (1975)
   JAMES J. PARSONS   3

2. Carl Sauer and His Critics
   KENT MATHEWSON   9

3. Thoughts on Bibliographic Citations to and by Carl Sauer
   DANIEL W. GADE   29

4. Bibliography of Commentaries on the Life and Work of Carl Sauer
   COMPILED BY WILLIAM M. DENEVAN   53

II. EARLY EFFORTS

5. Introduction
   WILLIAM M. DENEVAN   89

6. Exploration of the Kaiserin Augusta River in New Guinea, 1912–1913 (1915)   95

7. Preface to *The Geography of the Ozark Highland of Missouri* (1920 [1915])   99

8. Man's Influence Upon the Earth (1916)   103

9. Notes on the Geographic Significance of Soils: A Neglected Side of Geography (1922)   105

III. TOWARD MATURITY

10. Introduction
    GEOFFREY J. MARTIN   111

11. The Field of Geography (1927)   119

12. Cultural Geography (1931)   136

13. Correspondence [on Physical Geography in Regional Works] (1932)   144

IV. ECONOMY/ECONOMICS

14. Introduction
    KENT MATHEWSON, MARTIN S. KENZER, AND GEOFFREY J. MARTIN   149

15. Abstract: Geography as Regional Economics (1921)   160

16. Regional Reality in Economy (1984 [1936])   162

17. Economic Prospects of the Caribbean (1954)   173

V. CULTIVATED PLANTS

18. Introduction
    DANIEL W. GADE   185

19. Age and Area of American Cultivated Plants (1959)   198

20. Maize into Europe (1962)   213

VI. MAN IN NATURE

21. Introduction
    WILLIAM W. SPETH   231

22. Soil Conservation (1936)   242

23. Destructive Exploitation in Modern Colonial Expansion (1938)   245

24. The Relation of Man to Nature in the Southwest (1945)   252

25. Grassland Climax, Fire, and Man (1950)   261

VII. HISTORICAL GEOGRAPHY

26. Introduction
    W. GEORGE LOVELL   271

27. The Prospect for Redistribution of Population (1937)   277

28. About Nature and Indians (1939)   292

29. Middle America as a Culture Historical Location (1959)   296

30. *Terra Firma: Orbis Novus* (1962)   306

31. Chart of My Course (1980)   324

32. Decline of Indian Population (1980)   330

33. The End of the Century (1980)   338

VIII. CARL SAUER ON GEOGRAPHERS AND OTHER SCHOLARS

34. Introduction
    EDWARD T. PRICE   345

35. Ruliff S. Holway, 1857–1927 (1929)   352

36. Oskar Peschel, 1826–1875 (1934)   354

37. Friedrich Ratzel, 1844–1904 (1934)   356

38. Carl Ritter, 1779–1859 (1934)   358

39. Ellen Churchill Semple, 1863–1932 (1934)   360

40. Herbert Eugene Bolton, 1870–1953 (1954)   362

41. Homer LeRoy Shantz, 1876–1958 (1959)   367

42. Erhard Rostlund, 1900–1961 (1962)   371

43. Richard J. Russell, 1895–1971 (1967)   375

44. David I. Blumenstock, 1913–1963 (1968)   377

IX. INFORMAL REMARKS

45. Introduction
    PHILIP L. WAGNER   383

46. Letter to *Landscape* [on Past and Present American Culture] (1960)   390

47. The Seminar as Exploration (1976 [1948])   392

48. The Quality of Geography (1970)   398

49. Casual Remarks (1976)   407

Appendix: Doctoral Dissertations Supervised by Carl Sauer   417

Bibliography of Publications by Carl Ortwin Sauer
COMPILED BY WILLIAM M. DENEVAN   421

Contributors   433

Index   437

Foreword

> What we need more perhaps is an ethic and aesthetic under which man, practicing the qualities of prudence and moderation, may indeed pass on to posterity a good earth.
>
> Carl O. Sauer, "The Agency of Man on the Earth," 1956

In 1927 Carl Sauer had a disagreement with his mother over the value of doing the committee and administration work that had come his way in Berkeley. Whereas she thought that "service" to others was important, he wasn't sure that regulating other folks' affairs helped much and didn't think "carrying on studies [was] work worthwhile."[1] Administration had very little to offer; he was a Jeffersonian at heart and looked at all government and regulation as an annoyance and wished that life might be as unconfined as possible. "I took these administrative services because I thought it would help us build up the work in geography. And here they are becoming daily more in amount." Therefore, he concluded: "I am quite content . . . to be a geographer as I might be a botanist or philologist or something else that has nothing to do with *Weltverbesserung* [betterment of the world]. It is safe and yields the pleasure of a contemplative life. Perhaps that is why the old monks left so much that is fine and enduring. Be assured that I shall be quite content if I may add a few pages that people may occasionally read after me and say it was well done."[2] In this volume we have a few of his pages that we can read more than occasionally, and hopefully we can easily say that they are "well done."

This collection brings together some choice Sauer publications not selected for the two previous anthologies of his work (1963a, 1981a). It also does the service of making accessible essays that appeared originally in irregular and somewhat obscure publications. Examples include "Age and Area of

American Cultivated Plants" (1959c), "Middle America as Culture Historical Location" (1959a), and "Maize into Europe" (1962c) from the 33rd and 34th *Proceedings of the International Congress of Americanists*. Most welcome is his pathbreaking essay on "Destructive Exploitation in Modern Colonial Expansion" from the *International Congress of Geographers* (1938a). This forceful critique and condemnation of colonial land-use practices is one of the few instances of its kind appearing anywhere between about 1900 and 1970. Another piece of interest is Sauer's contribution to the 1962 *festschrift* for Hermann von Wissmann, "*Terra Firma: Orbis Novus.*" Other little-known papers in this volume include "Man's Influence Upon the Earth" (1916b), which can be considered an early preview of *Man's Role in Changing the Face of the Earth* (1956a); "Soil Conservation" (1936b); and "Casual Remarks" (1976b). These three are not listed in previous bibliographies of Sauer's works.

If this were not enough, William Denevan's meticulously compiled "Bibliography of Commentaries on the Life and Work of Carl Sauer" (chapter 4), which was, I know, an original impetus for the formulation of this volume, is justification enough for its publication. It is an impressive list of over five hundred articles and other entries that attest to the continuing interest and relevance of Sauer as a scholar and original thinker. Although the commentaries peaked during the mid-1970s and 1980s around the time of his death, the continuing flow of work suggests that his influences and ideas are anything but dead. Over one hundred commentaries after 2000 (see table 1 in chapter 4) give a very affirmative and definite answer to Bret Wallach's 1999 query, "Will Carl Sauer Make It across That Great Bridge to the Next Millennium?"

Some of the commentaries bring to mind an essay by one of Sauer's former students, Marvin Mikesell, titled "Sauer and 'Sauerology'" (1987). He sounded a cautionary note; many commentators miss the essential points of Sauer's personality, thought, and achievements, and concentrate too much on the programmatic statements contained in, for example, "The Morphology of Landscape" (1925), "Foreword to Historical Geography" (1941b), and "The Education of a Geographer" (1956d). These were less important than the values he cherished—of knowledge for its own sake, an historical perspective, cultural relativism, and the primacy of diffusion over independent invention, and his negative thoughts about academic fads, bureaucracy, social engineering, and destructive exploitation. Consequently, thought Mikesell (1987), a tension can arise between those who interpreted him almost entirely on the basis of a few publications, and those who knew him or were taught by him. They knew him not only as a heroic, larger-than-life figure, whose contribution to twentieth-century American scholarship was "substantial and his reputation well deserved," but as a shrewd academic politician and entrepreneur as well. In other words, the "real" Sauer is in danger of being lost in the theorizing of what he thought, or what people think he thought. What is needed is more of what Hermione Lee (2005) in her thoughtful book *Body Parts* calls "life-writing," which draws attention to a person's peculiarities and achievements in order to produce an idea of what they were really like.

If we follow that approach, then it is a surprise to learn that Sauer was a reluctant geographer, and by a very fine margin almost did not become one. After two years at Northwestern and then Chicago under the fairly relentless pressure of Rollin Salisbury, whom he admired intensely but feared, he decided that he had simply "drifted" into being a geographer. In one of his most revealing letters to his parents, written in July 1910, he threw considerable light on his underlying attitudes about geography and set out his philosophy of learning. He felt that he had no realistic prospect of moving beyond an instructor if he stayed in the university:

> I have just drifted. . . . When I graduated you wanted me to go on with school and I went. Then I had the offer at Chicago—and you urged me to take it. That is all there is to it. It has been a whole lot harder for me to make up my mind to change than to let things go on; the latter course would have required no definite purpose, I have just drifted on and things would have been all right. . . . I am perfectly happy with geology and geography. It is the academic life as compared to the practical life. My upbringing has been for the first but my inclination has always been for the second. . . . I do not have the fiber of a specialist. Now I also lack the wish to specialize; it is a narrow life. When I observe the teachers in the university and I put myself in their shoes, I do not find it attractive.[3]

Simply, university teaching and research were too isolating and he wanted contact with people. He wanted to pull out of geography altogether to pursue his ambition to become an editor of a small-town country newspaper. "The very smell of a printing office lures me," he said. He knew this would bitterly disappoint his father, who had ambitions for his clever son.

Whereas Sauer's father had no option but to agree reluctantly, Salisbury didn't give up on his outstanding pupil so easily. He offered him the opportunity to write the report for the Geological Survey on the upper Illinois River Valley with a guaranteed publication at the end (1916b). It was too great a prize for Carl to decline. Then, when "Duke" Wellington Jones left Chicago to study for a year in Heidelberg under Hettner, Salisbury appealed to Sauer's sense of loyalty to the department and offered him Jones's teaching for a year. Carl accepted it; however, he wanted to make it clear to his parents that it was done out of loyalty and commitment to the school and not because he wanted to go on in academia. "This does *not, emphatically not* mean that I will be a teacher!" he wrote. Anyhow, he rationalized, it would give him time to finish the drafting of his Illinois report. In another statement to his parents he said one of the things he was most clearly going to contradict in later life: "I have never wanted to learn for learning's sake; I have realized some subjects are interesting and could be useful as providing change or as pastimes, but they do not seem to me worth the effort as a life time's occupation when the practical world holds endless challenges which give us all outlets for our best endeavours." A little earlier he had told his parents, "I believe that there is more to life, or there should be, than unconformities of pre-glacial valleys, and interglacial periods."[4]

Carl Sauer did try the "endless challenges" of the practical world by doing what he considered to be an apprenticeship for editing: working in the publication house of Rand McNally in Chicago for eighteen months. It was one of the most miserable times he had ever spent in his life. The petty backbiting of the office, the broken promises, and the dashing of his own over-trusting and optimistic hopes made him deeply despondent. It was "a crooked game," he said. Harlan Barrows and Salisbury, whom he saw frequently over the Illinois report, could see the disillusionment mounting, and it was Barrows who made the first advances to Sumner Cushing in the Salem Normal School suggesting Carl as a promising geography teacher.

In the meantime, Carl's relationship with that "dandy little girl," Lorena Schowengerdt, whom he had known since their days together in Central Wesleyan College at Warrenton, blossomed, and he fell wildly in love with her. But Franklin Schowengerdt was not going to let one of his daughters marry an impoverished would-be editor/academic. He had built up one of the most successful general stores in Warrenton and had the grandest house in town, and whoever married Lorena had to keep her in the style to which she had, truthfully, been accustomed. Carl saw that the Salem job offered the way out of his many dilemmas. In a rare moment of self-analysis, he admitted that his protracted objection to teaching "was largely obstinacy, I guess everyone expected me to do [it] and so I felt as though I ought to do something else." A strong-willed father had produced a strong-willed son. Above all, securing the Salem position produced an immense relief, as at long last by becoming a teacher he was giving in to the inevitable, and by November 1913 he was prepared to confess to Lorena: "I had come to the conclusion yesterday that the way for me to work out my salvation was in academic life but in a larger clearer view of it than I held in my previous period at the University. I have seen it coming for a long time and now I'm glad it is here."[5] I think we can all be glad that he came to the conclusion that the academic life was his salvation, and some of the results of that are contained in this rich collection for us to enjoy nearly a hundred years later.

Michael Williams

NOTES

1. Unless otherwise stated, passages quoted are with permission from letters in the possession of Sauer's daughter, Mrs. Elizabeth FitzSimmons, Berkeley, and form part of my forthcoming biographical study of Sauer entitled *"To Leave a Good Earth": The Life and Work of Carl Ortwin Sauer.*
2. Carl O. Sauer to Mother, September 20 and October 2, 1927.
3. Carl O. Sauer to Parents, July 24, 1910.
4. Carl O. Sauer to Parents, December 4 and September 3, 1910.
5. Carl O. Sauer to Lorena Sauer, November 24 and November 19, 1913.

REFERENCES

Lee, Hermione. 2005. *Body Parts: Essays in Life Writing.* Chatto and Windus, London.
Mikesell, Marvin. 1987. "Sauer and 'Sauerology': A Student's Perspective." In *Carl O. Sauer: A Tribute,* ed. Martin S. Kenzer, 144–150. Oregon State Univ. Press, Corvallis.
Sauer, Carl O. For references, see the Sauer bibliography at the end of this volume.
Wallach, Bret. 1999. "Commentary: Will Carl Sauer Make It across that Great Bridge to the Next Millennium?" *Yearbook of the Association of Pacific Coast Geographers* 61:129–136.

Preface

> Man's relation to his environment [is] the relation of habit [culture] and habitat [nature] . . . habitat is revalued or reinterpreted with every change in habit.
>
> Carl O. Sauer, "Foreword to Historical Geography," 1941

I first encountered Carl Sauer in the fall of 1951 when, as a junior at the University of California, Berkeley, I took his course on Middle America. I don't recall that I did very well, but the experience led me to major in geography. I took or audited other classes from him on domestication, conservation of natural resources, and South America, and later several seminars as a graduate student. As an undergraduate I only spoke with him once, during graduation week when I asked him to sign my copy of his *Agricultural Origins and Dispersals*. He was not my graduate advisor; I was too intimidated by him to even think of it, but he certainly was the motivation for me to study early human impacts on the environment and indigenous cultivation in Latin America. In September of 1961 he wrote me while I was struggling with dissertation fieldwork in Bolivian Amazonia: "I am much pleased that you are . . . finding pay dirt. . . . You've picked yourself a major problem." Simple words, but inspirational coming from him. He was on my dissertation committee, but I doubt that he did more than skim the draft. No matter. My final communication to him was to offer congratulations on receiving the Victoria Medal of the Royal Geographic Society, in a letter dated July 18, 1975, which turned out to be the day he died. Now, over thirty years later, I once more cross paths with Mr. Sauer as I reread many of his publications while I select and edit for this anthology, my personal tribute to him.

Two volumes of papers by Carl Sauer have been published previously,

Land and Life in 1963a (nineteen items), edited by John Leighly, and *Selected Essays* in 1981a (seventeen items), edited by Bob Callahan. Numerous other fine articles and short monographs by Sauer, of interest for the history of geography and for the historical geography of North America and Latin America, are not included in these books.

Carl Sauer (1889–1975) was one of the most important, as well as controversial, geographers of the twentieth century. He had both positive and negative influences on the development of academic geography. Probably more has been written about him than any other geographer in the United States, including three collections of essays (Kenzer, 1987; Speth, 1999a; Mathewson and Kenzer, 2003), two collections of correspondence (Sauer, 1979, 1982), a monograph concerning his fieldwork (West, 1979), several theses (Solot, 1983; Kenzer, 1986), a book-length biography by Michael Williams (forthcoming), and over five hundred published commentaries, critiques, and memorials (see chapter 4). His work continues to be cited and his impact is still felt. "Arguably no geographer had more influence on American geography in the twentieth century than Carl Sauer" (Duncan, 2000:45). Much of the writing about him is by historians, anthropologists, archaeologists, ecologists, historical demographers, and other nongeographers. Given this interest, there is some justification for making more of his work, published over the period 1915–1980, readily available to scholars and students. Hence this third anthology of his research articles, essays, and published talks.

It can be questioned whether it is worthwhile to reprint a new collection of Sauer's papers, some of which were minor in importance when published and/or are now outdated. The articles we have chosen to include are mostly widely dispersed, difficult to locate, and little known. However, Sauer's stature and importance in the development of geography in the United States suggest that such articles merit attention. They provide linkages in the evolution of his thinking and research. The preface to his 1915 dissertation (1920b, chapter 7 herein) touches on themes that appear throughout his life. "The Field of Geography" in 1924 (chapter 11 herein) anticipates "The Morphology of Landscape" in 1925. "Grassland Climax, Fire, and Man" in 1950 (chapter 25 herein) previews an important part of "The Agency of Man on the Earth" in 1956. "*Terra Firma: Orbis Novus*" in 1962 (chapter 30 herein) is a summary of *The Early Spanish Main* in 1966.

Furthermore, this volume is more than an anthology. Three essays about Sauer are included. The first, by Sauer's student and longtime Berkeley associate James Parsons, was written shortly after Sauer's death—an appreciation highlighting his life, scholarship, and impact. The second is a commentary by Kent Mathewson on some of the vast literature about Sauer, both favorable and unfavorable, dating from "Morphology of Landscape" (1925) to recent antagonistic views by "new" cultural geographers and others. The third is an essay by Daniel Gade discussing both Sauer's use of citations, often sketchy, and citations by others to Sauer, often perfunctory. In addition, I have compiled a "Bibliography of Commentaries on the Life and Work of Carl Sauer" (chapter 4), which although certainly not complete, should be useful to anyone interested in Sauer. Table 1 in that chapter gives the number of commentaries on Sauer, by decade. A list of reviews of books and monographs written by Sauer appears in table 2. Dissertations supervised by Sauer are listed in the appendix. Finally, as complete a listing as possible is provided of his published works in the Sauer bibliography at the end of this volume.

Sauer's articles collected here are divided into eight sections, with the earliest (1915–1932) being in Sections II and III. The next four sections are topical. Section VIII contains ten short biographies and memorials on geographers and other scholars. Section IX consists of four informal statements: two commentaries Sauer gave at conferences, one a letter to the journal *Landscape*, and his 1948 "The Seminar as Exploration," which he passed out in his graduate seminars. Each of these eight sections is preceded by an introductory essay. These help provide an intellectual and historical perspective on how certain important aspects of geographical knowledge and ideas have accrued and changed in the twentieth century in relation to Sauer's seminal role in that history.

The contributors to the essays and introductions are authorities on the topics concerned and are knowledgeable about Sauer. All have previously published on Sauer. Three received their Ph.D.s under Sauer (Parsons, Price, and Wagner); two did their dissertations under students of Sauer (Denevan and Speth); three are third-generation Sauerians (Gade, Lovell, and Mathewson); one wrote his doctoral dissertation on Sauer (Kenzer); and one is a historian of geographical thought (Martin).

The authors of the introductions were asked to comment on the context of the papers in their sections and on the papers themselves, with guidelines as to length. However, I gave them the freedom to follow their own inclinations. Some focus more on the context than on the specific papers. Some give considerable attention to certain papers and little to others. Some statements are brief, some lengthy. Kenzer's comment (in chapter 14) on "Regional Reality in Economy" (chapter 16) is a revision of his earlier published introduction to it (1984). Speth's introduction (chapter 21) is an original essay in itself on the origins of Sauer's ideas on human impacts on the environment. Wagner (chapter 45) gives us more of a personal statement about his relationship with Sauer. The introductions by Martin, Gade, and Speth (chapters 10, 18, and 21) are more critical commentaries than are most of the others.

An effort has been made to include more of Sauer's early papers (pre-1940) than are in the other anthologies, including his first professional article in 1915 on exploration in New Guinea (chapter 6). Also, here is the last chapter, "The End of the Century" (chapter 33), from Sauer's final (posthumous) publication in 1980. Several of the papers were initially oral presentations that had been turned into articles, in some cases from transcripts of tapings (chapters 16, 19, 20, 23, 24, 29, 47, 48, and 49). These in particular are only partly documented, and some are not documented at all. They do, however, give the flavor of the informal, spoken Sauer.

This anthology and the two previous ones do not give sufficient attention to Sauer's substantive field and historical research. Most of that work, however, is in lengthy monographs and books not appropriate for inclusion in an anthology. As Mikesell (1987:149) points out, "Too much attention is being paid to what he preached and not enough to what he practiced." Here are included some of his historical articles and chapters.

I have tried to select items that are mostly brief in length, thus permitting a large and diverse collection. Important but long articles such as "Recent Developments in Cultural Geography" (1927, 58 pages) and "Cultivated Plants of South and Central America" (1950, 56 pages) were considered but had to be deleted. Aspects of the former appear in "The Field of Geography" (1927e, chapter 11) and in "Cultural Geography" (1931a, chapter 12). Aspects of the latter appear in chapter 19 herein.

Two little-known, one-page statements are included that are indicative of Sauer's early pre–"Morphology of Landscape" thinking about the content of geography, one on "Man's Influence Upon the Earth" (1916b, chapter 8) and the other on "Geography as Regional Economics" (1921b, chapter 15).

By "Culture and Landscape" in our title, we refer in a general sense to Sauer's concerns with human-modified landscapes, the interaction between people and their environments, how people live in and from their land, and the processes of change involved. The two terms "culture" and "landscape" are used frequently by Sauer, both separately and in conjunction, not always in the same way, but always with the understanding that they are mutually interconnected.

Hopefully, this collection and accompanying essays and introductions will contribute to a greater awareness of Carl Sauer's position in the history of geographic thought and of his pioneering contributions to what are today major issues of concern—"a humane use of the earth," "man's role in changing the face of the earth," sustainable land use, the importance of diversity of both people and nature, and the worth of traditional peoples past and present.

NOTE ON EDITING AND CITATIONS

Only minor editing has been done on Sauer's writing, mainly changes in style and format and corrections or alterations of typographical errors, spelling, and punctuation. Word insertions for clarity are in brackets. Graphics, few in the original articles, have been deleted except in chapter 28. Sauer's citations are often incomplete or unclear, but are mostly left unchanged. As pointed out by Gade (chapter 3), Sauer poorly documented much of his work.

ACKNOWLEDGMENTS

The following colleagues have been especially helpful in the preparation of this volume: William Speth, Daniel Gade, and particularly my coeditor, Kent Mathewson.

A special thanks to our primary typist, Sharon Ruch in Madison, and also to Heather Nagel in Baton Rouge. Fernando González in Madison assisted with library searching and copying. LSU Press editors Joseph B. Powell and George Roupe have been valued advisors, and copy editor Derik Shelor contributed greatly to revising the manuscript. Linda Webster prepared the index.

REFERENCES

Duncan, James S. 2000. "Berkeley School." In *The Dictionary of Human Geography*, 4th ed., ed. R. J. Johnston, Derek Gregory, Geraldine Pratt, and Michael Watts, 45–46. Blackwell, Malden, Mass.

Gould, Peter. 1979. "Geography 1957–1977: The Augean Period." *Annals of the Association of American Geographers* 69:139–151.

Kenzer, Martin S. 1984. "Commentary" on "Regional Reality in Economy." *Yearbook of the Association of Pacific Coast Geographers* 46:35–37.

———. 1986. "The Making of Carl O. Sauer and the Berkeley School of (Historical) Geography." Ph.D. diss., McMaster University, Hamilton, Canada.

———, ed. 1987. *Carl O. Sauer: A Tribute.* Oregon State Univ. Press, Corvallis.

Mathewson, Kent, and Martin S. Kenzer, eds. 2003. *Culture, Land, and Legacy: Perspectives on Carl O. Sauer and Berkeley School Geography.* Geoscience and Man, vol. 37. Geoscience Publications, Department of Geography and Anthropology, Louisiana State University, Baton Rouge.

Mikesell, Marvin W. 1987. "Sauer and 'Sauerology': A Student's Perspective." In *Carl O. Sauer: A Tribute,* ed. Martin S. Kenzer, 144–50. Oregon State Univ. Press, Corvallis.

Sauer, Carl O. For references, see the Sauer bibliography at the end of this volume.

Solot, Michael S. 1983. "Carl Sauer and Cultural Evolution." Master's thesis, University of California, Los Angeles.

Speth, William W. 1999. *How It Came To Be: Carl O. Sauer, Franz Boas and the Meanings of Anthropogeography.* Ephemera Press, Ellensburg, Wash.

West, Robert C. 1979. *Carl Sauer's Fieldwork in Latin America.* Dellplain Latin American Studies, No. 3. University Microfilms International, Ann Arbor, Mich.

I

ABOUT CARL SAUER

It is safe to say that there has been no American geographer who has been so thoroughly dissected and analyzed as Carl Sauer.

Robert Hoffpauir, 2000

I

Carl Ortwin Sauer, 1889–1975 (1975)

James J. Parsons (1915–1997)

Carl Ortwin Sauer, who died on July 18, 1975, at the age of eighty-five, left his mark indelibly on many of the historical, social, and biological sciences. By temperament and in his love for maps he was a born geographer. This was the label he chose, and it pleased him, but his work far transcended the bounds of any one academic discipline. As a student of the interrelatedness of land and life, of people and places, both now and throughout the full course of human history, he pointed the way towards a land ethic, "a responsible stewardship of the sustaining earth," for this and future generations. He, as much as anyone, has given the environmental movement an esthetic rationale and an historical perspective.

Sauer was born of German ancestry in Warrenton, Missouri, some fifty miles west of St. Louis, on Christmas eve, 1889. This mid-western German background had a major influence on his life and work. He must have found early pleasure in nature and the out-of-doors. His naturally retrospective character led him to refer back frequently and with affection to his Missouri days. He recently observed: "By good fortune the chart of my course began in rural surroundings, the Ozark highland to the south, 'prairie plains' to the north, the diverse land forms providing instruction in earth history from Cambrian time through the Ice Age. Natural history was learned from the ample flora and fauna. An earlier generation remembered buffalo, elk, passenger pigeons and parakeets and a little of Indians. Of the original settlers a small number of families remained, largely in the backwoods. Settlement in strength began about 1830 and by 1860 had fully occupied the land" [letter to the Royal Geographic Society, June 9, 1975].

As a schoolboy he had been sent to southern Germany, in the charge of relatives, for three years of study. After graduating from the long since defunct Central Wesleyan College in Warrenton, where his father was a teacher, he enrolled as a graduate in geology at Northwestern University. In his first year there he learned of "goings on" across town at the then new University of Chicago, where a geography program had been initiated, and he was attracted to it, enrolling in 1910 under the tutelage of the physiographer Rollin D Salisbury. A few years later he had married Lorena Schowengerdt, a school friend from Warrenton, a devoted and supportive wife throughout their long and fulfilling lives. She died in Berkeley only a month before her husband. It is somehow quite impossible to imagine Carl Sauer without her. They shared every triumph and every disappointment.

In 1915, after a brief stint of teaching at Massachusetts State Normal School at Salem, Sauer accepted an appointment at the University of Michigan. In the same year he had filed his dissertation on *The Geography of the Ozark Highland of Missouri*, published in 1920, which still stands as a model for regional cultural geography. His advance was rapid at Ann Arbor. In a period of seven years he moved from the rank of instructor to that of full professor. While formulating the Michigan Land Economic Survey, working in the cutover lands of the north, his attention was attracted to the consequences of mismanagement of the land and the pervasiveness of the destructive exploitation of resources, a theme that was to concern him greatly in later years.

In 1923 he was called to the University of California, Berkeley, and it was there that he made his life and work. The move west opened up new horizons that he soon followed south with his students into Mexico and later into the further American tropics. For more than thirty years he was chairman of the Department of Geography at Berkeley, an unimaginably long time in these days of university bureaucracy and paperwork, but a charge he seemingly dispatched with almost effortless ease. His students came to form what was undoubtedly the most distinctive group of academic geographers in the country. His graduate seminars and field excursions especially spurred and sustained in them the spark of inquiry. He directed a total of thirty-seven Ph.D. dissertations [see appendix]; for other students, including many non-geographers, his influence was dominant. The "Berkeley School," with its humanistic, historical approach to geography and cultural history, came to be another way of saying the students of Carl Sauer.

He was a superb teacher, and he inspired the affection and loyalty of those around him. His inspiration and guidance was sought by many from outside of academic geography, especially from history, anthropology, and the life sciences. At times he seemed to us to know everything about everything, and to be able to discourse wisely on any aspect of nature or culture, anywhere, to the farthest reaches of human time. His was an intelligence of rare originality. In his lifetime he achieved the status of a legend.

The publication of his influential essay, "The Morphology of Landscape" (1925), brought him international attention. In it he had laid to rest with eloquence the specter of environmental determinism that had been haunting much of geography and stunting its development. Man, he said, largely determines his own destiny; the understanding of the agency of man on earth is a principal obligation and opportunity of geographic scholarship. This theme was explored explicitly and in detail in the Wenner-Gren-sponsored international symposium at Princeton in 1955, which he organized with the assistance of W. L. Thomas Jr. The conference proceedings, published as *Man's Role in Changing the Face of the Earth* (1956a), was a landmark volume that set the stage for many of the sharpened environmental concerns of the following decade.

Sauer's extensive publication list suggests the breadth of his interest: land-use planning, the evolution of desert land forms, prehistoric cultures of the Southwest and Mexico, the diffusion of mankind and man's cultural baggage including agricultural origins and dispersals, discovery and exploration of the New World, the broad sweep of Latin American cultural and demographic history, the settlement of his native Middle West, destructive exploitation of resources, and the philosophy of geography as a humane discipline. *Land and Life* [1963a], a selection of his papers edited by John Leighly, contains his complete bibliography to 1962. In retirement he wrote four books, all dealing in one way or another with the early European probings along the Atlantic and Caribbean fringes of the New World and the impact of European cultures on native societies. The most recent, *Seventeenth Century North America*, will appear posthumously [1980]. He never stopped working. In his last days he was sketching out the outline of a monograph on the geographical background of the American nation at its Bicentennial.

Sauer was much concerned with the direction in which much of the academy seemed to be drifting. He decided what he sensed was an increasing overemphasis on theory and method in the social sciences and called repeatedly for a renewal of earlier concerns for the study of man in comparative historical and ecological terms. The attachment of excessive merit to quantification, in particular, he thought was leading to the confusion of means with ends, industry with intellectual achievement. "The more social science confines itself to what may be measured," he wrote, "the more restricted the range of personalities and temperaments that will be attracted to it." He was especially dubious of the increasing dependence of scholars on granting institutions and concocted group projects, designed to meet the grantor's prescription. In the end, he said, "the able researcher will know best how he should employ his mind.... No groups can or should wish to be wise and farseeing enough to predetermine the quest for knowledge."

He wrote in a powerful and simple prose, with cadence and ring, and with a felicity of expression, that in later years attracted the attention of numerous modern writers and poets.[1] For evocative descriptions of people and places he drew on a lifetime of direct field observation as well as on wide reading and archival search. He was much attracted to field study and chided modern geographers for their apparent disinclination to "get their boots muddy." He himself greatly enjoyed studying out a landscape and the traces of its past human occupance, the identification of flora, fauna, and surface features by direct observation, and mingling with simple country folk, for whom he had such sympathy and respect, and from whom he learned so much. But he was above all an historical geographer, especially of the New World, reaching back all the way "to the Ice Age and its witnesses." He was convinced that the Americas were peopled early and numerously and that with his fire and stone tools primitive man had begun to dominate his environment much earlier than conventional wisdom held. Recent archaeologic investigations have tended to bear out his "hunches."

Sauer towered like a Chimborazo over the field of academic geography. He gave the impression of one for whom the experience of life had been rich beyond measure. Few, surely, have taken such complete advantage of its opportunities. His field of vision, encompassing equally man and nature, was uncommonly wide. He disliked the pace of modern life, its conspicuous and continuous innovation, its planned obsolescence, the mood of progress as common destiny. "My span," he wrote in 1960 [chapter 46 herein], "has covered seven decades that reach back into a greatly different world, one it was very good to have lived in." He did not especially relish the prospect which the next century seemed to offer.

His catholic interests led him to maintain contact with an extraordinarily wide circle of scholars, in many disciplines. His services and counsel were much sought. He served on the advisory boards, among others, of the Rockefeller Foundation and the John Simon Guggenheim Memorial Foundation, the U.S. Soil Conservation Service, the Conservation Foundation, the Social Science Research Council, the Committee on Latin American Studies of the American Council of Learned Scholars, the Office of Naval Research (Geography Branch), and the President's Science Advisory Board. He particularly enjoyed his long years of service on the committee charged with the selection of Guggenheim fellows, under his good friend Henry Allen Moe, which offered him the continuing opportunity to sense the pulse of the entire range of American scholarship.

Carl Sauer was much honored in his lifetime. Honorary degrees came from Heidelberg, Glasgow, Syracuse, and his own University of California at Berkeley. The American Philosophical Society elected him to membership in 1944; in recent years he had been its only geographer. He was awarded the American Geographical Society's Daly Medal, the Vega Medal of the Swedish Society for Anthropology and Geography, the Humboldt Medal of the Berlin Geographical Society, and, a month before his death, the Victoria Medal of the Royal Geographical Society (London). He was twice president of the Association of American Geographers.

Sauer's intellectual virtuosity was matched by his personal warmth and unassuming grace. He encouraged young scholars of ability by his own example. He was probably about as close to the ideal of the nineteenth-century "universal man" as the twentieth century could produce. We are not likely to see another quite like him.

NOTES

Reprinted with permission from *Yearbook of the American Philosophical Society*, 1975, 163–167. Copyright 1975, The American Philosophical Society, Philadelphia.

1. Volume editors' note: Parsons (1976:88) expanded on Sauer and writers in a longer memorial to Sauer:

> In his later years his writings attracted the attention of a group of younger poets and writers, apparently initially through the influence of the poet Charles Olson, founder of Black Mountain College in North Carolina, with whom he had an extensive correspondence in the late 1940s and early 1950s. Sauer's insistence on writing clearly and com-

pactly, his reflectiveness and his realism, and perhaps his interest in pre-Columbian contacts between the Old World and the New World led Olson to urge young writers to read his works. Out of this group has come Robert Callahan, founder of the Turtle Island Foundation [in Berkeley], an independent publishing venture that has reissued *Northern Mists* [1968a, 1973] and *Man in Nature* [1939, 1975] and has announced a program of additional Sauer publications. Several papers on Sauer and his influence on modern writers have recently appeared in the literary "little magazines," especially in New England and in the Bay Area, and more may be anticipated.

Parsons (1996) later published a full article on "'Mr. Sauer' and the Writers." Also see Meinig (1983:319–320).

Poet Bob Callahan not only wrote a poem about Sauer (Parsons, 1996:24), but even turned Sauer's words and images into poetry or "song" in "Carl Sauer (The Migrations)" (1977:107–108):

> The route of dispersal south
> along the eastern base of the Rockies, southeast
> into the forest in pursuit of old world mammals,
> musk ox, giant elk, mammoths, bison, burning ahead
> wooded areas turning into grasslands . . . (see Sauer, 1944a).

REFERENCES

Meinig, Donald. 1983. "Geography as an Art." *Transactions, Institute of British Geographers,* New Series, 8:314–28.

Parsons, James J. 1976. "Carl Ortwin Sauer, 1889–1975." *Geographical Review* 66:83–89.

———. 1996. "'Mr. Sauer' and the Writers." *Geographical Review* 86:22–41.

Sauer, Carl O. Correspondence (cited in text). Carl O. Sauer Papers, Bancroft Library, University of California, Berkeley.

———. For references, see the Sauer bibliography at the end of this volume.

Thomas, William L., Jr., ed. 1956. *Man's Role in Changing the Face of the Earth.* Univ. of Chicago Press, Chicago.

2

Carl Sauer and His Critics

Kent Mathewson

INTRODUCTION

Carl O. Sauer, and the Berkeley school he inspired, should be granted a central place in any formal history or informal accounting of American geography. On this, there should be little disagreement. Yet, if one consults the standard extended histories, Sauer, his students, and confreres in the shared Berkeley enterprise are treated more in refractory than integral ways.[1] Taking a step beyond, one might make the controversial claim that the work of Sauer, his associates, and his adherents constitutes American geography's premier accomplishment over the past century. While this would be a minority position, his stature as a major figure persists, and if anything, appears to be growing (see chapter 4, table 1). Of course, this claim would invite vigorous debate, involving revisiting many earlier debates within the development of American geography. This inquest and appraisal would necessarily put the concepts of "culture" and "landscape" in sharp relief. Here, I will not attempt this. Rather, I will sink some test pits to identify and clear some of the grounds that must be prepared in order to build the case for placing the importance of Sauer and his school within twentieth-century American geography. Here, I only discuss some of his main critics, and the contexts out of which they mount their criticisms. A fuller treatment would necessarily engage the history of the development and deployment of a number of concepts in American geography, especially culture and landscape.

Not surprisingly, many of Sauer's chief critics also turn out to be representative figures of American geography's main paradigmatic moments over the past century. For simplicity's sake, this time frame can be divided into five temporal spans, each with its recognized orthodoxies or epistemic expressions, some of which are at odds with one another, as one may note. The phases with representative critics are: (1) 1900–1925, environmentalism and Davisian landform studies, few if any critics; (2) 1925–1955, static-synchronic chorology, Davis himself, some environmentalists, though muted, and famously Richard Hartshorne; (3) 1955–1970, spatial analytic and systems science, Peter Gould and various lesser positivist critics; (4) 1970–1980, structuralist and humanist approaches, seemingly few critics; and (5) 1980–present, post-structuralist perspectives and neo-positivist regroupings (especially around GIS), James Duncan and most New Cultural geographers almost as a rite of passage. This highly simplified scheme obviously requires caveats—for example: the rigid dating is obviously debatable, physical geography gets short shrift, and unpublished sources may contain more material than published critiques. It is useful, however, in putting his critics in changing disciplinary contexts. (For a substantial bibliography of commentaries on Sauer over the course of these phases, see chapter 4 herein.)

BACKGROUND: "THE MORPHOLOGY OF LANDSCAPE"

Throughout much of his professional life, Carl Sauer was at odds with mainstream American geography. While still in graduate school (1909–1915) he began to develop critiques of major currents within the mainstream. These departures or dissents were well informed and grounded in the philosophy and history of geography. Initially these came from close readings of the German and French geographical literature. Save for his advisor Rollin Salisbury, he was not impressed with the geography faculty at the University of Chicago, and even less taken by the cause-and-effect geographical determinism then enjoying paradigmatic dominance. He later remarked that by 1912 he had begun to distance himself from the Chicago program and spent his evenings reading the continental literature (Sauer, 1999b). It was not until after 1923, when he moved from the University of Michigan to Berkeley, that he began to publish critiques of mainstream currents and propose alternative approaches (Hooson, 1981). His major statement during this period, "The Morphology of Landscape," was published in 1925.[2]

Penn and Lukermann (2003) and other commentators on "The Morphology" agree that its immediate import was to offer American geographers a window on developments in European, especially German, geographic thought regarding the concept of not only "landscape," but also "culture" and "chorology." It also attacked environmental determinism head-on. It can be credited with discrediting and perhaps even derailing the environmentalist project within American geography. Of course, the environmentalist conceit did not disappear, but after 1925 it was moved to the margins, and within another two decades largely outside the bounds of respectable disciplinary discourse.[3]

IMMEDIATE RESPONSE TO "THE MORPHOLOGY"

Curiously, upon publication, "The Morphology" was met with little comment, at least in print. The only comment of note was by Charles R. Dryer in the 1926 *Geographical Review* (Lukermann, 1989:53–54). This review appeared under the subheading "The Nature of Geography."[4] Beyond Dryer's patronizing tone and his not so subtle ethnic slur on prolix German scholarship and by extension Sauer's own creation, Dryer does not find Sauer's attack on prevailing currents in American geography objectionable. He doubts that Sauer's landscape morphological method will yield much, but does commend the "young, competent, and ambitious" geographer for his seriousness.

Beyond the journals, "The Morphology" made for considerable commentary, or at least that is what Sauer's younger contemporaries, especially in the Midwest, have said in retrospect. Preston James (1929:85) and others put some of Sauer's ideas and methodology to work during the late 1920s and into the 1930s. The main adherents, however, were Sauer's own Berkeley students and the visiting German geographers that he favored for his early hires in the reconstructed Berkeley geography department (Speth, 1981).

RICHARD HARTSHORNE

The main critic of Sauer's concept of landscape, or his advocacy of putting landscape at geography's core, was Richard Hartshorne. In 1939 Hartshorne published his *The Nature of Geography*. It was subtitled "A Critical Survey of Current Thought in the Light of the Past." One of the principle targets was Sauer and his emerging Berkeley school. Two years earlier, John Leighly (1937), Sauer's close colleague at Berkeley, published an article modestly entitled "Some Comments on Contemporary Geographic Methods." He followed with "Methodologic Controversy in Nineteenth Century Geography" (1938). Leighly's interpretation of German geographers provoked Hartshorne to respond—first in an article-length paper, and then an extended paper that eventually was published as two entire numbers of the *Annals of the Association of American Geographers* (and later in various book editions and translations). Hartshorne found much of Sauer's concept of landscape objectionable—especially the emphasis on material culture and its tolerance for humanistic perspectives—and not within the proper bounds of geography. He also felt that both Leighly's and Sauer's advocacy of a genetic or historical method would be taking geography down misguided if not errant paths. Furthermore, for Hartshorne, a product of the Chicago department that Sauer had spurned, physical geography was best marginalized if not demobilized altogether. Finally, like many of Sauer's subsequent critics, Hartshorne saw little value in culture historical themes, such as plant and animal domestications, prehistoric human-environment interactions, material culture diffusions, and the dispossession and demise of indigenous and local peoples in the wake of European colonial global expansion.

Sauer's interdisciplinary foraging in the domains of history, anthropology, and the natural sciences, together with his disdain for positivistic social science focused on narrow political and economic concerns in the here-and-now, branded him as a maverick and possibly a subversive (at least in a disciplinary sense). For self-appointed paradigm policemen such as Richard Hartshorne, Sauer and his followers were beyond the pale and needed to be given disciplinary citations or censures, and were not widely cited in the geographic literature.

Sauer's response followed the next year. The Association of American Geographers held its annual meeting at Louisiana State University in 1940, and Sauer was the association's president. His presidential address, "Foreword to Historical Geography" (1941b), was directed squarely at Hartshorne and mainstream American geography that found landscape-as-concept confusing and landscape studies as geography's central focus deviant (Livingstone, 1992:260). Sauer (1941b:4–5) prescribed three remedies for the "pernicious anemia" that he took to be geography's current condition. He offered a "three-point underpinning" for geography: (1) that the history of geography not only be a foundational element in graduate education, but also a touchstone through a geographer's career; (2) that "American geography cannot dissociate itself from the great fields of physical geography"; and (3) that "the human geographer should be well based on the sister discipline of anthropology." Not surprisingly, his former colleague Richard Russell and former student Fred Kniffen (both had arrived at LSU from Berkeley in the late 1920s to found the ideal Sauerian department) were the hosts of his presidential meeting. Moreover, they were well on their way to building a program based on his three-point design, along with many other Sauerian features. Within the address, and within the LSU program, the study and reconstruction of cultural landscapes were at the core of what Sauer prescribed for a healthy discipline of geography.

After this exchange, Sauer did not expend much further effort in debating the eastern and midwestern custodians of geographic orthodoxy. Removed as he was beyond the western mountains in Berkeley, he and his students thrived in semi-isolation from the rest of American geography. His scholarly exchanges were more likely to be with anthropologists, historians, botanists, agronomists, and field-oriented natural scientists of all descriptions than with geographers who found increasingly less value in historical approaches to questions of culture and landscape. Thus, by the late 1950s, when Hartshorne and the unreconstructed regionalists came under fire from the spatial science "insurgents," Sauer and his associates were mostly ignored and largely unaffected.

PETER GOULD AND ALLAN PRED

During the 1960s, however, when the spatial positivists achieved what they assumed was an hegemony within American geography, Sauer and similar landscape enthusiasts were dismissed as irrelevant relics, or derided as retrograde and even reactionary. Peter Gould best expressed this sentiment in his retrospective essay "Geography 1957–1977: The Augean Period" (1979). Gould quotes Sauer as saying, "We may leave enumeration to census takers . . . to my mind we are concerned with processes that are largely nonrecurrent and involve time spans beyond the short runs available to enumeration." He finds Sauer's debunking of increasing quantification in the social sciences as "shabby, parochial, and unintelligent," and accuses Sauer (and by implication many of his contemporaries) of "bumbling amateurism and antiquarianism" (140).

Similarly, Allan Pred in a 1983 retrospective essay on the "quantitative revolution" remembered Sauer at Berkeley in the early 1960s as being antiurban and racist in his views of rural black migration to cities. Pred was understandably disturbed by what he imagined Sauer's views on the black civil rights movement to be. Pred paraphrased Sauer's position as "Negroes are simple, happy folk whose natural place is close to the soil. If only they hadn't been driven from the countryside into the cities we would have none of these problems" (1983:92–93). Rhetorical simplification aside, this may be a reasonably accurate rendering of Sauer's sentiments. Sauer's position on the plight of rural peoples in general, and indigenous and tribal populations in particular, is well documented. He was an outspoken advocate on rights of rural folk to defend and extend their traditional ties to the land, especially in the face of "development" and modernization. Pred's mid-century modernist outlook clashed with Sauer's earlier antimodernist convictions (Mathewson, 1986).

Later, a postmodernist Pred read Sauerian-inspired landscape studies the riot act in a review of James Duncan's 1990 work *The City as Text* (1991:115–116). Pred proclaimed that with Duncan's intervention: "The heavy ballast of Sauer-influenced landscape study is to be cast overboard. Fully. No more avowedly atheoretical undertakings. No more innocent reading of the superficial and the artefactual. No more satisfied, naïve claims that what you see is what you have. No more shunning of the human agency associated with landscape production. No more denial of the social processes and power relations with which built landscapes are inescapably interwoven. No further reliance on a notion of culture that is superorganic, unproblematic, divorced from the experiences of everyday life, devoid of the actively constructed and contested" (116).

Though both appraisals were honestly held by these critics, they say as much about their own epistemological perspectives (spatial positivist for Gould, and post-positivist for Pred) as about Sauer's own outlook and practice. Pred's latter comments may accurately describe the practices and perspectives of aspects of traditional cultural geography, but they cannot be applied in blanket fashion to Sauer's work.

JAMES BLAUT

During the decade of the 1970s, the phase I have associated with both structuralist and humanist currents in human geography, Sauer and the Sauerians once again did not elicit much critique. During this decade, cultural geography in general and landscape studies in particular generated little innovation, but still attracted enthusiasts. In various ways, traditional landscape studies offered some geographers a refuge or quiet backwater removed from the theoretical ferment and methodological challenges that both the spatial analysts and their structuralist or radical opponents represented. This condition was not to last for long, however. In 1980 (perhaps significantly coinciding with the onset of the Reagan-Thatcher era), two quite different critiques of cultural geography were published. James Blaut's "Radical Critique of Cultural Geography" appeared in *Antipode,* and James Duncan's "The Superorganic in American Cultural Geography" appeared in the *Annals of the Association of American Geographers.* Among other things, Blaut took aim at cultural geography's tentative moves toward the soft positivism implicit in much of the behaviorial geography of the time. He praised traditional cultural geography for its unwavering embrace of its historical and materialist groundings. But he challenged it to also take up a radical ethno-class perspective in looking at the world, whether past, present, or future. Not inconsequently, Blaut's doctoral work was with Fred Kniffen, a founder of the LSU program, and one of Sauer's early students (Mathewson and Stea, 2003).[5]

JAMES DUNCAN

Duncan had studied with one of Sauer's students—David Sopher, a specialist on South Asian landscapes and religions. But unlike Blaut, Duncan faulted the Sauerians specifically for their alleged embrace of the superorganic concept. The term superorganic was coined by Herbert Spencer, and adapted by Sauer's Berkeley anthropologist colleague Alfred Kroeber for a general cultural theory. Kroeber (1917) posited four separate levels of reality: the inorganic, the organic, the psychological, and finally, at the top, the social or cultural level (Duncan, 1980:184–185). According to Kroeber, each level is autonomous, and the cultural or "superorganic" level has separate ontological status and causative power. As such, culture is an entity above humans, not reducible to actions by individuals, and following its own laws. Duncan argues that Sauer and his students uniformly adopted Kroeber's superorganic concept, and hence American cultural geography reified culture, assumed internal homogeneity within cultures, and accepted a form of cultural determinism in their landscape studies.

Other than a few scattered quotes and Sauer's famous epigram from "The Morphology" that stated: "The cultural landscape is fashioned from a natural landscape by a culture group. Culture is the agent, the natural area is the medium, the cultural landscape the result" (1925:46), there is little actual evidence that Sauer, or most of his students for that matter, did accept Kroeber's superorganic concept (Mathewson, 1998). The main exception is Wilbur Zelinsky, an extremely eclectic and prolific cultural geographer and one of Sauer's students from the 1950s. Zelinsky (1973:40–41) did explicitly accept and employ the superorganic cultural concept in his writings. Beyond identifying Zelinsky's affinities for the superorganic, and Sauer's supposed close association with Kroeber, Duncan himself reifies Sauer and Berkeley school geography in attributing a quasi-superorganic control or direction over the Sauerians' perspectives and practice. Even though Duncan missed the mark in some of his critiques of Berkeley school geographers, there is no question that his article helped catalyze an emerging dissatisfaction with cultural geography in the traditional mode. It also generated both immediate response (Duncan, 1981; Richardson, 1981; Symanski, 1981) and protracted comment and debate that continue (Duncan, 1993; Mathewson, 1998).

THE "NEW CULTURAL GEOGRAPHY"

By the end of the 1980s the "New Cultural Geography" was rapidly emerging, especially in Britain (Cosgrove and Jackson, 1987). Cultural Marxists such as Denis Cosgrove (1983, 1984, 1985) and Stephen Daniels (1993) put historical eyes on elite culture and past rural landscapes. Others, such as Peter Jackson (1989), with comparable cultural materialist views, surveyed contemporary urban popular culture scenes and situations. Starting from this initial base, during the 1990s the New Cultural Geography exploded into multiple directions and modes, but most proponents were united in indicting traditional cultural geography, past and present, as irrelevant at best, and reactionary at worst. Duncan, Cosgrove, Jackson, and others, such as Derek Gregory (1989) and David Ley (1981, 1982), produced a cannon of criticism that caricatured the traditional cultural geographers as single-mindedly focused on mapping the distribution of material artifacts such as houses, barns, fences, and gasoline stations. Rarely does this kind of criticism admit, or even apparently see, that the core focus of Sauer's work, and that of most of his followers, was on ecological analysis and historical interpretation of cultural landscapes.

This oversight or, more accurately, ignorance on the part of many of the instigators of the New Cultural Geography has not gone without challenge. Perhaps the best rebuttal to date is Marie Price and Martin Lewis's article "The Reinvention of Cultural Geography" (1993). Beyond stressing the Sauerians' past and continuing emphasis on cultural landscape construction in its environmental and historical dimensions, they point to Sauer's own radical environmentalist stance, his defense of indigenous peoples and local folk in the face of capitalist development, and his profound skepticism toward positivist social science in refuting the notion that Sauer and his style of landscape studies are necessarily both irrelevant and reactionary.[6]

MICHAEL SOLOT

Along with generally uninformed critiques by the new cultural geographers during the past two decades, several well-informed geographers and anthropologists have offered critical opinions on Sauer's scholarship. In 1986, Michael Solot, then a geography graduate student at the University of Wisconsin, Madison, examined Sauer's rejection of cultural evolutionism and his championing of culture history as an alternative course for cultural geography. The crux of Solot's argument is that Sauer rejected cultural evolutionism, as Franz Boas had earlier, chiefly because of its associations with environmental determinism, and its "rationalistic" commitment to explaining culture change and transformation in unilinear and often providential terms. Solot suggests that Sauer offered as an alternative culture historical excavation of patterns of past landscape change with an emphasis on the visible, material elements of landscapes. Both of these appraisals are basically correct. Solot goes on to argue that in doing so, Sauer eschewed examination or explication of the processes involved. Again, this may be accurate for the first decade or so of Sauer's program and perspectives. This position is harder to support or demonstrate if one considers Sauer's last two or three decades of work (ca. 1950–1975). Sauer becomes increasingly concerned about the destructive nature not only of earlier colonial patterns *and* processes, but of modern life and civilization itself.

DICKSON, HARRIS, AND GADE ON
AGRICULTURAL ORIGINS AND DISPERSALS

One of the main texts from which Solot draws his conclusions is Sauer's *Agricultural Origins and Dispersals* (1952a). Several scholars sympathetic to Sauer's overall oeuvre have put critical eyes to aspects of his theory of agricultural origins, and his championing of non-reoccurrent histories of agriculture's diffusions. Anthropologist D. Bruce Dickson (2003) considered Sauer's domestication theories in light of subsequent theories. First he rehearsed Sauer's basic arguments, aspects of which were original and ran counter to conventional wisdom. Sauer hypothesized that agriculture was spawned in conditions of leisure and abundance, not undue toil and scarcity. He posited key cultural and environmental preconditions, and deemed the optimum locations to be riverine settings with diverse relief features in the humid tropics. His prime candidate was Southeast Asia. He also argued that root crops rather than seed crops were the first cultigens. Finally, he proposed that epidemic-style diffusion out of single hearths through cultural contact was the mode of dispersal. While Sauer's ideas were generally well received in geography, outside they met with muted response, or in some cases were rejected outright. Dickson traces this reception, and then turns to recent appraisals that have been more congenial. He shows that poststructural-functionalist approaches can accommodate Sauer's insights and conjectures. Specifically, both non-equilibrium development (or historical growth models) and evolutionary interpretations of domestication as naturally selected cases of mutualism between plants and/or animals and humans are not incompatible with Sauer's ideas.

Two geographers with ties to the Berkeley school have also reevaluated Sauer's speculations and prospectings involving the domestication process. David Harris (2002), archaeologist and Berkeley geography Ph.D., drawing on his own and others' grounded work (Near East and Southeast Asia/Australasia) corrects and counters much of Sauer's empirics (or lack of them) regarding Old World plant and animal domestication. Though Sauer comes up short in the test of time on these origins, Harris nevertheless gives him high marks on his deeper speculations. Sauer's prescient projections of early human migrations and dispersals and the importance of fire as perhaps humankind's primordial domestication continue to be corroborated by accumulating evidence. Daniel Gade, a cultural geographer in the Sauer mold, revisited Sauer's writings on New World crop diversity (1999:184–213). Much of Sauer's fieldwork on crop diversity was done as part of his Rockefeller Foundation-funded travel in 1942 to South America (1982). His principle publication on the topic is "Cultivated Plants of South and Central America" (1950a). Gade surveys Sauer's conjectures and conclusions regarding crop origins, chronologies, diffusions, and distributions. Much of what Sauer proposed has been shown to be accurate. In other cases, however, he was far from correct, and in one case—that of the coconut being an Old World rather than a New World domesticate, he declined to be corrected during his lifetime despite accumulating evidence to the contrary. Despite Sauer's streak of Missouri-style stubbornness, Gade credits Sauer's sharp insights and skillful syntheses with helping to put the question of crop diversity and agricultural origins before a broad spectrum of subsequent specialists. Perhaps even more important, Gade sees Sauer's moral defense of biodiversity as being a legacy that lives on in a dissenting academic tradition and resonates with the struggles of ordinary farmers to defend their agro-cultural patrimony in the face of a destructive modernization.

DAVID STODDART

Sauer's earliest graduate training at Northwestern University was in geology, and his dissertation advisor was Chicago's foremost physiographer, Rollin D Salisbury. Sauer continued to keep a foot (if not a particularly active hand) in physical geography for the first half of his career. However, he is little remembered for his efforts in this side of the field. David Stoddart, former Berkeley chair of geography and distinguished coastal geomorphologist and historian of science, exhumed and evaluated portions of Sauer's record as a geomorphologist (Stoddart, 1997). What he found was that Sauer was quite engaged at certain times in the 1920s and 1930s in planning and promoting research in geomorphology, climatology, and soil studies. Soon after arriving in Berkeley in 1923, Sauer launched a study of California's Peninsular Range (1929a). Stoddart sees this venture more as establishing a beachhead to advance geography's claim to geomorphic study turf within the university than as sustaining a commitment to geomorphic studies per se. At the same time, Sauer saw potential in applying the new German approach to land form studies pioneered by Walther Penck (1924). Sauer and his close colleagues John Leighly and John Kesseli saw great utility in Penck's analytical method as a means to challenge and deflate Harvard's William Morris Davis's reigning "cycle of erosion" concept and method. Although Sauer showed great élan and ingenuity in his analyses, few of his novel interpretations were born out in subsequent work. Stoddart's own appraisal suggests that some of Sauer's explanations were simply an inversion of the Davisian explanations.

A few years later Sauer found himself in southeast Arizona, deflected from his intended Mexican fieldwork by political troubles south of the border. There he decided to re-examine "some commonly accepted concepts concerning basin-range features" (Sauer, 1930a). The founding work had been carried out by Davis, and by Sauer's Berkeley geology colleagues Andrew Lawson and George Louderback (upon whose territory he had trespassed in his Peninsular Range studies). Kirk Bryan, a Harvard geologist and geographer, had carried out the most recent studies. Sauer advanced a provocative Penckian explanation for the processes at work. Little came of this venture, save for engendering the career-long enmity of Bryan.

Sauer's final foray into the physical geography arena was more successful. During the Depression, Isaiah Bowman (a member of Roosevelt's Science Advisory Board) named Sauer to the S.A.B.'s Committee on Land-Use. Sauer played a key role in fostering research on soil erosion. He drew effectively from the ranks of his growing cadre of Berkeley-trained geographers. The principal sites chosen were in New Mexico and the piedmont of the Southeast looking at arroyo and gully formation. Although Sauer by this time was not directly involved in the fieldwork, it did spark interest among geographers (primarily Berkeley-associated) in these questions for several decades subsequently. Among the main things that Sauer took away from this experience was the destructive potential of human agency to alter landscapes under colonial or subsequent exploitative conditions, and his own lack of enthusiasm for further involvement with bureaucratic agencies and organizations. This mid-Depression service also helped set the stage for Sauer's resolute turn from the 1940s on toward historical geography.

RICHARD SYMANSKI

Over time, Sauer's persona, his position as the key figure in a scholarly school, and his approach to geography have all generated critical comments, but to date the most sustained attempt at critique has been by Richard Symanski. Symanski, a geographer teaching in the ecology program at the University California, Irvine, has in recent years turned to Web-based and self-published broadsides aimed at selected geographers and tendencies within the discipline. He has singled out Sauer for extended criticism. A short essay titled "Coconuts on a Lava Flow in the Chiricahua Mountains" appears in a 2002 collection of Symanski's essays. He revisits the case of the coconut, and Sauer's relations with Henry Bruman, his former doctoral student who championed an Old World origin for its domestication. Symanski also unearths a field episode from the 1920s in which Sauer overrode a student's interpretation of local geology, thus putatively demonstrating disregard for the evidence as well as pulling rank with a subordinate. Among the conclusions Symanski draws from these episodes, along with several other recounted lapses and errors in the field, is that Sauer was neither much of a scholar nor a field-worker nor at times a gentlemen. Given the temptations toward hagiography that some of the literature on Sauer exhibits (Mikesell, 1987), Symanski's observations are a useful corrective. But also given Symanski's well-known, and well-honed, penchant for invective and ridicule, his motives must be questioned.[7]

DON MITCHELL

In contrast to Symanski's *ad hominen, ad infinitum* attacks on an expanding gallery of geographers, Don Mitchell has set his sights on de- and reconstructing landscape studies. In important ways, Mitchell (2000) picks up the Marxist critique of traditional cultural geography that Cosgrove, Daniels, and Jackson initiated but have not sustained. In addition, Mitchell's approach to landscape is more direct and his adherence to historical materialism more orthodox. For Mitchell, landscape is the product of human labor, and must be understood as such. Questions of representation, "reading the landscape as text," and similar approaches deployed by the first wave of new cultural geographers are deemed useful, but they are not at the core of his concerns. Unmasking the social relations at work in landscape construction and destruction are his concern. Unlike most critics of the Sauer and the Berkeley school, Mitchell has never singled out Sauer as the source of cultural geography's perceived problems and retrograde agendas. For the most part, Mitchell (2000) sees Sauer's cultural landscape concept quite accurately, especially Sauer's materialist and historical orientation, and his "Herderian" ethnopluralism in the face of Eurocentrism. In Mitchells's estimation, Sauer's main failing was in not adequately theorizing the place of labor in culture and landscape, leaving an ill-defined "culture" to do this work.

In a *Progress in Human Geography* report on current cultural landscapes studies, Mitchell (2003:787) asks the question: "Just landscapes or landscapes of justice?" He singles out Kenneth Olwig's excavations of the landscape concept for favorable comment, especially the connections Olwig (2002) makes between the construction of modern Atlantic imperial polities and the construction, or the production, of both cultural and political landscapes. Following on this, Mitchell argues for putting empire and imperial polities center stage in landscape studies. As he suggests: "It is doubly important now, as . . . the landscape of empire is every bit as much a landscape of destruction as it is a landscape of production" (2003:788). He goes on to comment on landscape destruction as a defining characteristic of our current times (lower Manhattan on 9/11, Afghanistan in its wake, Palestinian towns before, during, and after, and then Iraq) and to propose that landscape studies become laboratories, or at least incubators, of theory and practice directed at the propagation of landscapes of justice.

LANDSCAPES OF DESTRUCTION

If this refocusing of critical landscape studies on landscapes of destruction is to be taken seriously, then it must also take seriously its precursors. In nineteenth-century and early twentieth-century geography, the work of George Perkins Marsh (1864), Elisée Reclus (1905–1908), Jean Bruhnes (1910), and a number of other chroniclers of imperial and colonial "destructive exploitation" (to use the apt term of the times) provides the foundations for any contemporary study of landscapes of destruction. If the focal point is to be grounded in cultural landscape study, then Carl Sauer and his work indisputably need to be the starting point. Surveying his career, starting with his dissertation in 1915 and continuing until his death in 1975 (some sixty years of fieldwork, research, and publication), one will find that, for Sauer, landscape construction and destruction were central organizing concepts and were often conjoined to produce powerful and sometimes polemical critiques of European colonial expansion from the late Middle Ages onward.[8]

CONCLUSION

An adequate account has not yet been written of Sauer's contributions to American geography as realized and expressed through the collective production and directions taken by his associates, students, and those inspired by the Berkeley school approach. If and when it is, it will encompass the work of several generations of scholars, whose numbers now total several hundred at a minimum. Sauer's direct progeny within the Latin American branch of his "academic genealogical" tree (his advisees and their advisees, et cetera, that wrote dissertations on Latin American topics) numbered over 150 by 2000 (Brown and Mathewson, 1999). Sauer's works and those of some of his students have been or are currently being translated into Spanish and Portugese and a whole new generation of Latin American geographers and students are being introduced to the Berkeley school, many for the first time.[9] Nor has anyone yet attempted to tabulate the published contributions of Sauer, along with his associates and several generations of his legatees, to the multiple questions that have engaged this large group of like-minded scholars. I think it is safe to say, however, that this literature comprises some hundreds of books and monographs, and several thousand articles and lesser publications. Viewed collectively, this corpus amounts to one of the larger bodies of published work in North American geography. It has not previously been implicitly recognized or acknowledged in these terms.

I think it is also safe to say, despite retorts to the contrary, that Sauer's legacy is alive and well and is likely to persist as long as geographers and kindred scholars continue to take an interest in questions of culture and landscape and the history of humans' agency on earth.[10] The quality and quantity of the Sauerian oeuvre taken as a whole—or even in parts—is large and complex enough to ensure both continuing criticism and enduring admiration (see chapter 4, tables 1 and 2). For example, Don Mitchell's 2003 call for landscape scholars to put questions of landscape destruction front and center provides not only a fulcrum to redirect cultural geography, but also an appropriate lens to reassess Sauer's and his adherents' contributions. Mitchell's call could be the grounds for a critical survey of past thought, in light of the future. And, this is not likely to be the last opportunity either.

NOTES

1. In the standard text on the history of geography by Martin (2005), *All Possible Worlds*, Sauer appears as a leader of the younger field-oriented midwestern geographers after World War I, as the introducer and practitioner of the landscape approach in American geography, as an early advocate of historical geography, and later as an organizer of the monumental International Symposium on Man's Role in Changing the Face of Earth in 1955 (Thomas, 1956). Martin makes only fleeting references to the Berkeley school itself. Johnston and Sidaway (2004), *Geography and Geographers*, index the Berkeley school, but the references are to critical and reifying comments made by new cultural geographers. Johnston and Sidaway do accord Sauer significant roles in pre-1960s cultural and historical geography, and the debates over Hartshorne's (1939) *The Nature of Geography*. Lesser histories by Unwin (1992) and Holt-Jensen (1999) devote a few lines to Sauer mainly in relation to his debate with Hartshorne, but also as a possibilist and advocate of landscape studies. By far the most sympathetic and sophisticated treatment of Sauer in these histories is Livingstone's (1992). He devotes a major section to Sauer, especially his relations to Boasian anthropology. Here one is allowed to glimpse the larger significance of Sauer and his school for not only geography, but in the context of cognate fields. Yet, not even a synoptic accounting of the school is offered by any of these histories.

2. While much has been written on Sauer's "Morphology of Landscape" (1925), there has been little close reading. Penn and Lukermann's (2003) "Chorology and Landscape" is the ex-

ception. See them for a fuller understanding of the issues, implications, and legacy of Sauer's treatise.

3. This is not to say that the residua of environmental determinism has not been reconstituted and resurrected even in our times, but these revivals come almost entirely from beyond geography's borders. Examples range from Jared Diamond's (1999) well-intentioned but seriously flawed *Guns, Germs, and Steel* to the global development designs advanced by various economists and planners who have recently "discovered" geography in its most banal forms (see Sluyter, 2003).

4. Perhaps significantly, Hartshorne does not mention this, or cite Dryer (other than his 1919 presidential address to the AAG) in his own 1939 magnum opus entitled *The Nature of Geography*.

5. Blaut (1993), until his death in 2000, was proud of this lineage, and even more pleased to self-identify as a "Kniffenite/Sauerian-Marxist"—a seeming contradiction in terms for many geographers (Mathewson, 2005).

6. Also, see the commentaries on "The Reinvention of Cultural Geography" by Cosgrove, Duncan, and Jackson, with a response by Price and Lewis, *Annals of the Association of American Geographers* 83 (1993): 515–522).

7. To date, Gade (2004) has provided the most penetrating (psycho)analysis of Symanski's crusade to unmask the foibles, follies, and falsities of American geographers, ordinary and otherwise.

8. Much of Sauer's writing can be subsumed in this category, from his 1920s reports on the "cut-over" lands of northern Michigan, to his 1930s broadsides on colonialism and destructive exploitation, to his organizing the 1955 Man's Role in Changing the Face of the Earth symposium, and on to his cultural and historical studies of the European conquest of the Americas throughout his career.

9. Environmental historian Guillermo Castro H. based at the City of Knowledge Foundation, Panama, is currently translating Sauer's methodological papers for Web distribution. James Parsons's and Robert West's Colombian studies have been translated and published in Colombia under several auspices. Mexican geographer Narciso Barrera-Bassols is overseeing the translation and publication of geographical classics on Michoacán, Mexico, including studies by Sauer students Donald Brand, Dan Stanislawski, and Robert West. William Denevan, a student of Parsons, has published or republished three of his own monographs in Spanish plus several articles in Spanish or Portuguese. Brazilian geographers Roberto Lobato and Zeny Rosenthal are publishing some of Sauer's methodological writings in Portuguese. It will be interesting to see to what extent this foreign exposure generates new work along old lines, or if hybrid forms develop.

10. In what attempts to pass for a jocular marker of the extinction of Sauerian-inspired cultural geography, the lead illustration of the *Handbook of Cultural Geography* (Anderson et al., 2003) depicts an above-ground tomb in New Orleans with the inscription in Gothic script "Here Lies Cultural Geography, Born 1925, Died 2002. In Loving Memory." At least memory is indicated here, but Sauer and the Berkeley school are largely elided from this 580-page reference work. The only place they make a serious appearance is in Jane Jacobs's "Introduction: After Empire" (2003:348–350). And this is because Andrew Sluyter (1997, 2002) has made a cogent case for the potential of Sauerian-style geography for postcolonial analysis. Sympathetically, Bret

Wallach (1999) asked whether Sauer would make it "across that Great Bridge" to the next millennium? He forecast a decline "in the near term," but held out for his later "resurrection."

REFERENCES

Anderson, Kay, Mona Domosh, Steve Pile, and Nigel Thrift, eds. 2003. *Handbook of Cultural Geography.* Sage Publications, London.

Blaut, James M. 1980. "A Radical Critique of Cultural Geography." *Antipode* 12:25–30.

———. 1993. "Mind and Matter in Cultural Geography." In *Culture, Form, and Place: Essays in Cultural and Historical Geography,* ed. Kent Mathewson, 345–356. Geoscience and Man, vol. 32. Geoscience Publications, Department of Geography and Anthropology, Louisiana State University, Baton Rouge.

Brown, Scott S., and Kent Mathewson. 1999. "Sauer's Descent? Or Berkeley Roots Forever." *Yearbook of the Association of Pacific Coast Geographers* 61:137–157.

Bruhnes, Jean. 1910. *La Géographie Humaine.* Armand Colin, Paris.

Cosgrove, Denis E. 1983. "Towards a Radical Cultural Geography: Problems of Theory." *Antipode* 15:1–11.

———. 1984. *Social Formation and Symbolic Landscape.* Croom Helm, London.

———. 1985. "Prospect, Perspective and the Evolution of the Landscape Idea." *Transactions, Institute of British Geographers,* New Series, 10:45–62.

Cosgrove, Denis E., and Peter Jackson. 1987. "New Directions in Cultural Geography." *Area* 19:95–101.

Daniels, Stephen. 1993. *Fields of Vision: Landscape Imagery and National Identity in England and the United States.* Princeton Univ. Press, Princeton.

Diamond, Jared. 1999. *Guns, Germs, and Steel: The Fates of Human Societies.* Norton, New York.

Dickson, D. Bruce. 2003. "Origins and Dispersals: Carl Sauer's Impact on Anthropological Explanations of the Development of Food Production." In *Culture, Land, and Legacy: Perspectives on Carl O. Sauer and Berkeley School Geography,* ed. Kent Mathewson and Martin S. Kenzer, 117–134. Geoscience and Man, vol. 37. Geoscience Publications, Department of Geography and Anthropology, Louisiana State University, Baton Rouge.

Dryer, Charles R. 1926. "The Nature of Geography." Review of "The Morphology of Landscape," by Carl O. Sauer. *Geographical Review* 16:348–350.

Duncan, James S. 1980. "The Superorganic in American Cultural Geography." *Annals of the Association of American Geographers* 70:181–198.

———. 1981. "Comment in Reply" [to Miles Richardson and Richard Symanski on "The Superorganic"]. *Annals of the Association of American Geographers* 71:289–291.

———. 1993. "Commentary on 'The Reinvention of Cultural Geography' by Price and Lewis." *Annals of the Association of American Geographers* 83:517–519.

Gade, Daniel W. 1999. "Carl Sauer and the Andean Nexus in New World Crop Diversity." In *Nature and Culture in the Andes,* by D. W. Gade, 184–213. Univ. of Wisconsin Press, Madison.

———. 2004. Review of *Geography Inside Out,* by Richard Symanski. *Historical Geography* 32:181–185.

Gould, Peter. 1979. "Geography 1957–1977: The Augean Period." *Annals of the Association of American Geographers* 69:139–151.

Gregory, Derek. 1989. "Areal Differentiation and Post-Modern Geography." In *Horizons in Human Geography,* ed. D. Gregory and R. Walford, 67–96. Rowman and Littlefield, Lanham, Md.

Harris, David R. 2002. "The Farther Reaches of Human Time: Retrospect on Carl Sauer as Prehistorian." *Geographical Review* 92:526–544.

Hartshorne, Richard. 1939. *The Nature of Geography: A Critical Survey of Current Thought in the Light of the Past.* Association of American Geographers, Lancaster, Pa.

Holt-Jensen, Arild. 1999. *Geography: History and Concepts,* 3rd ed. Sage, London.

Hooson, David. 1981. "Carl O. Sauer." In *The Origins of Academic Geography in the United States,* ed. Brian W. Blouet, 165–174. Archon Books, Hamden, Conn.

Jackson, Peter. 1989. *Maps of Meaning.* Unwin Hyman, London.

Jacobs, Jane M. 2003. "Introduction: After Empire?" In *Handbook of Cultural Geography,* ed. Kay Anderson et al., 345–353. Sage, London.

James, Preston E. 1929. "The Blackstone Valley: A Study in Chorography in Southern New England." *Annals of the Association of American Geographers* 19:67–109.

Johnston, R. J., and J. D. Sidaway. 2004. *Geography and Geographers: Anglo-American Human Geography since 1945,* 6th ed. Edward Arnold, London.

Kroeber, Alfred Louis. 1917. "The Superorganic." *American Anthropologist* 19:163–213.

Leighly, John. 1937. "Some Comments on Contemporary Geographical Methods." *Annals of the Association of American Geographers* 27:125–141.

———. 1938. "Methodologic Controversy in Nineteenth Century German Geography." *Annals of the Association of American Geographers* 28:238–258.

Ley, David. 1981. "Cultural/Humanistic Geography." *Progress in Human Geography* 5:249–257.

———. 1982. "Rediscovering Man's Place." *Transactions, Institute of British Geographers,* New Series, 7:248–253.

Livingston, David N. 1992. *The Geographic Tradition: Episodes in the History of a Contested Enterprise.* Blackwell, Oxford.

Lukermann, Fred. 1989. "The Nature of Geography: Post Hoc, Ergo Propter Hoc?" In *Reflections on Richard Hartshorne's The Nature of Geography,* ed. J. N. Entrikin and S. D. Brunn, 53–68. Association of American Geographers, Washington, D.C.

Marsh, George P. 1864. *Man and Nature: Physical Geography as Modified by Human Action.* Scribner's, New York.

Martin, Geoffrey J. 2005. *All Possible Worlds: A History of Geographical Ideas*, 4th ed. Oxford Univ. Press, New York.

Mathewson, Kent 1986. "Sauer South by Southwest: Antimodernism and the Austral Impulse." In *Carl O. Sauer: A Tribute*, ed. Martin S. Kenzer, 90–111. Oregon State Univ. Press, Corvallis.

———. 1998. "Classics in Human Geography Revisited: J. S. Duncan's 'The Superorganic in American Cultural Geography.'" *Progress in Human Geography* 22:569–571.

———. 2005. "Jim Blaut: Radical Cultural Geographer." *Antipode* 37:911–926.

Mathewson, Kent, and David Stea. 2003. "James M. Blaut (1927–2000)." *Annals of the Association of American Geographers* 93:214–222.

Mikesell, Marvin W. 1987. "Sauer and 'Sauerology': A Student's Perspective." In *Carl O. Sauer: A Tribute*, ed. Martin S. Kenzer, 144–150. Oregon State Univ. Press, Corvallis.

Mitchell, Don. 2000. *Cultural Geography: A Critical Introduction*. Blackwell, Oxford.

———. 2003. "Cultural Landscapes: Just Landscapes or Landscapes of Justice?" *Progress in Human Geography* 27:787–796.

Olwig, Kenneth R. 2002. *Landscape, Nature, and the Body Politic: From Britain's Renaissance to America's New World*. Univ. of Minnesota Press, Minneapolis.

Penck, Walther. 1924. *Die Morphologische Analyse: Ein Kapitel der Physikalischen Geographie*. J. Englehorns Nachforschung, Stuttgart.

Penn, Misha, and Fred Lukermann. 2003. "Chorology and Landscape: An Internalist Reading of 'The Morphology of Landscape.'" In *Culture, Land, and Legacy: Perspectives on Carl O. Sauer and Berkeley School Geography*, ed. Kent Mathewson and Martin S. Kenzer, 233–259. Geoscience and Man, vol. 37. Geoscience Publications, Department of Geography and Anthropology, Louisiana State University, Baton Rouge.

Pred, Allan. 1983. "From Here and Now to There and Then: Some Notes on Diffusions, Defusions, and Disillusions." In *Recollections of a Revolution: Geography as Spatial Science*, ed. Mark Billinge, Derek Gregory, and Ron Martin, 86–95. St. Martin's Press, New York.

———. 1991. Review of *The City as Text: The Politics of Landscape Interpretation in the Kandyan Kingdom*, by James S. Duncan. *Journal of Historical Geography* 17:115–117.

Price, Marie, and Martin Lewis. 1993. "The Reinvention of Cultural Geography." *Annals of the Association of American Geographers* 83:1–17.

Reclus, Elisée. 1905–1908. *L'Homme et la Terre*. 6 vols. Librairie Universelle, Paris.

Richardson, Miles. 1981. "On 'A Critique of the Superorganic in American Cultural Geography.'" *Annals of the Association of American Geographers* 71:284–287.

Sauer, Carl O. For references, see the Sauer bibliography at the end of this volume.

Sluyter, Andrew. 1997. "Commentary: On Excavating and Burying Epistemologies." *Annals of the Association of American Geographers* 87:700–702.

———. 2002. *Colonialism and Landscape: Postcolonial Theory and Applications.* Rowman and Littlefield, Lanham, Md.

———. 2003. "Neo-Environmental Determinism, Intellectual Damage Control, and Nature/Society Science." *Antipode* 35:813–817.

Solot, Michael. 1986. "Carl Sauer and Cultural Evolution." *Annals of the Association of American Geographers* 76:508–520.

Speth, William W. 1981. "Berkeley Geography, 1923–33." In *The Origins of Academic Geography in the United States,* ed. Brian W. Blouet, 221–244. Archon Books, Hamden, Conn.

———. 1999. *How It Came To Be: Carl O. Sauer, Franz Boas and the Meanings of Anthropogeography.* Ephemera Press, Ellensburg, Wash.

Stoddart, David R. 1997. "Carl Sauer: Geomorphologist." In *Process and Form in Geomorphology,* ed. D. R. Stoddart, 340–379. Routledge, London.

Symanski, Richard. 1981. "A Critique of the Superorganic in American Cultural Geography." *Annals of the Association of American Geographers* 71:287–289.

Symanski, Richard, and Korski. 2002. "Coconuts on a Lava Flow in the Chiricahua Mountains." In *Geography Inside Out,* 103–114. Syracuse Univ. Press, Syracuse.

Thomas, William L., ed. 1956. *Man's Role in Changing the Face of the Earth.* Univ. of Chicago Press, Chicago.

Unwin, Tim. 1992. *The Place of Geography.* Longman, London.

Wallach, Bret. 1999. "Commentary: Will Carl Sauer Make It across that Great Bridge to the Next Millennium?" *Yearbook of the Association of Pacific Coast Geographers* 61:129–136.

Zelinsky, Wilbur. 1973. *The Cultural Geography of the United States.* Prentice-Hall, Englewood Cliffs, N.J.

3

Thoughts on Bibliographic Citations to and by Carl Sauer

Daniel W. Gade

Citations in the literature of geography and cognate fields form an unexamined aspect of the uncommon attention that Carl O. Sauer has received since his death in 1975. Unlike most intellectual ancestors of disciplinary knowledge, Sauer has not become a ghost. His ideas continue to get attention and, perhaps more, his inspiration as a scholarly exemplar live on. Though he himself was reserved in his own citation practice, others have heavily referenced his writings. Bibliographic attention in both directions tells us something about this luminary, but also evokes a broader reflection about referencing patterns and how they have changed. Citation study has centered on stratifying the world of scholarship through the use of a numerically based hierarchy, based on an assumption that statistical measurement is the ultimate template of scholarly significance.[1] The sociologist Robert K. Merton (1973:17) qualified this approach as hiding "speculative interpretations behind methodological fiats." A full understanding of this phenomenon deserves a wider scope in order to bring out cross-influences, sources of inspiration, and shifting trends. Referencing practice as it relates to Carl Sauer also conveys thoughts about American geography and the social sciences of which it is a part.

In its broadest context, citation practice can be seen as one very particular aspect of what Philip Wagner (1996:12) has called *Geltung*, defined as "competitive communicative display." This German word, which has no equivalent in English, combines the ideas of respect, esteem, reputation, acceptance, credibility, and authority. Though *Geltung* permeates the human brain as a survival mechanism of the species, scholars form an occupational group that is self-consciously concerned with both its implementation and outcomes. These people have a heavy ego investment in their work, for they operate in an intangible form of communication as their stock-in-trade. Citations received to publications produced help to validate scholarly efforts through prestige motivation because, unlike journalists or novelists, scholars rarely derive their livelihood from their publications. Citations, even when not complimentary, are welcome attention to the person who receives them, for work acknowledged is normally preferred to work ignored. A dyadic logic of association holds that, by implication, those who dispense these references are themselves good scholars. Self-citation forms a separate kind of referencing behavior that thrusts one's own previous work to the reader's attention. Whether seen as a form of ego enhancement or as a rhetorical device, self-citation is the least subtle expression of *Geltung* in the life of a scholar. Scrutiny of out-citation, in-citation, and self-citation can together form an entrance to another level of biographical understanding.

CARL SAUER, GEOGRAPHER

Carl Sauer was a consistently productive scholar (see the Sauer bibliography at the end of this volume) in spite of a heavy responsibility over three decades as departmental chair and supervisor of most of the dissertations written in the Berkeley department during that time (see appendix). When obligatory retirement in 1957 released him from time-consuming departmental duties, he became even more productive. Sauer published four books and twenty articles in the nearly twenty years between retirement and death. As people have written, his career as a geographer took a major turn when he moved to Berkeley (Speth, 1981). Research and writing were directed to the recuperation of European ideas about land and life, but even more in a southward gaze toward another way of life in Latin America (Mathewson, 1986). The inventory approach to land use that a youthful Sauer had gotten involved in at the University of Michigan led him to a fundamental realization: scholarship is best pursued as a result of compelling personal curiosity than as part of an applied agenda set by bureaucrats.

"Following one's bliss" unleashed a series of intellectual discoveries that demonstrate how, for him, education was an open-ended process without set goals. With no missionary zeal to save the world, this approach opened Sauer to the value of intellectual speculation. He was not, however, a polemicist and pointedly ignored most criticisms directed at his published work. Another aspect of Sauer's personality that provides clues to the citation phenomenon was a Romantic temperament that believed in the concept of cultural diversity (though he certainly would have eschewed postmodern identity politics). He was interested in landscape description and in reconstructing the distinctive life of people remote in space and time. In his thinking, he took issue with the conventional wisdom (e.g., he did not believe that necessity was the mother of invention), and he cultivated his own peculiarities.

The appeal of his geography had a good deal to do with the topics he chose. But one might also posit that much of that appeal had to do with two other elements. One of them was an unusual capacity to grasp a wide range of universal knowledge, and another was to impart intellectual wisdom. Those acquisitions came from a concern for self-education and a desire to express thoughts in an inimitable way. The push in that direction came from the fact that Sauer's parents were educated German immigrants who had absorbed the intellectual achievements of high German culture. Sauer kept Goethe's works as bedtime reading. He had absorbed the notion of *Kulturgeschichte* (culture history) first elaborated by Johann Christoph Adelug (1732–1806), which constituted a way of looking at the past that in North America was rare. The careful attention he gave the written word resulted in an uncanny ability to express complex ideas in disarmingly simple language free of jargon. Words conveyed his messages almost to the exclusion of pictures. He was chary in using iconographic representations; few of his articles contained halftone photos, and those that were included were taken by someone else.

Carl Sauer was productive as measured by number of publications, though many geographers have surpassed him on that score. Only a few of his published writings appear to have been truly refereed, which excludes all of his in-house publications at Berkeley and many invited addresses. He did not seek to fractionalize his work into a series of overlapping articles, nor did he himself follow a now common practice of including in a book articles previously published.[2] With a few exceptions, information in one publication did not duplicate the content found in another. Many of Sauer's studies have now been reprinted, but always at the instigation of others. If Sauer's work avoided redundancy, he was also not a recensionist. He did not keep a tight handle on the literature of his discipline as a way of formally evaluating its contributions to knowledge.³ Few book reviews came from his pen, mostly early on. Over a sixty-year period, Sauer published only eighteen of them, and most had been oriented to his research interest at the time. Unlike two of his Ph.D. students who became stellar scholars, James Parsons (1915–1997) and Wilbur Zelinsky (1921–), Sauer did little to chronicle the advance of geographical knowledge among his contemporaries. In his own articles and books, he provided only modest numbers of citations. He rarely referenced himself. Some of the maestro's idiosyncrasies in this regard can be better appreciated by general commentary on scholarly practice as it stands in the early twenty-first century.

No sharper example of a school of thought has occurred in American geography than the Berkeley tradition. Yet identifying overarching ideas is bound to come up short in characterizing them; all such generalizations have their caveats and exceptions. Fieldwork in remote settings does not define everyone in the group. No line of thinking fits all of them, though, generally, Sauerians have often professed a concordance of broad, but often unarticulated, views about the past, geography, and scholarship. The glue that binds the Sauerian perspective follows lines of thinking that came out of the research experience. Sauerians have disputed the triumphalist Eurocentrism implicit in a certain kind of thinking emanating not only from Europe but a host of transatlantic-looking scholars in North America whose working assumptions have been that the benighted New World was "civilized" only beginning in 1492. Included as a corollary to this mind set is the oft-repeated gloss that the Western Hemisphere was virgin territory until Europeans subdued and transformed it. Individuals in the Berkeley tradition have also contested the simplistic notion of economic development and modern technology as beneficent to the advancement of nations.

Already in the 1930s Sauer had discussed his reservations about the notion of progress as it related to Mexican agriculture; in the 1980s, aid organizations, faced with a surge of resistance to their modernist view of economic development, forced a shift of their own objectives to continue operations. Thus, the Berkeley tradition in geography has not stood in opposition to those aspects of the postmodernist project that have privileged culture and ethnicity and concern for the environment. That prescience of the Sauerians has not, however, extended to the disregard of the empirical fact favored by social theorists, some of whom call themselves cultural geographers. Finally the Sauerian circle of affinity has refused the notion that geographic research is most justified when it is applied to solving societal problems. Sauer was resolute in insisting that curiosity is the driving force behind doing good geography. He would have been the first to warn of the danger of the "relevancy orientation" recently promulgated by Lynn Staeheli and Don Mitchell (2005). Sauer's perspective, if laudable from a scholarly point of view, has had some consequences. The Sauerians have rarely dealt with the burning issues of the day, nor have they concentrated on getting geography into the schools or in projects to promote the discipline. Although Sauer had conservative attitudes on some, though not all, questions, it would be incorrect to generalize Sauerians in any politically or socially consistent way.

GENERAL THOUGHTS ABOUT CITATION PRACTICE

Citations convey something about both the person referenced ("in-citation") and the author who made the citation to the work of others ("out-citation") (Hyland, 2003). Both categories acknowledge the building block nature of scholarly work, a principle that has separated academic writing from popular books and articles. Footnotes and citations in the non-scholarly realm are considered an encumbrance or unnecessary affectation. Scholars, however, are sensitive to the cumulative nature of knowledge, which makes the presence of a bibliographical list indispensable in evaluating the publication. So basic is referencing as a process of scholarly ratification that an uncanny psychological effect takes over in transforming belief into knowledge. By referral to the list, the informed reader can determine the depth or shallowness of what the author has absorbed on the topic. Normally, however, space limitations confine the list to only those sources that the author has mentioned in the text. To those intimately familiar with the literature of a particular topic, some citations included can seem arbitrary. Omissions are more serious. Those who work on similar topics make judgments about who properly belongs on a reference list, and if names are omitted it raises questions about the text.

Authors usually assume that the citation of their work is a positive illustration of its intellectual value. But citation can also be decidedly noncomplimentary when it disputes the version of truth presented or the accuracy of a fact. Citation in several subject areas of geography now also tends to be monolingual. Except on topics dealing with foreign areas, referencing work in foreign languages is now uncommon.

Citation analysis in geography has not included discussion of who cites whom and why some works are cited and others not. Some patterns seem fairly clear. Theory and methodology, which cut across all other aspects of a discipline and frequently deal with fashionable or trendy issues, snare more general attention than other systematic or regional geography. Not surprisingly, geographers who work primarily with concepts or techniques receive the largest numbers of citations.[4] Whether bibliographic attribution is a product of the social structure—who the author knows or wants to know—or the intellectual structure, that is, the other authors working on the same or closely related topics, is more elusive. Although editors may intervene in some cases, citation attribution is a discretionary decision of the author or authors. An excellent article, well researched, analyzed, and written, does not by itself elicit many citations. Much depends on timing. For example, the quantitative movement of the 1960s and the postmodernist vogue of the 1990s generated hyperactive bibliographic search behavior. Citation can become a feeding frenzy, as during the early years of the "quantitative revolution" in geography when authors sought corroboration from like-minded contemporaries to validate their positions.

Building on Sauer's broad canvas of the New World past, William Denevan's 1992 "Pristine Myth" article exemplifies the importance of an idea that has fit into the spirit of the times. Although it avoids the trendy language of postmodernism, this article was clearly revisionist in its intent to offer a contesting perspective to the conventional wisdom that the Europeans in 1492 came to a virginal New World of primary forest and cultural primitiveness. Many ecologically minded individuals were predisposed to hear that message. Plant ecology had undergone a paradigmatic change in thinking about its once-sacred theme of stability, to a point of view that espoused the diametric opposite, that stability was a total chimera. Scholars concerned especially with culture were, even before that quincentennial year, open to the distortions of Eurocentrism. Denevan's article had two contributing aspects that have helped to explain its success as a widely disseminated piece of writing. One is its clear, understated prose that recalls Sauer's own style of writing. The other is its disarmingly short and concise title; "Pristine Myth" has now begun to enter dictionaries of allusions as a sideways reference in the same way as "tragedy of the commons." Thomas Headland (2004:1023) has attributed "pristine myth" as having been "created and popularized by U.S. geographer William M. Denevan."

Denevan's article became one of the most heavily cited non-methodological and non-theoretical pieces of writing by an American geographer in the period after it was published. From 1992 through 2007, it received over 260 citations in the Web of Science (Social Science, Science, and Arts and Humanities Indexes) and possibly as many others not recovered in any index. Nongeographers, who may never have previously taken the opportunity to pursue the *Annals of the Association of American Geographers*, discovered its special relevance to their thinking. Citations to Denevan (1992) have occurred in journals as diverse as the *Transactions of the American Fisheries Society* and the *Public Historian*. This example is instructive in seeing citation as a reflection of a convergence of factors that include the fact that Denevan had established his bona fides over more than three decades of productive fieldwork in South America on precisely this theme.

CARL SAUER'S USE OF THE CITATION

Sauer's writing contains modest but often telling attributions to the work of others. Six well-known publications chosen here provide a basis for understanding in more detail the peculiarities of his referencing pattern. Three methodological pieces and three substantive works at different regional scales are dissected for this purpose.

"The Morphology of Landscape" (1925)

As the earliest of the writings considered here, the long article "The Morphology of Landscape," sometimes catalogued as a book, reveals as much about Sauer as it does about the landscape concept. His referencing betrays this work as a prime statement of his *Bildung*. That untranslatable German term refers to self-education beyond the formal acquisition of knowledge. Implicit in *Bildung* is converging being and knowing that comes from a personal sense of possessing the living powers of cognition that incorporate the demand for self-discovery and a reflected relationship between reason and sensibility. In the *Bildung* tradition, wisdom comes by seeing oneself in a state of becoming rather than as a finished product. It helps to explain Sauer's lifelong quest of scholarship. One of his accolades was to refer to someone as "never having stopped learning." To place oneself in a state of becoming and toward self-actualization requires a diachronic perspective that leads seamlessly to a strong historical sense. Sauer's own *Bildung* had an accommodation with the natural sciences, but one that repudiated the physical environment as the central fact of human life on earth. This particular article suggests that Sauer's early training in landform morphology and process had started to also encompass biotic phenomena. In a long footnote, Sauer (1925:44–45) quoted Alfred Hettner (1859–1941) as a way of validating the inclusion of plants and animals into American geographical thinking about landscape.

Sauer cited in this lengthy article a dozen thinkers, many of them German, who in fact, had little or nothing to do with landscape per se. That group included Johann Wolfgang von Goethe (1749–1832), the greatest figure of German Romanticism, whose own *Bildung* has been the great model through the centuries of human self-awareness. Other thinkers referenced were the phenomenologist H. G. von Keyserling (1880–1946), the Kantian philosopher Hans Vaihinger (1852–1933), and especially the historian Oswald Spengler (1880–1936).[5] Sauer read and cited Spengler's key work, *Der Untergang des Abendlandes* (1922), in the German original.[6] No English translation of this work, known as *The Decline of the West*, was published until after "Morphology of Landscape" had appeared (vol. 1 in 1926 and vol. 2 in 1928).

Spengler influenced Sauer in four ways. The former wrote of *Morphologie der Weltgeschichte* (*Morphology of World History*), an idea transposed from Goethe's definition of morphology as the science of external forms. German geographers Carl Ritter (1779–1859), Albrecht Penck (1858–1945), Siegfried Passarge (1867–1958), and Karl Sapper (1866–1945) all used morphology in that Goethean sense of "world-as-nature." Sauer cited them together eight times in his paper to emphasize the importance of the trained eye to register forms and patterns. But Sauer, following Spengler's "world-as-history," applied the morphology idea to the humanized and historical landscape. Sauer was also taken with Spengler's organic idea of world cultures that rejected the linear model of history. Spengler's description of the senescence and death of cultures appealed to Sauer's historicism and may have contributed to his anti-modernism. Sauer applied that cyclical idea to geography by highlighting the notion of "landscape-as-history" rather than the "landscape-as-nature" approach of Penck, Passarge, or Sapper. Sauer transferred Spengler's line of thought about the cyclical nature of civilizations to the landscape. In that line of thinking, each place (i.e., landscape) formed a stage on which a succession of different cultures through time had left subtle traces on the land as in a palimpsest. Previously, Otto Schlüter (1872–1959) had developed the idea of landscape as a visible imprint (*Kulturlandschaft*) of a culture that showed the work of human action on the *Urlandschaft* (original landscape). Curiously, however, Sauer did not cite Schlüter in this paper. Perhaps as John Leighly had claimed, Sauer did not then know Schlüter's work

(Martin, 2003:33). In a paper published shortly thereafter, Sauer (1927) did cite Schlüter.

On a third score, Sauer seems also to have been influenced by Spengler's writing that "the history of the landscape, which included the plant cover and weathered mantle and the stage on which human history had played out for the past 5,000 years; this landscape has been so tied up with human history that life, soul, and thought cannot be understood without it" (Spengler, 1922: 2:45, my translation). Sauer, who had substantial training in the earth sciences, could relate to the idea of the fundament. Fourth, Spengler's argument that Western civilization had since reached its peak and was in a downward spiral must be seen as part of a deep malaise after the collapse of Germany between 1918–1923. A loss of faith in the idea of historical progress made Spengler's book, which reflected the pessimism of the period rather than many lasting truths, immensely popular when it was published. From reading Sauer's correspondence, it is not unreasonable to deduce that his "austral turn" to Mexico's Indian-dominated rural way of life was a result of disenchantment with the direction of Euro-American civilization.

Several non-German thinkers also contributed to Sauer's sense of *Bildung* conveyed in this article. They included Flinders Petrie (1853–1952), who wrote about historical reconstruction; Thomas Huxley (1825–1895), the agnostic biologist who promoted Darwinism; Henry Adams (1838–1918), an American philosopher who sought meaning in chaos; and Benedetto Croce (1866–1952), the philosophical idealist who argued that although everything was history, historical facts do not point to permanent truths. Sauer used Croce's idea to draw a parallel that even in geography, personal judgment is implicit in reading the landscape. Sauer's references to works in literature and philosophy indicated his concern for enduring values beyond the affairs of the day and reflected an educational ideal above any political tendency. Anthropologists cited in this paper were his colleague Alfred Kroeber (1876–1960), who developed his own cyclical ideas of history and of culture; Clark Wissler (1870–1947), who advanced the notion of culture area; and Leo Frobenius (1873–1938), the leading adherent of the *Kulturkreis* circle that provided a theoretical basis for understanding diffusion. Sauer cited several European scholars, notably Lucien Febvre (1878–1956), Paul Vidal de la Blache (1845–1918), and Samuel Van Valkenburg (1891–1976), for their views that countered the crude environmental determinism prevalent in the United States. Sauer integrated the writings of this diverse group of scholars with his concern for *Bildung* at a turning point in his life as a scholar. In his mid-thirties, but already a full professor with tenure and the chair of a department, he was determined to follow his own intellectual path rather than meet the expectations of others.

"Foreword to Historical Geography" (1941b)

Originally delivered as a presidential address, "Foreword to Historical Geography" sought to establish a vision for all geography as historical. Its thirty-five citations can be assembled into five cohort groups. One was Sauer's American contemporaries in geography, notably Ellen Churchill Semple (1863–1932), Richard Hartshorne (1899–1992), Harlan Barrows (1877–1960), and William Morris Davis (1850–1934), the middle two of whom were called to task as representing a "Great Retreat." A second category of citations went to sociologists Howard Odum (1884–1954) (Sauer's favorite regionalist); Robert E. Park (1864–1944); Ernest W. Burgess (1886–1966); and Vilfredo Pareto (1848–1923), as well as archaeologists James A. Ford (1911–1968) and Harold Sellers Colton (1881–1970) who had excavated southwestern sites.

The third group included citations to two people in Sauer's own academic circle: one to Rollin Salisbury (1858–1922), his *Doktorvater* at Chicago, cited for his 1907 textbook (a form of writing that Sauer largely disdained but for whose author Sauer had great respect); the other to Fred Kniffen (1900–1993), one of his early Ph.D. students at Berkeley whom Sauer had first met as a geology undergraduate at the University of Michigan two decades earlier. European geographers formed a fourth cited group, and included Eduard Hahn (1856–1928), Friedrich Ratzel (1849–1904), August Meitzen (1822–1910), and Carl Schott (1905–1990), who wrote in German; the Frenchman Jean Bruhnes (1869–1930); the Swede Sten De Geer (1881–1933); and the Britons Vaughn Cornish (1862–1948) and Herbert Fleure (1877–1969). A disparate collection of historical figures from Cluverius (1588–1622), Juan de Torquemada (1557–1604), Alexander von Humboldt (1769–1859), and George Perkins Marsh (1801–1882) formed the last citation category in this article.

"The Education of a Geographer" (1956d)

The methodological "Education of a Geographer" has become another classic statement in American geography. Half a century after it was first published, the amiable essay still resonates with quite a few budding scholars seeking to confirm their career choice. Its origin as a presidential address in 1956 helps to explain the absence of citations. Sauer did, however, evoke in the text the names of twenty-two worthies, from Humboldt to Robert Louis Stevenson (1850–1894) and Marjorie Rawlings (1896–1953), the last two of whom he commended for their evocative place-imbued fiction. A telling mention was his invocation of Henry Adams, whose life reflected a quest for *Bildung* and who saw his world, as Sauer did, through the eyes of an historian caught, as Sauer was, between the forces of rationalism and romanticism. Sauer's remarks in this essay reflected his decades of experience with graduate students, the scholarly values that had solidified in his mind, and the *Bildung* that was part of his own self-realization.

"The Personality of Mexico" (1941c)

Greater bibliographic reserve appears in three other works chosen here for comment. In "The Personality of Mexico," Sauer established a fundamental distinction between the civilized south and uncivilized north without a bibliographical foundation. Only two references can be found in this essay. One was to Cyril Fox's (1882–1969) work on the "personality of Britain," from which Sauer had gotten the idea for this anthropomorphic projection into place (Dunbar, 1974). "The Personality of Mexico" demonstrates Sauer's intense concern for a deep geographical past, one that goes beyond the nineteenth century of many historians into prehistory. He saw the regional differences in Mexico as a reflection of the strong pre-Hispanic differences between north and south and also the effects of the sixteenth-century conquest. More than half a century later, that regional difference is less sharp, replaced by a still strong contrast between city and countryside.

Agricultural Origins and Dispersals (1952a)

Agricultural Origins and Dispersals, for decades one of geography's classics, had a remarkably brief list of citations, given the enormous breadth of the topic. Sauer neither retrieved plant remains as an archaeologist would nor made systematic plant collections as botanists do. His contribution was to synthesize several different strands into a coherent and convincing story. Moreover, Sauer at that time had never been to the Near East, Eastern Asia, or Africa, or even the Mediterranean, which forced dependency on other authors. As in some other works, chary attribution came in part from the fact that this book had originated as a series of lectures delivered to a general, rather than specialist, audience. For example, the useful weed hypothesis of agricultural beginnings was a remarkable insight that began with Theodor Engelbrecht (1916), who was not cited. However, Sauer did cite in this connection André Haudricourt and Louis Hédin (1943:153) as suggesting that rice was originally a weed in taro fields. His use of self-citation (1936a; 1950a) was a legitimate rhetorical strategy to strengthen knowledge claims on this topic where American geographers had not tread before.

Sauer also cited eight German scholars, either as authorities or as sources of specific ideas, such as Eduard Hahn's reasoning for a religious motive to cattle domestication. Included in that category was the archaeologist Frederick Zeuner (1905–1963). Americans cited were the polyvalent tropical zoologist Marston Bates (1906–1974); George Gaylord Simpson (1902–1984) on prehistoric South American environments; John Leighly (1895–1985), a valued departmental colleague; and George Carter (1912–2004), an irrepressible Ph.D. student during the early post-war period. Sauer also acknowledged the help of Fred Simoons, then a Berkeley graduate student, whose off-beat ethnographic imagination led him to construct a highly original map of the non-milking area of the Old World (Simoons, 1970). Beyond direct citation, Sauer placed another list in that slim volume of twenty-six "selected general references" that subliminally conveyed a European dominance of this subject: seventeen of them were in German, two were written in French, and one in Dutch; the remaining six were in English, including a translation of Nikolai Vavilov from the original Russian.

The Early Spanish Main (1966a)

The Early Spanish Main was a display of Sauerian erudition on the first century of impact after European arrival on Hispaniola. It formed his primary meditation on the European Conquest and focused on an account of land use, resources, and native peoples. More memorable for many readers was the strong criticism of Columbus that foreshadowed the sharp criticism of the European conquests that burst onto the public stage twenty-five years later (Starrs, 1992). More than 80 percent of the references in this book were to the printed sources of Bartholome de Las Casas (1474–1566), Gonzalo Fernández de Oviedo (1478–1557), Christopher Columbus (1451–1506), and Peter Martyr d'Anghiera (1457–1526). Contrary to Donkin's (1997:249) assertion that in "Early Spanish Main we have more formal scholarship drawing on colonial archives," all of the references Sauer used were to printed sources. If Sauer had sought out archival documents, their inclusion would have enriched this study and gotten more respect from historians. Nor did many twentieth-century scholars receive attribution; among those who did were the Columbus authority Samuel Eliot Morison (1887–1976) and two Berkeley Ph.D.s (Charles Alexander [1916–1987] and Burton L. Gordon [1920–]), who received mention on specific details. Sauer's impressive synthesis, based on wide-ranging cogitation through the decades about aboriginal material culture in the Caribbean and the colonial Spanish mind, presented a natural history perspective unusual in historical treatments. It did not, however, come sufficiently to grips with the reasons for aboriginal decline.

CITATIONS TO CARL SAUER

Sauer received high numbers of citations to his publications in the second half of the twentieth century. This reference-rich period covered his life as a mature scholar and a twenty-five-year period after he died. Some of these citations were retrieved in the two pertinent data banks, the Social Science Research Index and the Science Research Index. In J. W. R. Whitehand's (1985: 228) analysis of geographers in the SSRI between 1966–1970, Sauer came in third. In the period 1971–1975, Sauer came in tenth out of thirty-five, but far below the first place holder, the quantitative geographer Brian Berry. Andrew Bodman (1991:26–27), who used both the SCI and SSCI as data sources for the period from 1981 to 1985, placed Sauer twentieth among the "centurions"— geographers who received at least one hundred citations for that period. However, the 290 citations that Sauer received were still only one-fourth the number of citations that Brian Berry received in the same period. Sheer tallies, however, are only one aspect of citation sociology. The remarkable array of subjects and disciplines that his citers represent, now electronically available for examination on the "Web of Science," testifies to his wide scholarly influence. A study of disciplinary affiliation of citers would reveal that geographers, not surprisingly, have most persistently referenced Sauer's work. Anthropologists and historians have also cited his work, whereas biologists have been more reticent to do so. North American scholars have cited him by far the most, which also reflects his modest influence in continental Europe and Latin America, two parts of the world with which he was involved.

Sauer's students and subsequent generations of intellectual descendants have generously cited him (Brown and Mathewson, 1999). That is one measure of the acceptance of his ideas; few descendants have reacted against his influence. These offspring have often worked on topics that had similarities with those of Sauer. Many of their attributions, however, have made an affinity connection with the maestro rather than acknowledging a source of specific information. Citation in those cases has constituted a dyadic relationship between the cited and the citer as a way to validate one's style of research or broad thematic concerns. Mentors have, of course, often gotten citation and acknowledgment from their students. Sauerians include more than Sauer's own students, and as a group they often have received inspiration from a revered professor they never knew. By the turn of the century, most self-identified adherents of the "Berkeley school" had not attended UC-Berkeley as students. At the same time, most Berkeley geography Ph.D.s, especially those who came through the department from the 1970s onward, have not claimed to be part of this particular tradition. Nor is California still the center of gravity for the loosely defined, nationally dispersed assemblage of Sauerian practitioners.

In the history of American geography, "The Morphology of Landscape" is frequently presented as the key anti-determinist manifesto that led the discipline away from the environmentalist trap. Remarkably, however, it was an idea that Sauer never implemented; he made no empirical study using the model he laid out. It was a method suitable only to detailed studies of small areas, and by the 1930s Sauer had gotten interested in domestication and diffusion. This piece of writing has received extraordinary attention—it is one of his two most cited publications—even though it was prepared as an unrefereed in-house publication that took him only "several weeks" to write. By modern standards, "Morphology of Landscape" offers an unsophisticated view of time and space in landscapes compared to that, say, presented in Claude and Georges Bertrand's 2002 study. Of course, Sauer said nothing about how the landscape has reflected social domination, concepts of territory, and environmental conflict that energized many younger geographers in the late twentieth century.

In contrast to his methodological papers, Sauer's publications on Latin America have received far fewer citations. Many methodologists and historians of geography think they know Carl Sauer without having read his work on Latin America. Their own references to Sauer have suggested that most have concentrated on just a few of his writings. In the world of anglophone geography, Latin America has rarely been brought to the foreground to demonstrate larger concepts (Sluyter and Mathewson, 2007). Selective use of Sauer's writings and misinterpretation of them became a way to establish a "new cultural geography," seen, for example, in the assertion of James Duncan (1980) about Carl Sauer's putative view of culture as superorganic, which

Price and Lewis (1993), Mathewson (1998), Penn and Lukermann (2003), and Price (2003) have all called into question.

CARL SAUER AND CHANGES IN CITATION PRACTICE

Scrutiny of the bibliographic practice of an individual geographer provides a comparative template with which to more broadly reflect on practitioners, the discipline they represent, and the periods in which certain kinds of scholarship have been carried out. Sauer's reticence at self-citation contrasts with present practice in scholarly circles that makes abundant reference to one's own work.[7] This practice sometimes extends to listing one's oral presentations and unfinished manuscripts that cannot be retrieved from any library or Web page. Personality and the period may explain this difference. Carl Sauer had a strong sense of his scholarly persona, but a comment he made to me in a 1965 graduate seminar in Madison suggested that he considered it self-serving to promote oneself through self-citation. He also did not like to brag about his students in public to elevate their qualities and thus himself. In his writings, Sauer rarely used the first-person pronoun. Part of his reluctance to self-reference went with the belief at the time that impersonality enhanced authority, but its roots may have been in the pietistic tradition of his Missouri youth. Self-reference also locked one into a position from which it was more difficult to disassociate oneself.

Wider acceptance of self-citation constitutes another change in American geography. With more than ten thousand self-described geographers in the United States, the profession is no longer the close-knit and decorous community that it was in Sauer's day. Applying Tönnies's (1957) categories, a *Gesellschaft* of rational self-interest replaced the previous *Gemeinschaft* of professional solidarity. In 1950 most geographers in the United States either knew or knew about each other. An imbedded status-conscious hierarchy voted on membership dossiers and made decisions about meetings, editors, and directions without much consultation. In the 1960s, as college and university systems throughout North America expanded, the number of geographers also increased. Between 1950 and 1975 the Association of American Geographers doubled in membership. That growth also broadened geography from the WASP-dominated and heavily male discipline that it had been to a more diverse and loose-knit association. An important early contingent of U.S. geography professors had grown up in rural communities in the Middle West and set the tone. Those born in cities and especially suburbs became more dominant after the 1950s. Women geographers, for many years a token and marginalized group, surged in the 1990s. Foreign-born geographers, not just from Europe but especially now from Asia and even some from Africa, became visible in the 1980s and after.

Graduate student apprentices, even without a major piece of research yet approved by any committee, were allotted fifteen minutes of fame and even allowed to form sections at conferences. Growth in the variety and number of members, credentialed or not, and in geography departments or not, made depersonalization of this open guild unavoidable. New equal-opportunity-based hiring protocols shunted aside the old-boy networks. Greater anonymity arguably also made academic life more overtly competitive. Jostling for jobs, grants, and journal space accelerated. The practice of publishing "work in progress" rather than the finished product did not stop with a favorable tenure decision. The case for subsequent teaching load reductions, research leaves, and subventions came from possession of a well-developed bio-bibliography.

The greater emphasis on coauthorship constitutes another change since Sauer's days as a professor. He shared his billing in only seven publications over a lifetime, once with Wellington Jones (1886–1957), a fellow graduate student friend at Chicago, once with Peveril Meigs (1903–1979) and three times with Donald Brand (1905–1984), who were at the time his Ph.D. students, on a monograph with [Gilbert H.] Cady and [Henry C.] Cowles, and on a report with [C. K.] Leith and others, all prior to 1935. He did not attach his name to projects carried out principally by students, nor did he engage in collaborative research projects with colleagues. This pattern was less a peculiarity of Sauer, perhaps, than a reflection of the time in which he did his work. In the two volumes of the *Annals of the Association of American Geographers* for 1953 and 1954, only 8 percent of the articles were coauthored, compared with 42 percent for the two years 2003 and 2004. Research attribution has changed because project collaboration increasingly reflects shared authorship. However, shared authorship is not always symmetrical with shared responsibility and effort, especially if a name is added to jump-start the publication list of a budding young scholar.

Sources of scholarly inspiration in American geography have also changed. In his earlier work, Sauer had a strong connection with continental European geography. Family heritage and a boarding school education in Württemberg made him fluent in German, although at that time other American geographers, among them E. C. Semple and W. M. Davis and later R. Hartshorne, also had strong skills in that direction. Most Ph.D.s in American geography had to pass a requirement to read German. When, after 1933, geographical scholarship in Germany largely became an ideological vehicle of the state, the transfer of geographical inspiration from there to North America decelerated greatly. With the post-war demise of a German language requirement in American graduate departments, German geography lost influence, and the work of Germans found little transatlantic resonance even when geographers from the German-speaking countries published in English. Much of the same decline of European influence was seen with French geography. Influential early in the twentieth century, francophone geographers have provided little intellectual nourishment for their American counterparts since World War II. In contrast, French social theorists, most notably M. Foucault (1926–1984) and J. Derrida (1930–2004), have had a strong impact on American geography and social science in general, but only after their works were translated into English.

How or if Sauer will be cited in the year 2020 cannot be known in 2009. Geography as a discipline has been marked by bandwagon mentalities and themes that have siphoned off younger practitioners anxious to make their mark in the world of scholarship by fitting their research program into the trend line. Positioned not too assuredly in the academic world that has sustained it, geographers are like cows moving in tandem to graze new intellectual pastures, many of which lie in the back forty of cognate disciplines. A good deal of collective cud-chewing centers on certain hot topics that are swallowed, ruminated, digested, and excreted. Today's lush green fascinations become tomorrow's dried-up stubble: superannuated idea, stale method, or conceptual wasteland. Given the vagaries of academic fashion, reference to Sauerian geography could expand as certain topics become once again of wide interest. If this resurgence does not occur, citations to Sauer may have reached their acme. Some attribution will continue, but not necessarily in the way that Whitehand (2002:517) opined when he wrote that continuation of citations to Sauer depends on "others working on topics that have a connection with those on which he worked." A careful reading of Sauer's in-citation record would show that more than a dozen of Sauer's writings have hardly been referenced at all. Sauer has received fewer citations for all of his work on Latin America put together than for the three methodological and philosophical statements discussed earlier.

Sauer's work on the culture history of agriculture still contains some provocative kernels for discussion, but they are no longer near the center stage of debatable ideas. New hypothesis and many new facts have appeared to deal with a knowledge base about agricultural origins and the world crop and animal inventory that in 2005 had increased more than three times since 1945. Citations to Sauer on this subject have slowed to a tiny trickle. Peter Bellwood (2005), who has provided the most ambitious synthesis in the new millennium on agricultural origins, cites Sauer twice in passing. A decade earlier, Bruce Smith's 1995 tome did not mention Sauer once. Many who now write on this subject ignore Sauer's work even though his speculations were one of the key twentieth-century points of discussion. Some archaeologists did not favor his book, for Sauer placed agricultural beginnings in the hot humid tropics devoid of prehistoric plant remains. Moreover, diffusionary extrapolations got top-billing, which came under heavy criticism. From a present perspective, Sauer's notions about the invention of agriculture as having occurred among sedentary fishermen with the time for leisurely experimentation have to compete with now well-reasoned arguments focused on population, environmental change, or coevolution that have at least as much explanatory power. Sauer's view that the idea of agriculture was invented no more than twice in the history of the world seems counterintuitive. However, new techniques or interpretations could trigger a return to diffusion as a way to unravel the puzzle of enigmatic distributions. The history of Alfred Wegener's continental drift theory is instructive about the need to keep an open mind about diffusion even when available evidence seems tenuous. Until the 1960s, most scientists rejected Wegener's theory as lacking in proof.

As indicated by citation retrieval on the Web of Science, Sauer's work has continued to be referenced year in and year out (see chapter 4, table 1). Some decline can be noted; for example, in the late 1980s survey of American geography, the twelve citations Sauer received contrast with the three citations made in a comparable work published fourteen years later (Gaile and Willmott, 1989; 2003). He is cited now as a foundational figure for his roles in cultural geography (Foote, Hugill, Mathewson, and Smith, 1994), historical geography (Baker, 2003), Latin American geography (Gade, 2002), and some aspects of biogeography (Veblen, 2003). Perhaps the most embedded single idea attributed to Sauer is the origination of the cultural landscape theme (e.g., Matthews and Herbert, 2004). Failure to read German may explain why so many geographers have not given Otto Schlüter the credit he deserves for this idea.

CONCLUSION

A study of citations dispensed and received by Carl Sauer coaxes another level of biographical and philosophical understanding not heretofore elaborated in the abundant writings about him. Sauer's referencing behavior points to his extensive borrowing of ideas, the transatlantic source of most of his inspiration, and a philosophical self-awareness in his trajectory as a scholar. His own use of citation was often to acknowledge personages in the history of geography, which Speth (1993) saw as a device to promote a "time-dependent-view" of the discipline. The bibliographic apparatus in his scholarly work was less than comprehensive, yet the slender referencing may actually have made his writings more, not less, inviting to a variety of readers. His spare out-citation behavior contrasts with the voluminous references he has received from others. From one perspective, these attributions reflect the appeal of his work: timeless themes, broad sweep of erudition, bold speculations, and a fetching, understated prose style. More in-citations have come his way as the godfather of a compelling school of thought than as the source of a piece of factual information.

At another level, bibliographical attribution suggests a different and more unconventional perspective about the "Sauer phenomenon" in American geography. The citations to the works of Carl Sauer that pepper the literature of geography have established him as a *gran eminencia* of the discipline. Analysis of these attributions shows that their intent was less to credit him for specific content than to reaffirm his importance as a geographer and a thinker. The conventional wisdom that Carl Sauer "made" cultural-historical geography in the United States should more accurately be turned around: geographers made "Mr. Sauer" and the circle of affinity called cultural-historical geography. Scholarly enterprises need luminaries to serve as beacons and sages, for a human exemplar is always a more compelling rallying point than a bloodless idea. These special individuals found in most every academic discipline are elevated to heroic status.

Geography has for several decades lionized Carl Sauer out of a collective need to define a higher purpose. To some he has been seen as a cultic figure above criticism and as a source of oracular wisdom. This is suggested by disputes about what Sauer "really said" about culture, history, or the environment (Willems-Braun, 1997). His publications so frequently cited have been infrequently scrutinized because for this purpose, content is less important than representation. In the same way that environmentally concerned people have selected Francis of Assisi as a "patron saint of ecology," quite a few geographers have subconsciously canonized Carl Sauer as a two-pronged symbol for their discipline. One aspect is that he has embodied the nobility of pure scholarly purpose. He was not interested in technique and he was nonchalant about building his department into an academic empire. Sauer was not a narrow careerist, but one whose self-directed life came from reading, questioning, and reflection. His professorial life gave him the autonomy and freedom to pursue *Bildung* and his interests, which at the time were not popular among geographers in the Midwest.

The second thrust of Sauer as a symbol provides an example of an environmental moralist who refused to accept society's prevailing view of progress at a time when it was heresy to buck that trend. Many would argue that the agribusiness, sprawl, and Wal-Mart-type commerce we have today have impoverished, not enriched, our lives. The human role that has changed the face of the earth has gone too far. Carl Sauer, the true scholar and non-modernist visionary, whose *Bildung* imparted intellectual and moral autonomy, is seen as subliminally addressing these two concerns, and that is why he has been cited so much since his death. By elevating Carl Sauer to disciplinary sainthood, his place in the canonical history of the discipline and the pantheon of notables is assured as long as geography in the United States remains a recognizable academic discipline.

NOTES

1. Quantifying citations as an index of scholarly influence is fraught with problems. The Web of Science, which incorporates the Science Research Index (SRI) and the Social Science Research Index (SSRI), tallies only a portion of citations received in the scholarly literature. Thousands of relevant journals are not listed in their register. Moreover, no evaluation of the cited reference is made in this data bank as to whether it is simply a perfunctory inclusion or a critical winnowing that builds on or challenges the findings vetted in the text.

2. It was John Leighly's idea to assemble Sauer's writings into an anthology (Sauer, 1963a). The impact of this assemblage must be considered in the remarkable attention that Sauer has gotten over the past half century. It simplified retrieval of the nineteen articles, some published in fairly obscure journals, that were reprinted in that book. The anthology grouped the themes of Sauer's work, put them in the larger context of his published corpus between 1915 and 1962, and provided the editor's graceful overview of Sauer's writings. Several of Sauer's major articles are now available freely on the Internet, which facilitates access to his thinking by a worldwide audience.

3. Letters written by Sauer indicate that at least before World War II he was negative about the published work of his fellow geographers. He did not, however, wish to make critical comments about them in print.

4. Eight of the top ten "centurions" in Bodman's (1991) citation analysis were geographers of British origin who work mainly on theoretical issues, a form of scholarship in which brief articles constantly raise ephemeral controversies that elicit retort. Distinctly different is research based on empirical knowledge derived from fieldwork. Only handfuls of people work on similar topics whose results are generated in that way, so that publication and citation are not so facile. More recently, Yeung (2002) has corroborated the profile in geography in which methodology, philosophy, and theory evoke the most citations.

5. The attention to German scholarship would not have occurred eight to ten years earlier when, in the guise of patriotism, strong anti-German wartime sentiment penetrated even the universities. At the University of Michigan, the chairman of Sauer's department, William H. Hobbs, vented his Germanophobic sentiments (Martin, 2003:32), one reason that spurred Sauer in 1923 to depart for California. Baker (2003:74) errs in stating that Sauer was born in Germany.

6. Sauer's German was fluent, yet he did not publish in German even when it seemed appropriate to do so. He had learned to read French, and the fact that he cited Juan de Torquemada suggests that he could also read Spanish at a time before he had started his research in Mexico. Competency in reading other languages is not clear. His citation of works in Dutch probably meant that he could read that language, whereas dependence on an English translation of Croce suggests he did not read Italian.

7. Comparing half a century's difference in the *Annals of the Association of American Geographers,* only 45 percent of the articles in vol. 53 (1953) and vol. 54 (1954) used self-citation, in contrast with the 92 percent of the articles published in vol. 93 (2003) and vol. 94 (2004). Moreover in the latter two volumes, thirty-four articles had five or more self-citations; one even had twenty-two. More than simply a strategy to enhance scholarly credibility, such fulsome self-citation might be interpreted as an unsavory kind of academic egotism and dubious form of self-aggrandizement. More to the point is that, with so much evidence of prior publications on closely related topics, one wonders how much new information is actually provided in the article one is about to read. However, as a discipline, geography may engage in less self-citation than the hard sciences, where between 10 to 20 percent of all citations are to the authors themselves (Rousseau, 1999). More multi-authored works (some with a half dozen or more authors) offer a greater likelihood that any given citation will be to one of those authors' projects.

REFERENCES

Baker, Alan R. H. 2003. *Geography and History: Bridging the Divide.* Cambridge Univ. Press, New York.

Bellwood, Peter S. 2005. *First Farmers: The Origins of Agricultural Societies.* Routledge, New York.

Bertrand, Claude, and Georges Bertrand. 2002. *Une géographie traversière: l'environnement à travers territoires et temporalités.* Appéditions Arguments, Paris.

Bodman, Andrew R. 1991. "Weavers of Influence: The Structure of Contemporary Geographic Research." *Transactions, Institute of British Geographers*, New Series, 16:21–37.

Brown, Scott K., and Kent Mathewson. 1999. "Sauer's Descent? Or Berkeley Roots Forever?" *Yearbook of the Association of Pacific Coast Geographers* 61:137–157.

Denevan, William M. 1992. "The Pristine Myth: The Landscape of the Americas in 1492." *Annals of the Association of American Geographers* 82:369–385.

Donkin, Robin A. 1997. "A 'Servant of Two Masters.'" *Journal of Historical Geography* 23:247–266.

Dunbar, Gary. 1974. "Geographical Personality." In *Man and Cultural Heritage: Papers in Honor of Fred B. Kniffen*, ed. H. Jesse Walker and William G. Haag, 25–33. Geoscience and Man, vol. 5. Geoscience Publications, Department of Geography and Anthropology, Louisiana State University, Baton Rouge.

Duncan, James S. 1980. "The Superorganic in American Cultural Geography." *Annals of the Association of American Geographers* 70:181–198. Baton Rouge.

Engelbrecht, Theodor. 1916. "Über die Entstehung einiger feldmässig angebauter Kulturpflanzen." *Geographische Zeitschrift* 22:328–334.

Foote, Kenneth E., Peter J. Hugill, Kent Mathewson, and Jonathan M. Smith, eds. 1994. *Re-Reading Cultural Geography*. Univ. of Texas Press, Austin.

Gade, Daniel W. 2002. "North American Reflections on Latin Americanist Geography." In *Latin America in the 21st Century: Challenges and Solutions*. Yearbook of the Conference of Latin Americanist Geographers 37:1–44.

Gaile, Gary L., and Curt J. Willmott, eds. 1989. *Geography in America*. Merrill Publishing, Columbus, Ohio.

———, eds. 2003. *Geography in America at the Dawn of the 21st Century*. Oxford Univ. Press, New York.

Haudricourt, André G., and Louis Hédin. 1943. *L'homme et les plantes cultivées*, 5th ed. Gallimard, Paris.

Headland, Thomas N. 2004. "Pristine Myth." In *Encyclopedia of World Environmental History*, ed. Stanley Krech III, John R. McNeill, and Carolyn Merchant, 3:1023–1024. Routledge, New York.

Hyland, Ken. 2003. "Self-Citation and Self-Reference: Credibility and Promotion in Academic Publication." *Journal of the American Society of Information Science and Technology* 54:251–259.

Leighly, John, ed. 1963. *Land and Life: A Selection from the Writings of Carl Ortwin Sauer*. Univ. of California Press, Berkeley.

Martin, Geoffrey J. 2003. "From the Cycle of Erosion to the 'Morphology of Landscape': Or Some Thought Concerning Geography as It Was in the Early Years of Carl Sauer." In *Culture, Land, and Legacy: Perspectives on Carl O. Sauer and Berkeley School Geography*, ed. Kent Mathewson and Martin S. Kenzer, 19–54.

Geoscience and Man, vol. 37. Geoscience Publications, Department of Geography and Anthropology, Louisiana State University, Baton Rouge.

Mathewson, Kent. 1986. "South by Southwest: Antimodernism and the Austral Impulse." In *Carl O. Sauer: A Tribute*, ed. Martin S. Kenzer, 90–112. Oregon State Univ. Press, Corvallis.

———. 1998. "Classics in Human Geography Revisited: J. S. Duncan's 'The Superorganic in American Cultural Geography.'" *Progress in Human Geography* 22:569–571.

Matthews, John A., and David T. Herbert. 2004. *Unifying Geography: Common Heritage, Shared Future*. Routledge, New York.

Merton, Robert K. 1973. *The Sociology of Science: Theoretical and Empirical Investigations*. Univ. of Chicago Press, Chicago.

Penn, Mischa, and Fred Lukermann. 2003. "Chorology and Landscape: An Internalist Reading of 'The Morphology of Landscape.'" In *Culture, Land, and Legacy: Perspectives on Carl O. Sauer and Berkeley School Geography*, ed. Kent Mathewson and Martin S. Kenzer, 233–260. Geoscience and Man, vol. 37. Geoscience Publications, Department of Geography and Anthropology, Louisiana State University, Baton Rouge.

Price, Edward T. 2003. "Crafting Cultural History." In *Culture, Land, and Legacy: Perspectives on Carl O. Sauer and Berkeley School Geography*, ed. Kent Mathewson and Martin S. Kenzer, 277–298. Geoscience and Man, vol. 37. Geoscience Publications, Department of Geography and Anthropology, Louisiana State University, Baton Rouge.

Price, Marie, and Martin Lewis. 1993. "The Reinvention of Cultural Geography." *Annals of the Association of American Geographers* 83:1–17.

Rousseau, Ronald. 1999. "Temporal Differences in Self-Citation Rates of Scientific Journals." *Scientometrics* 44:521–531.

Salisbury, Rollin D. 1907. *Physiography*. Henry Holt, New York.

Sauer, Carl O. For references, see the Sauer bibliography at the end of this volume.

Simoons, Frederick J. 1970. "The Traditional Limits of Milking and Milk Use in Southern Asia." *Anthropos* 65:547–593.

Sluyter, Andrew, and Kent Mathewson. 2007. "Intellectual Relations between Historical Geography and Latinamericanist Geography." *Journal of Latin American Geography* 6 (1): 25–41.

Smith, Bruce C. 1995. *The Emergence of Agriculture*. Scientific American Library, New York.

Spengler, Oswald. 1922. *Der Untergang des Abendlandes: Umrisse einer Morphologie der Weltgeschichte*, 2nd ed. 2 vols. C. H. Beck'sche Verlagsbuchhandlung, Munich.

Speth, William W. 1981. "Berkeley Geography, 1923–33." In *The Origins of Academic Geography in the United States*, ed. Brian W. Blouet, 221–244. Anchon Books, Hamden, Conn.

———. 1993. "Carl O. Sauer's Use of Geography's Past." *Yearbook of the Association of Pacific Coast Geographers* 55:37–65.
Starrs, Paul F. 1992. "Looking for Columbus." *Geographical Review* 82:367–374.
Staeheli, Lynn A., and Don Mitchell. 2005 "The Complex Politics of Relevance in Geography." *Annals of the Association of American Geographers* 95:357–372.
Tönnies, Ferdinand. 1957. *Community and Society*. Trans. Charles Loomis. Michigan State Univ. Press, East Lansing.
Veblen, Thomas T. 2003. "Carl O. Sauer and Geographical Biogeography." In *Culture, Land, and Legacy: Perspectives on Carl O. Sauer and Berkeley School Geography*, ed. Kent Mathewson and Martin S. Kenzer, 172–192. Geoscience and Man, vol. 37. Geoscience Publications, Department of Geography and Anthropology, Louisiana State University, Baton Rouge.
Wagner, Philip L. 1996. *Showing Off: The Geltung Hypothesis*. Univ. of Texas Press, Austin.
Whitehand, J. W. R. 1985. "Contributors to the Recent Development and Influence of Human Geography: What Citation Analysis Suggests." *Transactions, Institute of British Geographers*, New Series, 10:222–234.
———. 2002. "Classics in Human Geography Revisited: Author's Response." *Progress in Human Geography* 26:516–519.
Willems-Braun, Bruce. 1997. "Reply: On Cultural Politics, Sauer, and the Politics of Citation." *Annals of the Association of American Geographers* 87:703–708.
Yeung, Henry Wai-chung. 2002. "Deciphering Citations." *Environment and Planning A* 34:2093–2106.

4

Bibliography of Commentaries on the Life and Work of Carl Sauer

Compiled by William M. Denevan

Presented here is an incomplete bibliography of commentaries about Carl Sauer, his life, and his scholarship. These include critical and descriptive essays, obituaries and memorials, short biographies, and book reviews. Most items are articles and chapters specifically about Sauer and his published writings. However, also included are extracts about Sauer from articles, texts, and other books. There are many of these, and only a portion are included here, usually critical or descriptive statements of a few paragraphs or a few pages. Thus, this bibliography provides an historical perspective on thinking about Sauer during the course of his long career and after.

The commentaries are listed alphabetically by author. Notations are placed after the names of authors of commentaries who received their Ph.D.s under Sauer, who received their Ph.D.s during Sauer's tenure at Berkeley but not under him, or who were regular members of the geography faculty during Sauer's tenure there, 1923–1975. This distinguishes "insiders," who were particularly close to Sauer in time and place, from "outsiders," who may or may not have been strongly influenced by Sauer or biased toward him. Most of the major scholars of Sauer have not been his students or Berkeley Ph.D.s (Gade, Kenzer, Mathewson, Speth, M. Williams), the important exceptions being Leighly, Parsons, and West. Table 1 indicates the number of commentaries on Sauer, by decade, from the 1920s to the present. (Data for 1915–1949 are especially incomplete, counting primarily reviews of Sauer's works.)

Table 1. Number of Commentaries on Carl Sauer by Decade

1915–1919	3
1920–1929	5
1930–1939	11
1940–1949	4
1950–1959	25
1960–1969	54*
1970–1979	96†
1980–1989	122
1990–1999	88
2000–2007	104‡
Total:	514

*Includes the many commentaries on *Land and Life* (1963a) and *The Early Spanish Main* (1966a).
†Includes the numerous obituaries and memorials following Sauer's death in 1975.
‡Includes some commentaries in press at the time this list was compiled.

Reviews of Sauer's books have been difficult to track down, especially for his early books and his monographs in *University of California Publications in Geography* and in *Ibero Americana*. As seen in table 2, at least one review is included for each of his books after 1950, as well as for the collections of articles by Sauer edited by Leighly (Sauer, 1963a) and by Callahan (Sauer, 1981a). Sauer's most reviewed book is *The Early Spanish Main* (1966a), with twenty-one reviews.

Table 2. Reviews of Books and Monographs by Carl Sauer

Upper Illinois Valley, 1916a	Anonymous, 1917
Starved Rock State Park, 1918c	J. James, 1919
Geography of the Ozark Highland, 1920b	Unstead, 1922
	Violette, 1922
Morphology of Landscape, 1925	Dryer, 1926
Geography of the Pennyroyal, 1927a	Ekblaw, 1928
	Finch, 1931
	McMurry and Newman, 1928
Land Forms in the Peninsula Range, 1929c	Anonymous, 1930b
Prehistoric Settlements of Sonora, 1931b	Hoover, 1932
Aztatlán, 1932a	Anonymous, 1930a
Distribution of Aboriginal Tribes, 1934a	Spier, 1937

Aboriginal Population of Northwest Mexico, 1935a	Long, 1937
	Redfield, 1936
Man in Nature, 1939, 1975	Lovell, 2003
	Tucker, 1976
Colima of New Spain, 1948a	Aiton, 1949
Agricultural Origins and Dispersals, 1952a	Carter, 1954
	Cutler, 1953
	E.G.R.T., 1953
	Mangelsdorf, 1953
	West, 1953
Man's Role in Changing the Face of the Earth (co-collaborator), 1956a	Abel, 1960
	Cottrell, 1957
	Emory, 1957
	Glacken, 1956
	Green, 1957
	P. James, 1957
	Murphy and Hitchcock, 1955
	Nicholson, 1957
	Stamp, 1957
	Swanson, 1957
	Wagner, 1957
	Watson and Kleindienst, 1957
	M. Williams, 1987
Land and Life (Leighly, ed.), 1963a	Beals, 1965
	Carter, 1964
	Cornwall, 1965
	Denevan, 1964
	Evans, 1964
	Hart, 1964
	W. Jackson, 1964
	Knight, 1964
	Spate, 1965
	Stamp, 1966
Plant and Animal Exchanges, 1963d	Hindle, 1963
The Early Spanish Main, 1966a	Austin, 1992
	Blount, 1969
	Boxer, 1967
	Clark, 1967
	Crone, 1967
	Denevan, 1967
	Dodge, 1967

	Friede, 1971
	Hogan, 1966a, 1966b
	Madariaga, 1967
	McGann, 1968
	Melendez, 1967
	Navarro Garcia, 1967
	Niddrie, 1966
	Nowell, 1968
	Parry, 1966
	Penrose, 1968
	Rosenblat, 1976
	Sheridan, 1967
	Solnick, 1969
	Watts, 1967
Northern Mists, 1968a	Evans, 1969
	Pennington, 1969
	Quinn, 1969
	Washburn, 1968
Sixteenth Century North America, 1971b	Andrews, 1972
	Benninghoff, 1973
	Coker, 1972
	Detweiler, 1972
	Griffith, 1972
	Hawke, 1972
	M. Jackson, 1972
	Quinn, 1973
	Rowe, 1972
	Vorsey, 1973
	Washburn, 1973
	G. Williams, 1973
Seventeenth Century North America, 1980	McManis, 1981
	Meinig, 1981
	Nasatir, 1982
Selected Essays (Callahan, ed.), 1981a	Martin, 1984
Andean Reflections (West, ed.), 1982	Webb, 1984

Over five hundred items are listed, about 70 percent of which appeared after Sauer's death in 1975, which is indicative of his continuing importance. A total of 228 have "Sauer" in the title. I am responsible for decisions as to what is included and not included. Suggestions for additions have been received from several people, especially Kent Mathewson, and also William Speth and Daniel Gade. Undoubtedly, there are additional important commentaries that I am not aware of.

The size of this bibliography, the number and diversity of the authors, and the stature of many of the authors attest to Sauer's reputation, influence, and varied contributions to geography and other disciplines. Sauer may well be the most written about geographer of the twentieth century, with the only other contender being William Morris Davis (Martin, 2003b:554). As Robert Hoffpauir wrote, "It is safe to say that there has been no American geographer who has been so thoroughly dissected and analyzed as Carl Sauer" (2000:150).

Hopefully, this bibliography will be helpful to scholars of Sauer, as well as to others interested in this remarkable geographer and in the history of geography.

BIBLIOGRAPHY OF COMMENTARIES ON THE LIFE AND WORK OF CARL SAUER

Sauer's colleagues are designated in the list below as follows:

*Ph.D. under Carl Sauer.

†Ph.D. during Sauer's tenure, 1923–1975, in geography at Berkeley, but not under Sauer.

‡Regular faculty member in geography at Berkeley during Sauer's tenure, 1923–1975.

Abell, Dana L. 1960. Review of *Man's Role in Changing the Face of the Earth*, William L. Thomas Jr., ed. *Ecology* 41:400–401.

Aiton, Arthur S. 1949. Review of *Colima of New Spain in the Sixteenth Century*, by Carl O. Sauer. *Hispanic American Historical Review* 29:243–244.

Alpert, Barry. 1977. "Introduction." In *The Poet in the Imaginary Museum*, by Donald Davie, ed. B. Alpert, ix–xxi, esp. xviii–xix. Carcanet New Press, Manchester, U.K.

Anderson, Edgar. 1971 [1952]. *Plants and Man*, 3rd ed., esp. 142–144. Univ. of California Press, Berkeley.

Anderson, Kay. 1997. "A Walk on the Wild Side: A Critical Geography of Domestication." *Progress in Human Geography* 21:463–485, esp. 468–470.

Anderson, Kay, Mona Domosh, Steve Pile, and Nigel Thrift, eds. 2003. *Handbook of Cultural Geography*, esp. 20, 165, 171, 186, 187, 208, 238, 239, 270, 271, 455–456, 513. Sage, Thousand Oaks, Calif.

Andrews, K. R. 1972. Review of *Sixteenth Century North America: The Land and the People as Seen by the Europeans*, by Carl O. Sauer. *American Historical Review* 77:826–827.

Anonymous. 1916. "Addition of Geography to the Curriculum of the University of Michigan." *Geographical Review* 1:306–307.

Anonymous. 1917. Review of *Geography of the Upper Illinois Valley and History of Development*, by Carl O. Sauer. *Geographical Review* 4:497.

Anonymous. 1920. "The Reform of Political Divisions on a Geographical Basis." *Geographical Review* 10:185.

Anonymous. 1930a. Review of *Aztatlán: Prehistoric Mexican Frontier on the Pacific Coast*, by Carl O. Sauer. *Science* 72 (1868):403.

Anonymous. 1930b. Review of *Land Forms in the Peninsular Range of California as Developed about Warner's Hot Springs and Mesa Grande*, by Carl O. Sauer. *Geographical Review* 20:529–530.

Anonymous. 1935. "Honorary Corresponding Members." *Geographical Review* 25:486–487.

Anonymous. 1972. "A Taste for Human Geography." *California Monthly* (October).

Anonymous. 1974. "AAG Special Award in Recognition of a Career of Unusual Distinction: Carl Ortwin Sauer." *The Itinerant Geographer*, p. 2. Department of Geography, University of California, Berkeley. Originally in the "Association of American Geographers Newsletter" 9 (6) (1974): 1.

Anonymous. 1975a. "Dr. Carl O. Sauer Dies: Dean of Geographers, 85." *New York Times*, July 21.

Anonymous. 1975b. "Milestones—Died: Carl O. Sauer, 85, American Geographer." *Time*, August 4, 53.

Anonymous. 1975c. "Deaths—UC Geographer Carl Sauer." *San Francisco Examiner and Chronicle*, July 20, 5.

Anonymous. 1975d. "Carl Ortwin Sauer." *Geographical Journal* 141:516–517.

Anonymous. 1975e. "Carl Sauer: Human Student of the Earth." *California Monthly* (October), 23.

Anonymous. 1976a. "Sauer, Carl Ortwin." In *Who Was Who in America* 6:359. Marquis, Chicago.

Anonymous. 1976b. "Carl Ortwin Sauer, 1889–1975." *Contemporary Authors* 61–64:483–484.

Apffel-Marglin, Frédérique. 1996. "Introduction: Rationality and the World." In *Decolonizing Knowledge: From Development to Dialogue*, ed. F. Apffel-Marglin and Stephen A. Marglin, 1–40, esp. 24–25. Clarendon Press, Oxford.

Aschmann, Homer.* 1987. "Carl Sauer, A Self-Directed Career." In Kenzer, 1987a:137–143.

———. 1999. Letter to William W. Speth, November 18, 1976. In Speth, 1999a:212–213.

Austin, Daniel F. 1992. Review of *The Early Spanish Main*, by Carl O. Sauer. *Bulletin of the Torrey Botanical Club* 119:343.

Baker, Alan R. H. 2003. *Geography and History: Bridging the Divide*, esp. 37–38, 49, 74, 89–91, 110–111, 130, 164–166. Cambridge Univ. Press, Cambridge.

Ballas, Donald J. 1975. "Carl Ortwin Sauer's Life and Work." *Places* 11 (3): 24–33.

Barton, Thomas F., and Pradyumna P. Karan. 1992. "Carl Ortwin Sauer, December 24, 1889–July 18, 1975." In *Leaders in American Geography*, by T. F. Barton and Pradyumna P. Karan, 1:74–77. New Mexico Geographical Society, Mesilla, N.M.

Bassin, Mark, and Vincent Berdoulay. 2004. "Historical Geography: Locating Time in the Spaces of Modernity." In *Human Geography: A History of the 21st Century*, ed. Georges Benko and Ulf Strohmayer, 64–82, esp. 68–69. Edward Arnold, London.

Beals, Ralph. 1965. Review of *Land and Life: A Selection from the Writings of Carl Ortwin Sauer*, ed. John Leighly. *American Anthropologist* 67:205–206.

———. 1999. Letter to William W. Speth, February 22, 1977. In Speth, 1999a:213–217.

Belil, Mireia, and Isabel Clos. 1983. "Notes a l'entorn del pensament de Carl O. Sauer (1889–1975)." *Documents d'Analisi Geografica* 2:177–188.

Benninghoff, William S. 1973. Review of *Sixteenth Century North America: The Land and the People as Seen by the Europeans*, by Carl O. Sauer. *Ecology* 54:717–718.

Blaut, James M. 1993. "Mind and Matter in Cultural Geography." In *Culture, Form, and Place: Essays in Cultural and Historical Geography*, ed. Kent Mathewson, 345–356, esp. 345–346, 348. Geoscience and Man, vol. 32. Geoscience Publications, Department of Geography and Anthropology, Louisiana State University, Baton Rouge.

Blount, Stanley F. 1969. Review of *The Early Spanish Main*, by Carl O. Sauer. *Terrae Incognitae* 1:78.

Bolsi, Alfredo S. 1971. "Carl Ortwin Sauer." *Serie Traducciones* 3:5–8. Instituto de Geografía, Universidad Nacional del Nordeste, Resistencia, Argentina.

Bowden, Martyn J. 1980. "The Cognitive Renaissance in American Geography: The Intellectual History of a Movement." In *Les Écoles Geographiques*, ed. Jósef Babicz, 65–70. Polish Scientific Publishers, Warsaw.

———. 1988. Review of *Carl O. Sauer: A Tribute*, ed. Martin S. Kenzer. *Isis* 79:741–743.

Bowen, Dawn S. 1996. "Carl Sauer, Field Exploration, and the Development of American Geographic Thought." *Southeastern Geographer* 36:176–191.

Boxer, C. R. 1967. Review of *The Early Spanish Main*, by Carl Ortwin Sauer. *Man* 2:484.

Brand, Stewart. 1976. Introduction to the reprint of "Theme of Plant and Animal Destruction in Economic History," by Carl Sauer. *Co-Evolution Quarterly* 10:53.

Brookfield, Harold C. 1964. "Questions on the Human Frontiers of Geography." *Economic Geography* 40:283–303.

Brown, Judith K. 1970. "Subsistence Variables: A Comparison of Textor and Sauer." *Ethnology* 9:160–164.

Brown, Lawrence A., and E. G. Moore. 1969. "Diffusion Research in Geography." *Progress in Geography: International Reviews of Current Research*, ed. Christopher Board et al. 1:120–157, esp. 121–122. St. Martin's Press, New York.

Brown, Scott S., and Kent Mathewson. 1999. "Sauer's Descent? Or Berkeley Roots Forever." *Yearbook of the Association of Pacific Coast Geographers* 61:137–157.

Bruman, Henry J.* 1976. "In Memory of Carl Sauer." *Historical Geography Newsletter* 6 (1): 5–6.

———. 1987. "Carl Sauer in Midcareer: A Personal View by One of His Students." In Kenzer, 1987a:125–136.

———. 1996. "Recollections of Carl Sauer and Research in Latin America." *Geographical Review* 86:370–376.

Brunn, Stanley D. 2004. "The Midwesternization of Southern Geography Departments." In *The Role of the South in the Making of American Geography*, ed. James O. Wheeler and S. D. Brunn, 3–17, esp. 9. Bellwether, Columbia, Md.

Bulliet, Richard W. 2005. *Hunters, Herders, and Hamburgers: The Past and Future of Human-Animal Relationships*, esp. 90–91. Columbia Univ. Press, New York.

Butzer, Karl W. 1989. "Hartshorne, Hettner, and *The Nature of Geography*." In *Reflections on Richard Hartshorne's The Nature of Geography*, ed. J. N. Entrikin and S. D. Brunn, 35–52, esp. 35–36, 41–42. Association of American Geographers, Washington, D.C.

Callahan, Bob. 1976. "A Carl Sauer Checklist" [brief comments on Sauer's books]. *Co-Evolution Quarterly* 10:52–53.

———. 1977a. "Carl Sauer: (The Migrations)." *New World Journal* 1 (2–3):94–111, esp. 106–110.

———. 1977b. Introduction to "The Correspondences: Charles Olson and Carl Sauer." *New World Journal* 1 (4): 136–139, 168.

———. 1978. "On Turtle Island," an interview by Michael Helm. *City Miner* (Berkeley), 3 (3): 24–26, 33–39, esp. 25–26.

———. 1980. "For Parents and Teachers: Notes for a New Edition." In *Man in Nature: America before the Days of the White Man*, by Carl O. Sauer, i–ii. Turtle Island Foundation, Berkeley.

———. 1981. "Introduction." In *Selected Essays, 1963–1975*, by Carl O. Sauer, ed. Bob Callahan, xi–xv. Turtle Island Foundation, Berkeley.

Carter, George F.* 1954. Review of *Agricultural Origins and Dispersals*, by Carl O. Sauer. *American Journal of Archaeology* 58:260–262.

———. 1964. Review of *Land and Life: A Selection from the Writings of Carl Ortwin Sauer*, ed. John Leighly. *Quarterly Review of Biology* 39:118.

———. 2003. "Carl Sauer's Legacy: A Personal View." In Mathewson and Kenzer, 2003: 339–349.

Chaunu, Pierre. 1961. "Une histoire hispano-américaniste pilote: En marge de l'oeuvre de l'École de Berkeley." *Revue Historique* 124:339–368, esp. 353–355.

Checkovich, Alex, and Jeremy Vetter. 2001. Review of *How It Came To Be: Carl O. Sauer, Franz Boas and the Meanings of Anthropogeography*, by William W. Speth. *Journal of the History of the Behavioral Sciences* 37:404–405.

Clark, Andrew H.* 1954. "Historical Geography." In *American Geography: Inventory* *and Prospect,* ed. Preston E. James and Clarence F. Jones, 70–105, esp. 85–89. Syracuse Univ. Press, Syracuse.

———. 1967. Review of *The Early Spanish Main,* by Carl Ortwin Sauer. *Professional Geographer* 19:157.

Clark, Tom. 2002. *Edward Dorn: A World of Difference,* esp. 88–93, 250–251, 386. North Atlantic Books, Berkeley.

Claval, Paul, and J. Nicholas Entrikin. 2004. "Cultural Geography: Place and Landscape between Continuity and Change." In *Human Geography: A History of the 21st Century,* ed. Georges Benko and Ulf Strohmayer, 25–46, esp. 27, 29, 37–38. Edward Arnold, London.

Cloke, Paul J., Ian Cook, Philip Crang, Mark A. Goodwin, Joe M. Painter, and Chris Philo, eds. 2004. *Practising Human Geography,* esp. 3, 5–7, 8–11, 17–18, 123, 182, 322. Sage, London.

Coker, William S. 1972. Review of *Sixteenth Century North America: The Land and the People as Seen by the Europeans,* by Carl O. Sauer. *Hispanic American Historical Review* 52:468–469.

Constance, Lincoln. 1978. "Berkeley and the Latin American Connection," esp. 10–12, 27–28, 34. Twentieth Bernard Moses Memorial Lecture. Regents, University of California, Berkeley.

Conzen, Michael P. 1993. "The Historical Impulse in Geographical Writing about the United States, 1850–1990." In *A Scholar's Guide to Geographical Writing on the American and Canadian Past,* ed. M. P. Conzen, T. A. Rumney, and G. Wynn, 3–90, esp. 25–33; also 173–176. Univ. of Chicago Press, Chicago.

Cornwall, L. W. 1965. Review of *Land and Life: A Selection from the Writings of Carl Ortwin Sauer. Man* 65:168–169.

Cosgrove, Denis. 1984. *Social Formation and Symbolic Landscape,* esp. 16–17, 29–33. Croom Helm, London.

———. 1985. "Prospect, Perspective and the Evolution of the Landscape Idea." *Transactions, Institute of British Geographers,* New Series, 10:45–62, esp. 57.

———. 1993. "Commentary on 'The Reinvention of Cultural Geography' by Price and Lewis." *Annals of the Association of American Geographers* 83:515–517.

Cosgrove, Denis, and Peter Jackson. 1987. "New Directions in Cultural Geography." *Area* 19:95–101, esp. 95–96.

Cottrell, W. F. 1957. Review of *Man's Role in Changing the Face of the Earth,* William L. Thomas Jr., ed. *American Sociological Review* 22:115–116.

Crang, Mike. 1998. *Cultural Geography,* esp. 15–23. Routledge, London.

Cresswell, Tim. 2003. "Landscape and the Obliteration of Practice." In *Handbook of Cultural Geography,* ed. Kay Anderson, Mona Domosh, Steve Pile, and Nigel Thrift, 269–281, esp. 270–271. Sage, Thousand Oaks, Calif.

Crone, G. R. 1967. Review of *The Early Spanish Main,* by Carl O. Sauer. *Economic Geography* 43:277–278.

Crosby, Alfred W. 1995. "The Past and Present of Environmental History." *American Historical Review* 100:1177–1189, esp. 1183–1184, 1186.

Cutler, Hugh C. 1953. Review of *Agricultural Origins and Dispersals*, by Carl O. Sauer. *American Anthropologist* 55:434–436.

———. 1968. "Origins of Agriculture in the Americas." *Latin American Research Review* 3:3–21, esp. 3, 6–7.

Daniels, Stephen. 1991. "Landscape." In *Modern Geography: An Encyclopedic Survey*, ed. Gary S. Dunbar, 101–103, esp. 101–102. Garland, New York.

Davie, Donald. 1977. *The Poet in the Imaginary Museum*, ed. Barry Alpert, esp. 166–169. Carcanet New Press, Manchester, U.K.

Davies, Lawrence. 1957. "Professor of Human Geography, Dr. Sauer Maps People First, Land Is Merely a Setting." *Saturday Review* 40 (May 4): 64.

Demerrit, David. 1994. "The Nature of Metaphors in Cultural Geography and Environmental History." *Progress in Human Geography* 18:163–185, esp. 163–167.

Denevan, William M.† 1964. Review of *Land and Life: A Selection from the Writings of Carl Ortwin Sauer*, ed. John Leighly. *Hispanic American Historical Review* 44: 595–597.

———. 1967. Review of *The Early Spanish Main*, by C. O. Sauer. *Pacific Viewpoint* 8:208–209.

———. 1987. Introduction to "Observations on Trade and Gold in the Early Spanish Main." In Kenzer, 1987a, 164–165.

———. 1996. "Carl Sauer and Native American Population Size." *Geographical Review* 86:385–397. Also in Mathewson and Kenzer, 2003:157–172.

———. 2009a. "Preface." Herein.

———. 2009b. "Introduction" to Section II, Early Efforts. Chapter 5 herein.

Detweiler, Robert. 1972. Review of *Sixteenth Century North America: The Land and the People as Seen by the Europeans*, by Carl O. Sauer. *American Quarterly* 24:313–314.

Dicken, Samuel N.* 1959. "Baja California, 1926." *The Itinerant Geographer*, p. 1. Department of Geography, University of California, Berkeley.

———. 1988. *The Education of a Hillbilly: Sixty Years in Six Colleges*, esp. 30–43. Department of Geography, University of Oregon, Eugene.

Dickinson, Robert E. 1970. *Regional Ecology: The Study of Man's Environment*, esp. 26–27, 83, 91–92, 95, 180–181. John Wiley and Sons, New York.

———. 1976. "Carl Ortwin Sauer, 1889–1975." In *Regional Concept: The Anglo-American Leaders*, by R. E. Dickinson, 314–326. Routledge and Kegan Paul, London.

Dickson, D. Bruce. 1982. "Anthropological Utopias and Geographical Epidemics: Competing Models of Social Change and the Problem of the Origins of Agriculture." In *The Transfer and Transformation of Ideas and Material Culture*, ed. Peter J. Hugill and D. Bruce Dickson, 45–72, esp. 47–53. Texas A&M Univ. Press, College Station.

———. 2003. "Origins and Dispersals: Carl Sauer's Impact on Anthropological Explanations of the Development of Food Production." In Mathewson and Kenzer, 2003:117–134.

Dobyns, Henry F. 1966. "Estimating Aboriginal American Population: An Appraisal of Techniques with a New Hemispheric Estimate." *Current Anthropology* 7:395–449, esp. 398–399, 403–404.

Dodge, Ernest S. 1967. Review of *The Early Spanish Main*, by Carl O. Sauer. *American Historical Review* 72:749.

Doolittle, William E. 1988. *Pre-Hispanic Occupance in the Valley of Sonora, Mexico: Archaeological Confirmation of Early Spanish Reports.* Anthropological Papers, No. 48, esp. pp. 52–53. Univ. of Arizona Press, Tucson.

Donkin, R. A. 1997. "A 'Servant of Two Masters'?" *Journal of Historical Geography* 23:247–266, esp. 248–250, 264–265.

Dorn, Edward. 1980. "An Interview with Barry Alpert." In *Edward Dorn: Interviews*, ed. Donald Allen, 7–35, esp. 20–22. Four Seasons Foundation, Bolinas, Calif.

Dow, Maynard W. 1970–1988. "Mr. Sauer: Mentor and Disciples." In *Geographers on Film*. National Gallery of the Spoken Word, Michigan State University, East Lansing.

Dryer, Charles R. 1926. Review of *The Morphology of Landscape*, by Carl O. Sauer. *Geographical Review* 16:348–350.

Dunbar, Gary S. 1981. "Geography in the University of California (Berkeley and Los Angeles), 1868–1941," esp. 5–8. DeVoss and Co., Marina del Rey, Calif.

———, ed. 1991. *Modern Geography: An Encyclopedia Survey*, esp. 101–102, 160. Garland, New York.

———. 1996. *The History of Geography: Collected Essays*, esp. 69–71. Dodge-Graphic Press, Utica, N.Y.

Duncan, James S. 1977. "The Superorganic in American Cultural Geography." Ph.D. diss., Syracuse University, Syracuse.

———. 1980. "The Superorganic in American Cultural Geography." *Annals of the Association of American Geographers* 70:181–198.

———. 1981. "Comment in Reply" [to Miles Richardson and to Richard Symanski on "The Superorganic in American Cultural Geography"]. *Annals of the Association of American Geographers* 71:289–291.

———. 1993. "Commentary on 'The Reinvention of Cultural Geography' by Price and Lewis." *Annals of the Association of American Geographers* 83:517–519.

———. 1998. "Classics in Human Geography Revisited ('The Superorganic in American Cultural Geography'): Author's Response." *Progress in Human Geography* 22:571–573.

———. 2000. "Berkeley School." In *The Dictionary of Human Geography*, 4th ed., ed. R. J. Johnston, Derek Gregory, Geraldine Pratt, and Michael Watts, 45–46. Blackwell, Malden, Mass.

Effland, Anne B. W., and William R. Effland. 1992. "Soil Geomorphology Studies in the U.S. Soil Survey Program." *Agricultual History* 66:189–212, esp. 190–195.

E. G. R. T. 1953. Review of *Agricultural Origins and Dispersals*, by Carl O. Sauer. *Geographical Journal* 119:109–110.

Ekblaw, W. Elmer. 1928. Review of *Kentucky Geological Survey, Series VI* (includes *The Geography of the Pennyroyal*, by Carl Ortwin Sauer). *Economic Geography* 4:109–110.

Emory, S. T. 1957. Review of *Man's Role in Changing the Face of the Earth*, William L. Thomas Jr., ed. *Social Forces* 35:377–378.

Entrikin, J. Nicholas. 1982. "Sauer on Social Science: Howard Odum's Institute for Research in Social Science." *History of Geography Newsletter* 2:35–38.

———. 1984. "Carl O. Sauer, Philosopher in Spite of Himself." *Geographical Review* 74:387–408.

———. 1987. "Archival Research: Carl O. Sauer and Chorology." *Canadian Geographer* 31:77–79.

Evans, E. Estyn. 1964. Review of *Land and Life: A Selection from the Writings of Carl Ortwin Sauer. Geographical Review* 54:596–597.

———. 1969. Review of *Northern Mists*, by Carl O. Sauer. *Geographical Review* 59:297–298.

Finch, V. C. 1931. "The Influence of Geology and Physiography upon the Industry, Commerce, and Life of a People, as Described by Carl Ortwin Sauer and Others": Review of *The Geography of the Pennyroyal*. In *Methods in Social Science: A Case Book*, ed. Stuart A. Rice, 237–241. Univ. of Chicago Press, Chicago.

Ford, O. J. 1973–1974. "Charles Olson and Carl Sauer: Towards a Methodology of Knowing." *Boundary 2*, 2:145–150.

Fosberg, F. R., D. J. Blumenstock, P. W. Richards, P. H. Nesbitt, M. H. Sachet, and C. G. G. J. van Steenis. 1958. "Discussion," of Carl O. Sauer, "Man in the Ecology of Tropical America." *Proceedings of the Ninth Pacific Science Congress, 1957*, 20:110. Bangkok.

Freeman, T. W. 1961. *A Hundred Years of Geography*, esp. 167–168. Aldine, Chicago.

———. 1980. Review of *Carl Sauer's Fieldwork in Latin America*, by Robert G. West. *Geographical Journal* 146:447.

Friede, Juan. 1971. Review of *The Early Spanish Main*, by Carl Ortwin Sauer. *Ethnohistory* 18:67–68.

Fuson, Robert H. 1969. *A Geography of Geography*, esp. 105. Brown, Dubuque, Iowa.

Gade, Daniel W. 1976. "L'optique culturelle dans la géographie américaine." *Annales de Géographie (Paris)* 472:672–693.

———. 1988. "Cultural Geography as a Research Agenda for Peru." *Yearbook, Conference of Latin Americanist Geographers* 14:31–37, esp. 31–32, 35–36.

———. 1989. "Cultural Geography, Its Idiosyncrasies and Possibilities." In *Applied Geography*, ed. Martin S. Kenzer, 135–150, esp. 135–138. Kluwer, Dordrecht.

———. 1999a. "Reflections and Trajectories." In *Nature and Culture in the Andes*, by D. W. Gade, 3–30, esp. 16–21. Univ. of Wisconsin Press, Madison.

———. 1999b. "Carl Sauer and the Andean Nexus in New World Crop Diversity." In *Nature and Culture in the Andes*, by D. W. Gade, 184–213. Univ. of Wisconsin Press, Madison.

———. 2002. "North American Reflections on Latin Americanist Geography." *Yearbook of the Conference of Latin Americanist Geography* 27:1–44, esp. 5, 8–10.

———. 2003–2004. "Diffusion as a Theme in Cultural-Historical Geography." *Pre-Columbiana: A Journal of Long-Distance Contacts* 3 (1–3): 19–39, esp. 22–26, 36.

———. 2004a. "Espaço, tempo e cultura em convergência: Perspectivas americanas sobre a difusão histórico-cultural." *Espaço e Cultura* (Rio de Janeiro) 17–18:129–147, esp. 131–133, 141–142.

———. 2004b. Review of *Geography Inside Out*, by Richard Symanski [and Korski]. *Historical Geography* 32:181–185, esp. 182.

———. 2005. Review of *Culture, Land, and Legacy: Perspectives on Carl O. Sauer and Berkeley School Geography*, ed. Kent Mathewson and Martin Kenzer. *Annals of the Association of American Geographers* 95:481–483.

———. 2009a. "Introduction" to Section V, Cultivated Plants. Chapter 18 herein.

———. 2009b. "Thoughts on Bibliographic Citations to and by Carl Sauer." Chapter 3 herein.

———. 2008. "Irreverent Musings on the Dissertation in Latin Americanist Geography." In *Ethnographic Research in Latin America: Essays Honoring William V. Davidson*, ed. Peter Herlihy, Kent Mathewson, and C. S. Revels, 29–59. Geoscience and Man, vol. 40, Geoscience Publications, Department of Geography and Anthropology, Louisiana State University, Baton Rouge.

Gale, Fay. 2003. "Textbooks that Moved Generations: *Readings in Cultural Geography*," Philip L. Wagner and Marvin W. Mikesell, editors. *Progress in Human Geography* 27:233–236.

Germundsson, Tomas. 1996. "Människan, kulturen och landskapet: Speglingar av några teman hos Carl Sauer." *Svensk Geografisk Årsbok* (Lund) 72:84–92.

Glacken, Clarence J.*‡ 1956. "Man and the Earth." Review of the International Symposium on "Man's Role in Changing the Face of the Earth." *Landscape* 5 (6): 27–29.

———. 1983. "A Late Arrival in Academia." In *The Practice of Geography*, ed. Anne Buttimer, 20–34, esp. 28–29. Longman, London.

Glick, Thomas F. 1988. "History and Philosophy of Geography." *Progress in Human Geography* 12:441–450, esp. 446–447.

Gould, Peter. 1979. "Geography 1957–1977: The Augean Period." *Annals of the Association of American Geographers* 69:139–151, esp. 140.

Green, F. H. W. 1957. Review of *Man's Role in Changing the Face of the Earth*, William L. Thomas Jr., ed. *Geographical Journal* 123:111–112.

Gregory, Derek. 1978. *Ideology, Science and Human Geography,* esp. 28–29. St. Martin's Press, New York.
———. 1994. *Geographical Imaginations,* esp. 286–287. Blackwell, Oxford.
Griffith, William J. 1972. Review of *Sixteenth Century North America: The Land and the People as Seen by the Europeans,* by Carl O. Sauer. *American Anthropologist* 74:862–863.
Guelke, Leonard. 1982. *Historical Understanding in Geography: An Idealist Approach,* esp. 6–11. Cambridge Univ. Press, Cambridge.
———. 1997. "The Relations between Geography and History Reconsidered." *History and Theory* 36:216–234, esp. 217–220, 223.
Haggett, Peter. 1990. *The Geographer's Art,* esp. 25–28, 179–180. Blackwell, Oxford.
———. 1992. "Sauer's 'Origins and Dispersals': Its Implications for the Geography of Disease." *Transactions, Institute of British Geographers,* New Series, 17:387–398.
Harlan, Jack R. 1975. *Crops and Man,* esp. 46–47. American Society of Agronomy, Madison, Wisc.
———. 1995. *The Living Fields: Our Agricultural Heritage,* esp. 21–22. Cambridge Univ. Press, Cambridge.
Harmond, Richard, and Thomas J. Curran. 1999. "Sauer, Carl Ortwin." In *American National Biography,* ed. John A. Garraty and Mark C. Carnes, 302–304. Oxford Univ. Press, New York.
Harris, David R.† 1981. "Breaking Ground: Agricultural Origins and Archaeological Explanations." *Bulletin of the Institute of Archaeology* 18:1–20, esp. 12–14.
———. 1989. Review of *Carl O. Sauer: A Tribute,* ed. Martin S. Kenzer. *Geographical Journal* 155:422–423.
———. 1990. "Vavilov's Concept of Centres of Origin of Cultural Plants: Its Genesis and Its Influence on the Study of Agricultural Origins." *Biological Journal of the Linnean Society* 39:7–16, esp. 11–13.
———. 2002. "The Farther Reaches of Human Time: Retrospect on Carl Sauer as Prehistorian." *Geographical Review* 92:526–544.
Hart, John Fraser. 1964. Review of *Land and Life: A Selection from the Writings of Carl Ortwin Sauer,* ed. John Leighly. *Annals of the Association of American Geographers* 54:612–614.
———. 1983a. "Foreword." In *Evolution of Geographic Thought in America: A Kentucky Root,* ed. Wilford A. Bladen and Pradyumna P. Karan, ix–xi. Kendall/Hunt, Dubuque, Iowa.
———. 1983b. "Hart on Sauer." In *The Evolution of Geographic Thought in America: A Kentucky Root,* ed. Wilford A. Bladen and Pradyumna P. Karan, 113–114. Kendall/Hunt, Dubuque, Iowa.
Hartshorne, Richard. 1939. *The Nature of Geography: A Critical Survey of Current Thought in the Light of the Past,* esp. 155–156, 176–184. Association of American Geographers, Lancaster, Pa.

———. 1971. "Introduction." In *On Geography: Selected Writings of Preston E. James*, ed. Donald W. Meinig, ix–xiv, esp. xiii–xiv. Syracuse Univ. Press, Syracuse.

———. 1979. "Notes toward a Bibliobiography of 'The Nature of Geography.'" *Annals of the Association of American Geographers* 69:63–76, esp. 65, 67, 69–70.

Harvey, David. 1969. *Explanation in Geography*, esp. 111, 114–115, 135, 291, 408, 413. Edward Arnold, London.

Hawke, David F. 1972. Review of *Sixteenth Century North America: The Land and the People as Seen by the Europeans*, by Carl O. Sauer. *Journal of American History* 59:122–123.

Head, Lesley. 2004. "Landscape and Culture." In *Unifying Geography: Common Heritage, Shared Future*, ed. John A. Matthews and David T. Herbert, 240–255, esp. 240–244. Routledge, London.

Heiser, Charles B., Jr. 1985. *Of Plants and People*, esp. 211–215. Univ. of Oklahoma Press, Norman.

Henige, David. 1998. *Numbers from Nowhere: The American Indian Contact Population Debate*, esp. 27, 234. Univ. of Oklahoma Press, Norman.

Herbert, David T., and John A. Matthews. 2004. "Geography: Roots and Continuities." In *Unifying Geography: Common Heritage, Shared Future*, ed. John A. Matthews and David T. Herbert, 3–18, esp. 7–8. Routledge, London.

Hess, Westher. 1957. "Carl Sauer." *The Itinerant Geographer*, p. 1. Department of Geography, University of California, Berkeley.

Hewes, Leslie.* 1983. "Carl Sauer: A Personal View." *Journal of Geography* 82:140–147.

Hindle, Edward. 1963. Review of *Plant and Animal Exchanges between the Old and New Worlds*, by Carl O. Sauer. *Geographical Journal* 129:549–550.

Hoffpauir, Robert. 2000. Review of *How It Came To Be: Carl O. Sauer, Franz Boas and the Meanings of Anthropogeography*, by William W. Speth. *Yearbook of the Association of Pacific Coast Geographers* 62:147–151.

Hogan, William. 1966a. "The Author" [Carl O. Sauer, *The Early Spanish Main*]. *Saturday Review*, July 2, p. 25.

———. 1966b. "Genocide and the Columbus Myth." Review of *The Early Spanish Main*, by Carl O. Sauer. *San Francisco Chronicle*, July 17.

Holt-Jensen, Arild. 1999. *Geography: History and Concepts*, 3rd ed., esp. 52. Sage, London.

Hooson, David.‡ 1981. "Carl O. Sauer." In *The Origins of Academic Geography in the United States*, ed. Brian Blouet, 165–174. Archon, Hamden, Conn.

Hoover, J. W. 1932. Review of *Prehistoric Settlements of Sonora, with Special Reference to Cerros de Trincheras*, by Carl O. Sauer and Donald Brand. *Geographical Review* 22:510–511.

Horowitz, Helen L. 1997. "J. B. Jackson and the Discovery of the American Landscape." In *Landscape in Sight: Looking at America*, by John B. Jackson, ed. H. L. Horowitz, ix–xxxi, esp. xxvi–xxvii. Yale Univ. Press, New Haven, Conn.

Hugill, Peter J., and Kenneth E. Foote. 1994. "Re-Reading Cultural Geography." In *Re-Reading Cultural Geography*, ed. Kenneth E. Foote et al., 9–23. Univ. of Texas Press, Austin.

Humlum, Johannes. 1975. "Carl Ortwin Sauer." *Kulturgeografi* (Denmark) 125:236–237.

Hutchinson, George. 1982. "The Pleistocene in the Projective: Some of Olson's Origins." *American Literature* 54:81–96, esp. 84.

Isaac, Erich. 1970. *Geography of Domestication*, esp. 9–11, 27–28. Prentice-Hall, Englewood Cliffs, N.J.

Jackson, Melvin H. 1972. Review of *Sixteenth Century North America: The Land and the People as Seen by the Europeans*, by Carl O. Sauer. *William and Mary Quarterly*, Series 3, 29:660.

Jackson, Peter. 1989. *Maps of Meaning: An Introduction to Cultural Geography*, esp. 9–24. Unwin Hyman, London.

———. 1993. "Commentary on 'The Reinvention of Cultural Geography' by Price and Lewis: Berkeley and Beyond: Broadening the Horizons of Cultural Geography." *Annals of the Association of American Geographers* 83:519–520.

Jackson, W. A. Douglas. 1964. Review of *Land and Life: A Selection from the Writings of Carl Ortwin Sauer. Science* 143 (3609):945.

Jacobs, Jane M. 2003. "Introduction: After Empire?" In *Handbook of Cultural Geography*, ed. Kay Anderson et al., 345–353, esp. 348–350. Sage, Thousand Oaks, Calif.

Jacobs, Wilbur R. 1978. "The Great Despoliation: Environmental Themes in American Frontier History." *Pacific Historical Review* 47:1–26, esp. 21–22.

James, J. A. 1919. Review of *Starved Rock State Park and Its Environs*, by Carl O. Sauer, Gilbert H. Cady, and Henry C. Cowles. *Mississippi Valley Historical Review* 6:137–138.

James, Preston E. 1957. Review of *Man's Role in Changing the Face of the Earth*, William L. Thomas Jr., ed. *Economic Geography* 33:267–274.

———. 1980. "The Development of Professional Geography in the United States (1885–1940)." In *Les Écoles Geographiques*, ed. Jósef Babicz, 50–63, esp. 54–56, 67. Polish Scientific Publishers, Warsaw.

———. 1983. "The University of Michigan Field Station at Mill Springs, Kentucky, and Field Studies in American Geography." In *The Evolution of Geographic Thought in America: A Kentucky Root*, ed. Wilford A. Bladen and Pradyumna P. Karan, 59–83, esp. 61–71. Kendall/Hunt, Dubuque, Iowa.

James, Preston E., and Geoffrey J. Martin. 1972. *All Possible Worlds: A History of Geographical Ideas*, esp. 320–324. Wiley, New York.

———. 1978. *The Association of American Geographers: The First Seventy-Five Years, 1904–1979*, esp. 71–73. Association of American Geographers, Washington, D.C.

Janke, James. 1975. "Some Thoughts on the Impact of Sauer's 'Morphology of Landscape.'" *Kansas Geographer* 10:21–27.

Jarrell, Randall. 2000. *Raymond F. Dasmann: A Life in Conservation Biology*, esp. 38. Xlibris, Philadelphia.

Jennings, Bruce H. 1988. *Foundations of International Agricultural Research: Science and Politics in Mexican Agriculture*, esp. 50–57, 192. Westview Press, Boulder, Colo.

Jennings, Francis. 1975. *The Invasion of America: Indians, Colonialism, and the Cant of Conquest*, esp. 18. Univ. of North Carolina Press, Chapel Hill.

Johannessen, Carl L.* 1987. "Domestication Process: An Hypothesis for Its Origin." In Kenzer, 1987a:177–204, esp. 179–180, 184–186.

———. 2003. "Early Maize in India? A Case for 'Multiple-Working Hypotheses.'" In Mathewson and Kenzer, 2003:299–314, esp. 299–304.

Johnston, R. J., Derek Gregory, Geraldine Pratt, and Michael Watts, eds. 2000. *The Dictionary of Human Geography*, 4th ed., esp. 45–46, 134–135, 138–140, 283–284, 430. Blackwell, Malden, Mass.

Johnston, R. J., and J. D. Sidaway. 2004. *Geography and Geographers: Anglo-American Human Geography since 1945*, 6th ed., esp. 57–58, 106–107, 213–215. Edward Arnold, London.

Jordan, Terry G. 1989. "Preadaptation and European Colonization in Rural North America." *Annals of the Association of American Geographers* 79:489–500, esp. 489, 491, 495.

———. 1991. "Settlement Geography." In *Modern Geography: An Encyclopedic Survey*, ed. Gary S. Dunbar, 159–160, esp. 160. Garland, New York.

Karan, Pradyumma P. 1983. "Regional Studies in Kentucky and American Geography." In *The Evolution of Geographical Thought in America: A Kentucky Root*, ed. Wilford A. Braden and P. P. Karan, 87–111, esp. 95–98. Kendall/Hunt, Dubuque, Iowa.

———. 2004. "The South and the Development of Field Methods in American Geography." In *The Role of the South in the Making of American Geography: Centennial of the AAG, 2004*, ed. James O. Wheeler and Stanley D. Brunn, 35–45, esp. 38–44. Bellwether, Columbia, Md.

Kates, Robert W., B. L. Turner II, and William C. Clark. 1990. "The Great Transformation." In *The Earth as Transformed by Human Action*, ed. B. L. Turner II et al., 1–17, esp. 3–4. Cambridge Univ. Press, Cambridge.

Kearns, Gerry. 2004. "Environmental History." In *A Companion to Cultural Geography*, ed. James S. Duncan, Nuala C. Johnson, and Richard H. Schein, 194–208, esp. 196–197. Blackwell, Malden, Mass.

Kemper, Franz-Josef. 2003. "Landschaften, Texte, sozialew Praktiken-Weger der angelsachsischen Kulturgeographie." *Petermanns Geographische Mitteilungen* 147 (2): 6–15.

Kenzer, Martin S. 1984. "Commentary" on "Regional Reality in Economy," by Carl O. Sauer. *Yearbook of the Association of Pacific Coast Geographers* 46:35–37.

———. 1985a. "Carl O. Sauer: Nascent Human Geographer at Northwestern." *California Geographer* 25:1–11.

———. 1985b. "Milieu and the 'Intellectual Landscape': Carl O. Sauer's Undergraduate Heritage." *Annals of the Association of American Geographers* 75:258–270.

———. 1986a. "The Making of Carl O. Sauer and the Berkeley School of (Historical) Geography." Ph.D. diss., McMaster University, Hamilton, Canada.

———. 1986b. "Carl Sauer and the Carl Ortwin Sauer Papers." *History of Geography Newsletter* 5:1–9.

———, ed. 1987a. *Carl O. Sauer: A Tribute.* Oregon State Univ. Press, Corvallis.

———. 1987b. "Introduction." In Kenzer, 1987a:1–8.

———. 1987c. "Like Father, Like Son: William Albert and Carl Ortwin Sauer." In Kenzer, 1987a:40–65.

———. 1987d. "Tracking Sauer across Sour Terrain." *Annals of the Association of American Geographers* 77:469–474.

———. 1988. "Commentary on Carl O. Sauer" [on Leighly on Sauer]. *Professional Geographer* 40:333–336.

———. 1989. "Areal Differentiation: A Rejoinder [to David Stoddart]." *Annals of the Association of American Geographers* 79:617.

———. 1999. Abstract. "The *Other* Carl Sauer: Personal, Emotional, and Immature." *Yearbook of the Association of Pacific Coast Geographers* 61:236–237.

———. 2001. Review of *How It Came To Be: Carl O. Sauer, Franz Boas and the Meanings of Anthropogeography,* by William W. Speth. *Annals of the Association of American Geographers* 91:752–753.

———. 2003. "From 'Morphology' to 'Foreword': Toward a Clearer Understanding of Carl Sauer's Intellectual Growth." In Mathewson and Kenzer, 2003:55–79.

Kerns, Virginia. 2003. *Scenes from the High Desert: Julian Steward's Life and Theory,* esp. 88–89, 98–100, 112, 204, 233, 368. Univ. of Illinois Press, Urbana.

Kersten, Earl W. 1982. "Sauer and 'Geographic Influences.'" *Yearbook of the Association of Pacific Coast Geographers* 44:47–73.

Kniffen, Fred B.* 1979. "Why Folk Housing?" *Annals of the Association of American Geographers* 69:59–63, esp. 62–63.

———. 1985. "Fred B. Kniffen," interviewed by Anne Buttimer, 1982. Dialogue Project, Transcript Series G28, 17 pp., esp. 4. Lund, Sweden.

———. 2003. "Foreword." In Mathewson and Kenzer, 2003:5–9.

Knight, M. M. 1964. Review of *Land and Life: A Selection from the Writings of Carl Ortwin Sauer. Journal of Economic History* 24:404–405.

Kramer, Fritz L.* 1975. "Carl Ortwin Sauer, Geographer (1889–1975)." *Geopub Review* 1:337–346.

Krim, Arthur. 2003. "Carl Sauer and Southeast Asia: Archaeology of Early Culture." In Mathewson and Kenzer, 2003:135–156.

Kroeber, Alfred L. 1963 [1939]. *Cultural and Natural Areas of Native North America,* esp. 177–178. Univ. of California Press, Berkeley.

Kuhlken, Robert. 2003. Review of *How It Came To Be: Carl O. Sauer, Franz Boas and the Meanings of Anthropogeography,* by William W. Speth. *Historical Geography* 31:216–217.

Larkin, Robert P., and Gary L. Peters. 1993. "Carl Ortwin Sauer." In *Biographical Dictionary of Geography,* 259–264. Greenwood Press, Westport, Conn.

Lathrap, Donald W. 1974. "The Moist Tropics, the Arid Lands, and the Appearance of Great Art Styles in the New World." In *Art and Environment in Native America,* ed. M. E. King and I. R. Traylor Jr., 115–158, esp. 115–116. Special Publications of the Museum, No. 7, Texas Technical Univ. Press, Lubbock.

———. 1984. Review of *The Origins of Agriculture: An Evolutionary Perspective,* by David Rindos. *Economic Geography* 60:339–344, esp. 339.

Leighly, John.*‡ 1955. "What Has Happened to Physical Geography?" *Annals of the Association of American Geographers* 45:309–318, esp. 315–316.

———. 1963. "Introduction." In *Land and Life: A Selection from the Writings of Carl Ortwin Sauer,* ed. J. Leighly, 1–8. Univ. of California Press, Berkeley.

———. 1975. "Carl Ortwin Sauer, 1889–1975." *Newsletter, Association of Pacific Coast Geographers* (fall). Also in *Historical Geography Newsletter* 6 (1) (1976): 3–4.

———. 1976a. "Carl Ortwin Sauer, 1889–1975." *Annals of the Association of American Geographers* 66:337–348.

———. 1976b. Letter to William W. Speth, November 19, 1976. In Speth, 1999a: 203–205.

———. 1978a. "Carl Ortwin Sauer, 1889–1975." *Geographers Biobibliographical Studies,* 2:99–108. Mansell, London.

———. 1978b. "Scholar and Colleague: Homage to Carl Sauer." *Yearbook of the Association of Pacific Coast Geographers* 40:117–133.

———. 1979. "Drifting into Geography in the Twenties." *Annals of the Association of American Geographers* 69:4–9.

———. 1983. "Memory as Mirror." In *The Practice of Geography,* ed. Anne Buttimer, 80–89, esp. 82–83, 85, 87–88. Longman, London.

———. 1987 [1976]. "Ecology as Metaphor: Carl Sauer and Human Ecology," William M. Denevan, ed., *Professional Geographer* 39:405–412.

——— (with introduction and notes by William W. Speth). 1995 [1965–1969]. "The Emergence of Cultural Geography." *Yearbook of the Association of Pacific Coast Geographers* 57:158–180, esp. 173–174.

Lewis, Peirce. 1999. "The Monument and the Bungalow." *Geographical Review* 88: 507–527, esp. 509–511.

Livingstone, David N. 1992. *The Geographical Tradition: Episodes in the History of a Contested Enterprise,* esp. 260–262, 294–302. Blackwell, Oxford.

Long, Richard C. E. 1937. Review of *Aboriginal Population of Northwestern Mexico*, by Carl Sauer. *Man* 37:102.

Lopez, Barry H. 1988. *Crossing Open Ground*, esp. 95, 100. Scribner's, New York.

Lovell, W. George. 2003. "'A First Book in Geography': Carl Sauer and the Creation of *Man in Nature*." In Mathewson and Kenzer, 2003:323–338.

———. 2009. "Introduction" to Section VII, Historical Geography. Chapter 26 herein.

Lovell, W. George, and Christopher H. Lutz. 1996. "'A Dark Obverse': Maya Survival in Guatemala, 1520–1994." *Geographical Review* 86:398–407, esp. 398–399.

Lukermann, Fred. 1989. "*The Nature of Geography:* Post Hoc, Ergo Propter Hoc?" In *Reflections on Richard Hartshorne's The Nature of Geography*, ed. J. N. Entrikin and S. D. Brunn, 53–68. Association of American Geographers, Washington, D.C.

MacNeish, Richard S. 1992. *The Origins of Agriculture and Settled Life*, esp. 8, 274. Univ. of Oklahoma Press, Norman.

Macpherson, Anne.† 1987. "Preparing for the National Stage: Carl Sauer's First Ten Years at Berkeley." In Kenzer, 1987a:69–88.

Madariaga, Salvador de. 1967. "Passionate History." Review of *The Early Spanish Main*, by Carl Ortwin Sauer. *New York Review*, December 1, 1967.

Mangelsdorf, Paul C. 1953. Review of *Agricultural Origins and Dispersals*, by Carl O. Sauer. *American Antiquity* 19:87–90.

Mangelsdorf, Paul C., Richard S. MacNeish, and Gordon R. Willey. 1964. "Origins of Agriculture in Middle America." In *Handbook of Middle American Indians*, vol. 1, *Natural Environment and Early Cultures*, ed. Robert C. West, 427–445, esp. 429. Univ. of Texas Press, Austin.

Marglin, Stephen A. 1996. "Farmers, Seedsmen, and Scientists: Systems of Agriculture and Systems of Knowledge." In *Decolonizing Knowledge: From Development to Dialogue*, ed. Frédérique Apffel-Marglin and Stephen A. Marglin, 185–248, esp. 211–217, 221. Clarendon Press, Oxford.

Martin, Geoffrey J. 1980. *The Life and Thought of Isaiah Bowman*, esp. 206. Archon Books, Hamden, Conn.

———. 1984. Review of *Selected Essays, 1963–1975: Carl O. Sauer*, ed. Bob Callahan, *Professional Geographer* 36:253.

———. 1987. "Foreword." In Kenzer, 1987a:ix–xvi.

———. 1988. "Preston E. James, 1899–1986." *Annals of the Association of American Geographers* 78:164–175, esp. 170.

———. 2003a. "From the Cycle of Erosion to 'The Morphology of Landscape': Or Some Thoughts Concerning Geography as It Was in the Early Years of Carl Sauer." In Mathewson and Kenzer, 2003:19–53.

———. 2003b. "The History of Geography." In *Geography in America at the Dawn of the 21st Century*, ed. Gary L. Gaile and Cort J. Willmott, 550–561, esp. 554. Oxford Univ. Press, Oxford.

———. 2005. *All Possible Worlds: A History of Geographical Ideas*, 4th ed., esp. 385–388, 391, 457–458, 525. Oxford Univ. Press, Oxford.

———. 2009. "Introduction" to Section III, Toward Maturity. Chapter 10 herein.

Mather, Cotton. 1983. "Kentucky and American Geographic Thought." In *The Evolution of Geographic Thought in America: A Kentucky Root*, ed. Wilford A. Bladon and Pradyumna P. Karan, 1–11, esp. 5–6. Kendall/Hunt, Dubuque, Iowa.

Mather, John R., and Marie Sanderson. 1996. *The Genius of C. Warren Thornthwaite, Climatologist-Geographer*, esp. 9–16. Univ. of Oklahoma Press, Norman.

Mathewson, Kent. 1986. "Alexander von Humboldt and the Origins of Landscape Archaeology." *Journal of Geography* 85 (2): 50–56, esp. 50.

———. 1987a. "Humane Ecologist: Carl Sauer as Metaphor?" *Professional Geographer* 39:412–413.

———. 1987b. "Sauer South by Southwest: Antimodernism and the Austral Impulse." In Kenzer, 1987a:90–111.

———. 1988. "Response to Kenzer" [on Leighly on Sauer]. *Professional Geographer* 40:336–337.

———. 1996. "High/Low, Back/Center: Culture's Stages in Human Geography." In *Concepts in Human Geography*, ed. Carville Earle, Kent Mathewson, and Martin S. Kenzer, 97–125, esp. 103–107, 113. Rowman and Littlefield, Lanham, Md.

———. 1998. "Classics in Human Geography Revisited ('The Superorganic in American Cultural Geography'): Commentary 2." *Progress in Human Geography* 22: 569–571.

———. 1999. "Cultural Landscapes and Ecology II: Regions, Retrospects, Revivals." *Progress in Human Geography* 23:267–281, esp. 268–269.

———. 2002. Review of *How It Came To Be: Carl O. Sauer, Franz Boas and the Meanings of Anthropogeography*, by William W. Speth. *American Anthropologist* 104: 380–381.

———. 2003. "Introduction." In Mathewson and Kenzer, 2003:11–16.

———. 2004. "Sauer and the South: A Deferred Agenda." In *The Role of the South in the Making of American Geography: Centennial of the AAG, 2004*, ed. James O. Wheeler and Stanley D. Brunn, 353–363. Bellwether, Columbia, Md.

———. 2006. "Alexander von Humboldt's Image and Influence in North American Geography, 1804–2004." *Geographical Review* 96:416–438, esp. 429–430.

———. 2009. "Carl O. Sauer and His Critics." Chapter 2 herein.

Mathewson, Kent, and Martin S. Kenzer, eds. 2003. *Culture, Land, and Legacy: Perspectives on Carl O. Sauer and Berkeley School Geography*. Geoscience and Man, vol. 37. Geoscience Publications, Department of Geography and Anthropology, Louisiana State University, Baton Rouge.

Mathewson, Kent, Martin S. Kenzer, and Geoffrey J. Martin. 2009. "Introduction" to Section IV, Economy/Economics. Chapter 14 herein.

Mathewson, Kent, and Vincent J. Shoemaker. 2004. "Louisiana State University Geography at Seventy-Five: Berkeley on the Bayou and Beyond." In *The Role of the South in the Making of American Geography: Centennial of the AAG, 2004*, ed. James O. Wheeler and Stanley D. Brunn, 245–267. Bellwether, Columbia, Md.

May, J. A. 2003. "Some Remarks on the Implicit Philosophy of Carl Sauer." In Mathewson and Kenzer, 2003:261–276.

McDowell, Linda. 1994. "The Transformation of Cultural Geography." In *Human Geography: Society, Space, and Social Science*, ed. Derek Gregory, Ron Martin, and Graham Smith, 146–173, esp. 147–150. Univ. of Minnesota Press, Minneapolis.

McGann, Thomas F. 1968. Review of *The Early Spanish Main*, by Carl O. Sauer. *Geographical Review* 58:316–317.

McManis, Douglas R. 1981. Review of *Seventeenth Century North America*, by Carl O. Sauer. *Geographical Review* 71:492–493.

McPheron, William. 1988. *Edward Dorn*, esp. 12–13, 15. Boise State University Western Writers Series, No. 85. Boise.

Meinig, Donald W. 1981. Review of *Seventeenth Century North America*, by Carl O. Sauer. *Journal of Historical Geography* 7:432–434.

———. 1983. "Geography as an Art." *Transactions, Institute of British Geographers*, New Series, 8:314–328, esp. 319–320.

———. 2002. "The Life of Learning." In *Geographical Voices: Fourteen Autobiographical Essays*, ed. Peter Gould and Forrest R. Pitts, 188–210, esp. 202. Syracuse Univ. Press, Syracuse.

Melendez, Carlos. 1967. Review of *The Early Spanish Main*, by Carl Ortwin Sauer. *William and Mary Quarterly* 24:662–664.

Merrill, Elmer Drew. 1954. *The Botany of Cook's Voyages and Its Unexpected Significance in Relation to Anthropology, Biogeography and History.* Chronica Botanica, vol. 14, nos. 5/6, esp. 271–287. Chronica Botanica Co., Waltham, Mass.

Mikesell, Marvin W.* 1969. "The Borderlands of Geography as a Social Science." In *Interdisciplinary Relationships in the Social Sciences*, ed. M. Sherif and C. W. Sherif, 227–248, esp. 228–229. Aldine: Chicago.

———. 1978. "Tradition and Innovation in Cultural Geography." *Annals of the Association of American Geographers* 68:1–16, esp. 2–3.

———. 1983. Comment on "morale" at Sauer's Berkeley, in "The Environment of Graduate School," 66–79. In *The Practice of Geography*, ed. Anne Buttimer. Longman, London.

———. 1987. "Sauer and 'Sauerology': A Student's Perspective." In Kenzer, 1987a:144–150.

———. 1994. "Afterword: New Interests, Unsolved Problems, and Persisting Tasks." In *Re-Reading Cultural Geography*, ed. Kenneth E. Foote et al., 437–444, esp. 441. Univ. of Texas Press, Austin.

Miller, Vincent P., Jr. 1971. "Some Observations on the Science of Cultural Geography." *Journal of Geography* 70:27–35.

Mitchell, Don. 1996. *The Lie of the Land: Migrant Workers and the California Landscape,* esp. 24–25. Univ. of Minnesota Press, Minneapolis.

———. 2000. *Cultural Geography: A Critical Introduction,* esp. 20–29 on "Carl Sauer and Cultural Theory" and "The Morphology of Landscape." Blackwell, Oxford.

Morin, Karen M. 2003. "Landscape and Environment: Representing and Interpreting the World." In *Key Concepts in Geography,* ed. Sarah Holloway, Stephen P. Rice, and Gill Valentine, 319–334, esp. 321. Sage, London.

Murphy, Robert C., and Charles B. Hitchcock. 1955. "Conference on 'Man's Role in Changing the Face of the Earth.'" *Geographical Review* 45:583–586.

Myers, Garth A. 2003. Review of *Handbook of Cultural Geography,* ed. Kay Anderson et al. *Annals of the Association of American Geographers* 93:935–936.

Nadal i Piqué, Francesc. 1997. "Breus notes sobre 'The Morphology of Landscape' de Carl Ortwin Sauer." *Treballs de la Societat Catalana de Geografia* 43:151–154.

Nasatir, Abraham. 1982. Review of *Seventeenth Century North America,* by Carl O. Sauer. *Hispanic American Historical Review* 62:142.

Navarro Garcia, Luis. 1967. Review of *The Early Spanish Main,* by Carl Ortwin Sauer. *Caribbean Studies* 7 (3): 80–81.

Newcomb, Robert M. 1976. "Carl O. Sauer, Teacher." *Historical Geography Newsletter* 6 (1): 21–22, 30.

Newson, Linda. 1981. Review of *Carl Sauer's Fieldwork in Latin America,* by Robert C. West. *Journal of Latin American Studies* 13:214.

Nicholson, E. M. 1957. Review of *Man's Role in Changing the Face of the Earth,* William L. Thomas Jr., ed. *Journal of Ecology* 45:955–958.

Niddrie, David L. 1966. "A Geographer Explores the Caribbean." Review of *The Early Spanish Main,* by Carl Ortwin Sauer. *Landscape* 16 (2): 36–37.

Norton, William. 1989. *Explorations in the Understanding of Landscape: A Cultural Geography,* esp. 36–39, 97–98. Greenwood Press, New York.

———. 2000. *Cultural Geography: Themes, Concepts, Analyses,* esp. 70–75. Oxford Univ. Press, Oxford.

———. 2005. Review of *Culture, Land, and Legacy: Perspectives on Carl O. Sauer and Berkeley School Geography,* Kent Mathewson and Martin S. Kenzer, eds. *Journal of Cultural Geography* 23:140–143.

Nowell, Charles E. 1968. Review of *The Early Spanish Main,* by Carl Ortwin Sauer. *Hispanic American Historical Review* 48:691–693.

Oasa, Edmund K., and Bruce H. Jennings. 1982. "Science and Authority in International Agricultural Research." *Bulletin of Concerned Asian Scholars* 14:30–44, esp. 34–36.

Olson, Charles. 1974 [1964]. "A Bibliography on America for Ed Dorn." In *Additional Prose,* ed. George F. Butterick, 3–14. Four Seasons Foundation, Bolinas, Calif.

Olwig, Kenneth R. 1996. "Recovering the Substantive Nature of Landscape." *Annals of the Association of American Geographers* 86:630–653, esp. 643–644.

Padgen, Anthony. 1992. "Foreword." In *The Early Spanish Main*, by Carl Ortwin Sauer, 4th printing, vii–x. Univ. of California Press, Berkeley.

Parry, J. H. 1966. "Discoverers or Desperados?" Review of *The Early Spanish Main*, by Carl O. Sauer. *Saturday Review*, July 2, pp. 24–25.

Parsons, James J.*‡ 1964. "The Contribution of Geography to Latin American Studies." In *Social Science Research on Latin America*, ed. Charles Wagley, 33–85, esp. 52–53, 58. Columbia Univ. Press, New York.

———. 1967. "Geography." In *The Centennial Record of the University of California*, ed. Verne A. Stadtman, 86–87. University of California, Berkeley.

———. 1969. "Carl O. Sauer." In *International Encyclopedia of the Social Sciences* 8:17–19. Crowell, Collier, and Macmillan, New York.

———. 1975. "Carl Ortwin Sauer, 1889–1975." *Yearbook of the American Philosophical Society, 1975*, 163–167. Philadelphia. Also in *CoEvolution Quarterly* 10 (1976): 45–47. Chapter 1 herein.

———. 1976a. "Carl O. Sauer 1889–1975." *The Itinerant Geographer*, pp. 1–4. Department of Geography, University of California, Berkeley.

———. 1976b. "Carl Ortwin Sauer, 1889–1975." *Geographical Review* 66:83–89.

———. 1977b. "Saueriana." *The Itinerant Geographer*, pp. 12–16. Department of Geography, University of California, Berkeley.

———. 1979a. "The Later Sauer Years." *Annals of the Association of American Geographers* 69:9–17.

———. 1979b. "Saueriana." *The Itinerant Geographer*, pp. 10–12. Department of Geography, University of California, Berkeley.

———. 1981. Review of *Carl Sauer's Field Work in Latin America*, by Robert C. West. *Hispanic American Historical Review* 61:346–347.

———. 1985. "Sauerology." *The Itinerant Geographer*, pp. 7–8. Department of Geography, University of California, Berkeley.

———. 1986. "Sauerology." *The Itinerant Geographer*, pp. 11–12. Department of Geography, University of California, Berkeley.

———. 1987a. "On Sauer." *The Itinerant Geographer*, p. 15. Department of Geography, University of California, Berkeley.

———. 1987b. Introduction to "Now This Matter of Cultural Geography: Notes from Carl Sauer's Last Seminar at Berkeley." In Kenzer, 1987a, 153–154.

———. 1993. "On Sauer." *The Itinerant Geographer*, pp. 17–18. Department of Geography, University of California, Berkeley.

———. 1994. "Cultural Geography at Work." In *Re-Reading Cultural Geography*, ed. Kenneth E. Foote et al., 281–288, esp. 281–283, 285, 287. Univ. of Texas Press, Austin.

———. 1996a. "Carl Sauer's Vision of an Institute for Latin American Studies." *Geographical Review* 86:377–384.

———. 1996b. "'Mr. Sauer' and the Writers." *Geographical Review* 86:22–41. Also in Mathewson and Kenzer, 2003:193–216.

———. 1999. Letter to William W. Speth, January 4, 1977. In Speth, 1999a:209–211.
Parsons, James J., J. B. Leighly, W. Borah, and L. B. Simpson. 1977. "Carl Ortwin Sauer, 1889–1975." In *In Memoriam*, 205–206. University of California, Berkeley.
Parsons, James J., and Natalia Vonnegut. 1983. "Biographical Sketch: Carl Ortwin Sauer (1889–1975)." In *60 Years of Berkeley Geography, 1923–1983*, ed. J. J. Parsons and N. Vonnegut, 157–160. Department of Geography, University of California, Berkeley.
Paul, Sherman. 1989. "Charles Olson." In *Hewing to Experience: Essays and Reviews on Recent American Poetry and Poetics, Nature and Culture*, 181–234, esp. 218–220. Univ. of Iowa Press, Iowa City.
Pederson, Leland R.† 1985. "A Dedication to the Memory of Carl Ortwin Sauer, 1889–1975." *Arizona and the West* 27:304–308.
Peet, Richard.† 1998. "Cultural Geography." In *Modern Geographical Thought*, by R. Peet, 14–16. Blackwell, Oxford.
———. 2003. "Social Relations: The Missing Dimension in Carl Sauer's Theorization." In Mathewson and Kenzer, 2003:315–321.
Penn, Mischa, and Fred Lukermann. 2003. "Chorology and Landscape: An Internalist Reading of 'The Morphology of Landscape.'" In Mathewson and Kenzer, 2003: 233–259.
Pennington, Loren E. 1969. Review of *Northern Mists*, by Carl O. Sauer. *Hispanic American Historical Review* 49:324–325.
Penrose, Boies. 1968. Review of *The Early Spanish Main*, by Carl O. Sauer. *Renaissance Quarterly* 21:209–210.
Pfeifer, Gottfried.‡ 1938. "Enwicklungstendenzen in Theorie und Methode der regionalen Geographie in den Vereinigten Staaten nach dem Kriege." *Zeitschrift der Gesellschaft für Erdkunde zu Berlin*, 93–125, esp. 98–103. English translation by John Leighly, "Regional Geography in the United States since the War: A Review of Trends in Theory and Methods," mimeographed, 1938. American Geographical Society Library, Milwaukee.
———. 1965. "Carl Ortwin Sauer zum 75." *Geographische Zeitschrift* 53:1–9, 74–77.
———. 1975. "Carl Ortwin Sauer, 24.12.1889–18.7.1975." *Geographische Zeitschrift* 63: 161–169.
———. 1986. "Carl Ortwin Sauer, die Berkeleyer Schüle und das 'Morale' der Geographie." *Frankfurter Geographische* 55:423–437.
Philo, Chris. 1998. "Animals, Geography, and the City: Notes on Inclusions and Exclusions." In *Animal Geographies: Place, Politics, and Identity in the Nature-Culture Borderlands*, ed. Jennifer Wolch and Jody Emel, 51–71, esp. 56–58. Verso, London.
Piperno, Delores R., and Deborah M. Pearsall. 1998. *The Origins of Agriculture in the Lowland Neotropics*, esp. 18–21. Academic Press, San Diego.
Platt, Robert S. 1952. "The Rise of Cultural Geography in America." *Proceedings of the*

Seventeenth International Geographical Congress, 485–490. NAS-NRC, Washington, D.C. Also in Wagner and Mikesell, 1962:35–43, esp. 36.

Porter, Philip W. 1978. "Geography as Human Ecology: A Decade of Progress in a Quarter Century." *American Behavioral Scientist* 22:15–39, esp. 18.

———. 1987. "Ecology as Metaphor: Sauer and Human Ecology." *Professional Geographer* 39:414.

———. 1988. "Sauer, Archives, and Recollections" [on Leighly on Sauer]. *Professional Geographer* 40:337–339.

Pred, Allan.‡ 1983. "From Here to There and Then: Some Notes on Diffusions, Defusions and Disillusions." In *Recollections of a Revolution: Geography as Spatial Science,* ed. Mark Billinge, Derek Gregory, and Ron Martin, 86–103, esp. 90–95. St. Martin's Press, New York.

———. 1991. Review of *The City as Text: The Politics of Landscape Interpretation in the Kandyan Kingdom,* by James S. Duncan. *Journal of Historical Geography* 17:115–117, esp. 115–116.

Price, Edward T.* 1968. "Cultural Geography." In *International Encyclopedia of the Social Sciences* 6:129–134. Macmillan, New York.

———. 2003. "Crafting Culture History." In Mathewson and Kenzer, 2003:277–296.

———. 2009. "Introduction" to Section VIII, Carl Sauer on Geographers and Other Scholars. Chapter 34 herein.

Price, Marie, and Martin Lewis. 1993a. "The Reinvention of Cultural Geography." *Annals of the Association of American Geographers* 83:1–17.

———. 1993b. "Commentary on 'The Reinvention of Cultural Geography' by Price and Lewis. Reply: On Reading Cultural Geography." *Annals of the Association of American Geographers* 83:520–522.

Prince, Hugh. 1975. "Death of Carl Sauer, a Geographical Pioneer." *Geographical Magazine* 47:778.

Provinse, John H. 1937. Review of *Aboriginal Population of Northwestern Mexico,* by Carl O. Sauer. *American Anthropologist* 39:148–149.

Quinn, David B. 1969. Review of *Northern Mists,* by Carl O. Sauer. *William and Mary Quarterly,* 3rd series 26:623–624.

———. 1973. Review of *Sixteenth Century North America: The Land and the People as Seen by the Europeans,* by Carl Ortwin Sauer. *Annals of the Association of American Geographers* 63:252–253.

Redfield, Robert. 1936. Review of *Aboriginal Population of Northwestern Mexico,* by Carl Sauer. *American Journal of Sociology* 41:562–563.

Riess, Richard O. 1976. "Some Notes on Salem Normal School and Carl Sauer, 1914." *Proceedings of the New England-St. Lawrence Valley Division of the Association of American Geographers* 6:63–67.

Rindos, David. 1984. *The Origins of Agriculture: An Evolutionary Perspective,* esp. 84. Academic Press, Orlando.

Rival, Laura. 2006. "Amazonian Historical Ecologies." In *Ethnobiology and the Science of Humankind*, ed. Roy Ellen, 97–115, esp. 111–112. Journal of the Royal Anthropological Institute, Special Issue No. 1. Blackwell, Oxford.

Robbins, Paul. 2004. *Political Ecology: A Critical Introduction*, esp. 29–30. Blackwell, Oxford.

Robinson, David J. 1972. "Historical Geography in Latin America." In *Progress in Historical Geography*, ed. Alan R. H. Baker, 168–186, esp. 173–174. David and Charles, Newton Abbot, U.K.

———. 1980. "On Preston E. James and Latin America: A Biographical Sketch." In *Studying Latin America: Essays in Honor of Preston E. James*, ed. D. J. Robinson, 1–101, esp. 20–21. Dellplain Latin American Studies, No. 4. University Microfilms International, Ann Arbor.

Rodrigue, Christine M. 1992. "Can Religion Account for Early Animal Domestication?" *Professional Geographer* 44:417–430, esp. 417–419, 427–428.

Romero, Emilio. 1975. "La obra de Carl Sauer." *Boletín de la Sociedad Geográfica de Lima* 94:11.

Rosenblat, Ángel. 1976. "The Population of Hispaniola at the Time of Columbus." In *The Native Population of the Americas in 1492*, ed. William M. Denevan, 43–66, esp. 61–63. Univ. of Wisconsin Press, Madison.

Rowe, John. 1972. Review of *Sixteenth Century North America: The Land and the People as Seen by the Europeans*, by Carl Ortwin Sauer. *Geographical Journal* 138: 239–240.

Rowley, Virginia M. 1964. *J. Russell Smith: Geographer, Educator, and Conservationist*, esp. 37–38. Univ. of Pennsylvania Press, Philadelphia.

Rowntree, Lester B.† 1996. "The Cultural Landscape Concept in American Human Geography." In *Concepts in Human Geography*, ed. Carville Earle, Kent Mathewson, and Martin S. Kenzer, 127–159, esp. 130–134. Rowman and Littlefield, Lanham, Md.

Rucinque, Héctor F. 1987. "Introducción" to "La educación de un Geógrafo," by Carl O. Sauer, v–vii. GEOFUN, Bogota.

———. 1990. "Carl O. Sauer: Geógrafo y maestro *par excellence*." *Trimestre Geografico* (Bogotá) 14:3–19.

Sachs, Aaron. 2006. *The Humboldt Current: Nineteenth-Century Exploration and the Roots of American Environmentalism*, esp. 341. Viking, New York.

Sawatzky, N. L.* 1987. "Legacy and Stewardship." In Kenzer, 1987a:205–213, esp. 205–206.

Schein, Richard H. 2004. "Cultural Traditions." In *A Companion to Cultural Geography*, ed. James S. Duncan, Nuala C. Johnson, and Richard H. Schein, 11–23. Blackwell, Malden, Mass.

Sestini, Aldo. 1976. "Carl Ortwin Sauer." *Revista Geografica Italiana* 83:123–124.

Shaw, Denis J. B., and Jonathan D. Oldfield. 2007. "Landscape Science: A Russian Geographical Tradition." *Annals of the Association of American Geographers* 97:111–126, esp. 113–114.

Sheridan, Richard B. 1967. Review of *The Early Spanish Main*, by Carl O. Sauer. *Journal of Economic History* 27:265–266.

Siemens, Alfred H. N.d. "Sucesión de paisajes: Una conceptualización de la relación entre el ser humano y su ambiente natural en el transcuro del tiempo." In *Lugar, espacio y paisaje en arqueología*. VI Coloquio Pedro Bosch Gimpera, Ciudad de México, in press.

Sluyter, Andrew. 1997. Commentary. "On Excavating and Burying Epistemologies." *Annals of the Association of American Geographers* 87:700–702.

———. 2001. "Colonialism and Landscape in the Americas: Material/Conceptual Transformations and Continuing Consequences." *Annals of the Association of American Geographers* 91:410–428, esp. 412, 422.

———. 2002. *Colonialism and Landscape: Postcolonial Theory and Applications*, esp. 6–7, 24, 65, 147. Rowman and Littlefield, Lanham, Md.

———. 2005. "Blaut's Early Natural/Social Theorization, Cultural Ecology, and Political Ecology." *Antipode* 37:963–980, esp. 966–967.

Smith, Neil. 2003. *American Empire: Roosevelt's Geographer and the Prelude to Globalization*, esp. 226–228. Univ. of California Press, Berkeley.

Solnick, Bruce B. 1969. Review of *The Early Spanish Main*, by Carl O. Sauer. *Terrae Incognitae* 1:79.

Solot, Michael S. 1983. "Carl Sauer and Cultural Evolution." Master's thesis, University of California, Los Angeles.

———. 1986. "Carl Sauer and Cultural Evolution." *Annals of the Association of American Geographers* 76:508–520.

———. 1987. "Reply to Kenzer and Speth." *Annals of the Association of American Geographers* 77:476–478.

———. 1988. Review of *Carl O. Sauer: A Tribute*, ed. Martin S. Kenzer. *Journal of Historical Geography* 14:96–98.

Spate, O. H. K. 1965. Review of *Land and Life: A Selection from the Writings of Carl Ortwin Sauer*, ed. John Leighly. *Economic Geography* 41:93–94.

Speeding, Nick. 2003. "Landscape and Environment: Biophysical Processes, Biophysical Forms." In *Key Concepts in Geography*, ed. Sarah L. Holloway, Stephen P. Rice, and Gill Valentine, 281–303, esp. 289–292, 297. Sage, London.

Spencer, Joseph E.* 1975. "Carl Sauer: Memories about a Teacher." *California Geographer* 15:83–86.

———. 1976. "What's in a Name? 'The Berkeley School.'" *Historical Geography Newsletter* 6 (1): 7–11.

———. 1979. "Western and Southern Frontiers: A Geographer West of the Sierra Nevada." *Annals of the Association of American Geographers* 69:46–52, esp. 47–49.

———. 1999. Letter to William W. Speth, November 24, 1976. In Speth, 1999a:205–209.
Speth, William W. 1972. "Historicist Anthropogeography: Environment and Culture in American Anthropological Thought from 1890 to 1950," esp. 1–4, 32–36, 40, 194, 199, 203–239, 244–247. Ph.D. diss., University of Oregon, Eugene.
———. 1977. "Carl Ortwin Sauer on Destructive Exploitation." *Biological Conservation* 11:145–160. Also in Speth, 1999a:47–59.
———. 1981. "Berkeley Geography, 1923–33." In *The Origins of Academic Geography in the United States*, ed. Brian W. Blouet, 221–244. Archon Books, Hamden, Conn. Also in Speth, 1999a:61–79.
———. 1987a. "Historicism: The Disciplinary World View of Carl O. Sauer." In Kenzer, 1987a:11–39. Also in Speth, 1999a:5–30.
———. 1987b. "Against Presentist Histories: Solot's View of Sauer." *Annals of the Association of American Geographers* 77:474–475.
———. 1987c. "On the Discrimination of Anthropogeographies." *Canadian Geographer* 31:72–74. Also in Speth, 1999a:123–126.
———, ed. 1988. "The Berkeley School Questionnaire." *Proceedings, New England-St. Lawrence Valley Geographical Society* 18:26–30. (Comments by W. Speth, J. Leighly, H. Raup, J. Spencer, P. Wagner, D. Innis, M. Mikesell, L. Hughes, G. Carter, J. Parsons, D. Miller, C. Edwards, and E. Hammond.)
———. 1993. "Carl O. Sauer's Uses of Geography's Past." *Yearbook of the Association of Pacific Coast Geographers* 55:37–65. Also in Speth, 1999a:31–46.
———. 1995. "Introduction" to "The Emergence of Cultural Geography," by John Leighly. *Yearbook of the Association of Pacific Coast Geographers* 57:159–162.
———. 1999a. *How It Came To Be: Carl O. Sauer, Franz Boas and the Meanings of Anthropogeography*. Ephemera Press, Ellensburg, Wash.
———. 1999b. "Author's Note." In Speth, 1999a:vii–xiii.
———. 1999c. "A Sociological View of the Sauer School, 1923–44." In Speth, 1999a:89–118. Also in Mathewson and Kenzer, 2003:81–114.
———. 1999d. "Carl Sauer's Reinterpretation of Anthropogeography." In Speth, 1999a:175–195.
———, ed. 1999e. "Berkeley Geography: Selected Correspondence." In Speth, 1999a:197–217.
———. 2007. Abstract. "Carl O. Sauer's Philosophical Anthropology." *Yearbook of the Association of Pacific Coast Geographers* 69:236.
———. 2009. "Introduction" to Section VI, Man in Nature. Chapter 21 herein.
Spier, Leslie. 1937. Review of *The Distribution of Aboriginal Tribes and Languages in Northwestern Mexico*, by Carl O. Sauer. *American Anthropologist* 39:146–148.
Stamp, L. Dudley. 1957. Review of *Man's Role in Changing the Face of the Earth*, William L. Thomas Jr., ed. *Geographical Review* 47:597–600.
———. 1966. Review of *Land and Life: A Selection from the Writings of Carl Ortwin Sauer*, ed. John Leighly, *Geographical Journal* 132:145–146.

Stanislawski, Dan.*‡ 1975. "Carl Ortwin Sauer, 1889–1975." *Journal of Geography* 74: 548–554.
Starrs, Paul F. 1992. "Looking for Columbus." *Geographical Review* 4:367–374.
Stoddart, David R.‡ 1986. *On Geography and Its History*, esp. 3, 8, 146–147, 152, 206–207, 241. Blackwell, Oxford.
———. 1987. "To Claim the High Ground: Geography for the End of the Century." *Transactions, Institute of British Geographers*, New Series, 12:327–338, esp. 327, 333–334.
———. 1989. "Areal Differentiation" [on Solat and Kenzer on Sauer]. *Annals of the Association of American Geographers* 79:616.
———. 1991. "Carl Sauer: The Man and His Work." *Newsletter, Pacific Division, AAAS* 16:17–20.
———. 1997. "Carl Sauer: Geomorphologist." In *Process and Form in Geomorphology*, ed. D. R. Stoddart, 340–379. Routledge, London.
Sundberg, Juanita. 2003. "Masculinist Epistemologies and the Politics of Fieldwork in Latin Americanist Geography." *Professional Geographer* 55:180–190, esp. 182.
Sutton, Mark Q., and E. N. Anderson. 2004. *Introduction to Cultural Ecology*, esp. 27–28. Altamira Press, Walnut Creek, Calif.
Swanson, C. P. 1957. Review of *Man's Role in Changing the Face of the Earth*, William L. Thomas Jr., ed. *Quarterly Review of Biology* 32:319–320.
Symanski, Richard. 1981. "A Critique of 'The Superorganic in American Cultural Geography.'" *Annals of the Association of American Geographers* 71:287–289.
Symanski, Richard, and Korski. 2002. "Coconuts on a Lava Flow in the Chiricahua Mountains." In *Geography Inside Out*, by Richard Symanski and Korski, 103–114. Syracuse Univ. Press, Syracuse.
Tandarich, John P., Robert G. Darmody, and Leon R. Follmer. 1988. "The Development of Pedological Thought: Some People Involved." *Physical Geography* 9:162–174, esp. 165.
Thomas, Edwin S. 1987. "Reflections on 73 Years of Friendships and Geography at Cal, 1914–1987." *The Itinerant Geographer*, pp. 30–33. Department of Geography, University of California, Berkeley.
Troy, Tim. 2004. "Emil W. Haury and Carl O. Sauer in Sonora." *Journal of the Southwest* 46:113–116.
Tuan, Yi-Fu.† 1982. Interview by Clyde E. Browning. In *Conversations with Geographers: Career Pathways and Research Styles*, ed. Clyde E. Browning, interlocutor, 114–127, esp. 116–118. Studies in Geography No. 16, Department of Geography, University of North Carolina, Chapel Hill.
———. 1999. *Who Am I?: An Autobiography of Emotion, Mind, and Spirit*, esp. 29. Univ. of Wisconsin Press, Madison.
Tucker, Carll. 1976. Review of *Man in Nature*, by Carl O. Sauer. *Village Voice* (New York) 21 (31): 28–29.

Turner, B. L., II. 1987. "Comment on Leighly" [on Sauer and "Human Ecology"]. *Professional Geographer* 39:415–416.

———. 2002. "Contested Identities: Human-Environment Geography and Disciplinary Implications in a Restructuring Academy." *Annals of the Association of American Geographers* 92:52–74, esp. 58–59.

Unstead, J. F. 1922. Review of *The Geography of the Ozark Highland of Missouri*, by Carl Sauer. *Geographical Journal* 59:55–59.

Urquhart, Alvin W.† 1987. "Carl Sauer: Explorer of the Far Sides of Frontiers." In Kenzer, 1987a:217–224.

Unwin, Tim. 1992. *The Place of Geography*, esp. 101–103. Longman, Harlow.

Van Cleef, Eugene. 1940. "The Finns of the Pacific Coast of the United States, and Consideration of the Problem of Scientific Land Settlement." *Annals of the Association of American Geographers* 30:25–38, esp. 35–36.

Veblen, Thomas T.† 2003. "Carl O. Sauer and Geographical Biogeography." In Mathewson and Kenzer, 2003:173–192.

Violette, E. M. 1922. Review of *The Geography of the Ozark Highland of Missouri*, by Carl O. Sauer. *Mississippi Valley Historical Review* 9:90–92.

Vorsey, Louis de. 1973. Review of *Sixteenth Century North America: The Land and the People as Seen by the Europeans*, by Carl Ortwin Sauer. *Geographical Review* 63:412–413.

Wagner, Philip L.* 1957. Review of *Man's Role in Changing the Face of the Earth*, William L. Thomas Jr., ed. *Annals of the Association of American Geographers* 47:191–193.

———. 2000. "Each Particular Place: Culture and Geography." In *Cultural Encounters with the Environment*, ed. Alexander B. Murphy and Douglas L. Johnson, 311–322. Rowman and Littlefield, Lanham, Md.

———. 2009. "Introduction" to Section IX, Informal Remarks. Chapter 45 herein.

Wagner, Philip L., and Marvin W. Mikesell, eds. 1962. *Readings in Cultural Geography*, esp. 25–26, 60–61, 375. Univ. of Chicago Press, Chicago.

Wallach, Bret.† 1989. "Carl O. Sauer (1889–1975): In Memoriam." *The Itinerant Geographer*, pp. 3–7. Department of Geography, University of California, Berkeley.

———. 1991. *At Odds with Progress: Americans and Conservation*, esp. "Introduction," vii–xiv. Univ. of Arizona Press, Tucson.

———. 1999. "Commentary: Will Carl Sauer Make It across that Great Bridge to the Next Millennium?" *Yearbook of the Association of Pacific Coast Geographers* 61:129–136.

———. 2005. *Understanding the Cultural Landscape*, esp. 2–4, 33. Guilford Press, New York.

Warf, Barney, ed. 2006. *Encyclopedia of Human Geography*, esp. 10, 19, 71, 75–76, 79–80, 82–83, 212–213, 221, 391, 426. Sage, Thousand Oaks, Calif.

Washburn, Wilcomb E. 1968. Review of *Northern Mists*, by Carl O. Sauer. *American Historical Review* 74:698–699.

———. 1973. Review of *Sixteenth Century North America: The Land and the People as Seen by the Europeans*, by Carl O. Sauer. *Reviews in American History* 1:65–69.

Watson, Patty Jo, and Maxine R. Kleindienst. 1957. Review of *Man's Role in Changing the Face of the Earth*, William L. Thomas Jr., ed. *American Antiquity* 23:88–89.

Watts, David. 1967. Review of *The Early Spanish Main*, by Carl Ortwin Sauer. *Cahiers de Géographie de Québec* 24:590–591.

Webb, Kempton. 1984. Review of *Andean Reflections: Letters from Carl O. Sauer while on a South American Trip under a Grant from the Rockefeller Foundation, 1942*. *Journal of Historical Geography* 10:99–101.

Wedel, Waldo R. 1977. "The Education of a Plains Archeologist." *Plains Anthropologist* 75:1–11, esp. 5–6, 11.

West, Robert C.* 1953. Review of *Agricultural Origins and Dispersals*, by Carl O. Sauer. *Economic Geography* 29:371–372.

———. 1979. *Carl Sauer's Fieldwork in Latin America*. Dellplain Latin American Studies, No. 3. University Microfilms International, Ann Arbor. Pages 123–126 reprinted in *Journal of the Southwest* 46 (2004): 117–119.

———. 1980. "A Berkeley Perspective on the Study of Latin American Geography in the United States and Canada." In *Studying Latin America: Essays in Honor of Preston E. James*, ed. David J. Robinson, 135–175, esp. 143–158. Dellplain Latin American Studies, No. 4. Syracuse University, Syracuse.

———. 1981. "The Contribution of Carl Sauer to Latin American Geography." *Proceedings of the Conference of Latin Americanist Geographers* 8:8–21.

———. 1982. "Preface" and "Introduction to the Letters." In *Andean Reflections: Letters from Carl O. Sauer while on a South American Trip under a Grant from the Rockefeller Foundation, 1942*, ed. R. C. West, ix, 1–7. Dellplain Latin American Studies, No. 11. Westview Press, Boulder, Colo.

Whitaker, J. R. 1940. "World View of Destruction and Conservation of Natural Resources." *Annals of the Association of American Geographers* 30:143–162, esp. 158–161.

Willems-Braun, Bruce. 1997. "Reply: On Cultural Politics, Sauer, and the Politics of Citation." *Annals of the Association of American Geographers* 87:703–708.

Williams, Glyndwr. 1973. Review of *Sixteenth Century North America: The Land and the People as Seen by the Europeans*, by Carl O. Sauer. *English Historical Review* 88:629.

Williams, Michael. 1983. "The Apple of My Eye: Carl Sauer and Historical Geography." *Journal of Historical Geography* 9:1–28.

———. 1987. "Sauer and '*Man's Role in Changing the Face of the Earth*.'" *Geographical Review* 77:218–231.

———. 2001. "Sauer, Carl Ortwin (1889–1975)." *International Encyclopedia of the Social and Behavioral Sciences*, ed. Neil J. Smelser and Paul B. Baltes, 5:13,490–13,492. Elsevier, Amsterdam.

———. 2002. "Carl Sauer." In *Encyclopedia of Global Change: Environmental Change and Human Society*, ed. Andrew S. Goudie, 2:351–352. Oxford Univ. Press, Oxford.

———. 2003a. "Carl O. Sauer and the Legacy of *Man's Role*." In Mathewson and Kenzer, 2003:217–230.

———. 2003b. *Deforesting the Earth: From Prehistory to Global Crisis*, esp. 385. Univ. of Chicago Press, Chicago.

———. 2003c. "The Creation of the Humanized Landscape." In *A Century of British Geography*, ed. Ron Johnston and Michael Williams, 167–212, esp. 172–174. Oxford Univ. Press, Oxford.

———. 2005. "To Leave 'a Good Earth.'" *Environmental History* 10:765–767.

———. 2009. "Foreword." Herein.

———. N.d.(a). "The Berkeley School." In *Encyclopedia of Human Geography*, ed. Rob Kitchen and Nigel Thrift. Elsevier, Amsterdam, in press.

———. N.d.(b). "Carl O. Sauer." In *Encyclopedia of Human Geography*, ed. Rob Kitchen and Nigel Thrift. Elsevier, Amsterdam, in press.

———. N.d.(c). "Carl O. Sauer." In *Encyclopedia of Geography*, ed. Barney Warf. Sage, Thousand Oaks, Calif., in press.

———. N.d.(d). *To Leave a Good Earth: The Life and Work of Carl Ortwin Sauer*, in preparation.

Wilson, Leonard S. 1948. "Geographical Training for the Postwar World: A Proposal." *Geographical Review* 38:575–589, esp. 575–579, 581–582.

Wolch, Jennifer, and Jody Emel. 1998. "Preface." In *Animal Geographies: Place, Politics, and Identity in the Nature-Culture Borderlands*, ed. J. Wolch and J. Emel, xi–xx, esp. xiii. Verso, London.

Worster, Donald. 1977. *Nature's Economy: The Roots of Ecology*, esp. 245, Sierra Club Books, San Francisco.

Wright, Angus. 1984. "Innocents Abroad: American Agricultural Research in Mexico." In *Meeting the Expectations of the Land: Essays in Sustainable Agriculture and Stewardship*, ed. Wes Jackson, Wendell Berry, and Bruce Colman, esp. 139–140, 150–151. North Point Press, San Francisco.

Wycoff, William. 1979. "On the Louisiana School of Cultural Geography and the Case of the Upland South." Discussion Paper Series, 54, esp. p. 2. Department of Geography, Syracuse University, Syracuse.

Yoon, Hong-Key. 1991. Review of *Carl O. Sauer: A Tribute*, Martin S. Kenzer, ed. *New Zealand Geographer* 47:38.

Zelinsky, Wilbur.* 2000. "Remembering Berkeley." *The Itinerant Geographer: Cele-

brating 100 Years of Berkeley, pp. 253–254. Department of Geography, University of California, Berkeley.

Zimmerer, Karl. 1996. "Ecology as Cornerstone and Chimera in Human Geography." In *Concepts in Human Geography*, ed. Carville Earle, Kent Mathewson, and Martin S. Kenzer, 161–188, esp. 167–171. Rowman and Littlefield, Lanham, Md.

II

EARLY EFFORTS

5

Introduction

William M. Denevan

> The effects of man on the earth, although universally recognized, have received but occasional systematic attention.
>
> Carl O. Sauer, "Man's Influence Upon the Earth," 1916

Carl Sauer's apprenticeship as a published geographer extended, we could say, from 1915, when he completed his dissertation and received an appointment at the University of Michigan, until 1922, just before leaving Michigan for the University of California at Berkeley. His publications during these eight years were largely based on fieldwork as a graduate student in 1910 in the Upper Illinois Valley (1916a) and in 1914 in the Ozark Highlands for his doctoral dissertation (1915a, 1920a, 1920b), along with the coauthored monograph on Starved Rock State Park in Illinois (1918c). His *Pennyroyal* monograph, based on fieldwork with graduate students at Michigan in the early 1920s, was published later (1927a). His research and publications at Michigan mainly involved the development of agricultural and soil surveys. However, this was a period in which Sauer's interests were eclectic, and some articles are surprising and unrelated to what was to follow. His initial papers included such topics as New Guinea (1915b), high school geography (1917a), the gerrymander (1918a), and Niagara Falls (1919b).

Sauer's first publication was not on geography but rather on educational opportunities in Chicago in 1911, a piece that I have been unable to obtain. From 1914 to 1919, he published nine brief book reviews, mostly on physical geography topics, in the *Bulletin of the American Geographical Society* and the *Geographical Review*, eight of them on books written in German, including one by Friedrich Ratzel (1914a, b, c, d, e; 1915d, e; 1916c; 1919c). His first geography article was on the Kaiserin Augusta River in New Guinea in 1915. That same year, he was the second coauthor of "Outline for Field Work in Geography" with Wellington D. Jones (1915c). This is a detailed outline of procedures and topics for fieldwork, based on a seminar at the University of Chicago. Suggestions for it came from Harlan H. Barrows, Douglas W. Johnson, Wallace W. Atwood, Isaiah Bowman, William H. Hobbs, Mark Jefferson, and Bailey Willis. Young Sauer (age twenty-five), apparently, was the primary author.

"Exploration of the Kaiserin Augusta River in New Guinea, 1912–1913" (1915b, ch. 6 herein) is an "abstract" by Sauer of a report by geographer Walter Behrmann on a large German Colonial Office expedition to a little known region in Kaiser Wilhelm Land, which had been acquired by Germany in 1885–1886, and later by Australia. Sauer, of course, was not a participant, but he was certainly interested in German geographical research. He mainly describes the terrain, vegetation, and climate of the region examined by the Germans. The indigenous people are briefly mentioned. Those on the upper tributaries "had never heard of white men and were unacquainted with iron." Those in the lower reaches had "assembly houses, which are well designed and elaborately decorated with carvings, paintings, and woven goods . . . considerable aptitude in the production of cultivated crops" (343). A map was included but is omitted in the reprint here. This descriptive summary, long forgotten, presages Sauer's much later interest in the tropics, indigenous people, crops, and exploration.

Sauer's doctoral dissertation at Chicago, "The Geography of the Ozark Highland of Missouri" (1915a), was completed under Rollin D Salisbury. It was published in 1920 by the Geographical Society of Chicago and reprinted in 1968 by Greenwood Press. The preface is included here (ch. 7 herein). Sauer begins by stating, "This volume is a study in regional geography, the most urgent field of geographic inquiry." This Chicago (and European) approach to geography at the time was later rejected by Sauer (see Martin, ch. 10 herein). However, Sauer does go on to say that "the collection of facts" about a region, past and present, but "not with the enunciation of a theory," will make possible "an adequate explanation of the conditions of life in a given area . . . which will aid in working out fundamental principles" (1920b:vii, ch. 7 herein).

The Ozark study is in three parts: (1) the physical environment, (2) the historical background, and (3) the current economic conditions. Sauer says the "second part considers the influences of environment on the settlement and development," but he also says that for "three racial groups" there are "curious contrasts in their fortunes under the same environing conditions" (viii). Thus, Sauer's stance about geographic influences when he wrote his dissertation may seem ambiguous. However, Sauer here is referring to environmental "influences," not determinants. "Curious contrasts in their fortunes" indicates a recognition of the importance of culture (see p. 325 herein).

The 1920 volume contains numerous borrowed maps and maps based on existing data, rather than original mapping by Sauer, in line with later critics who say that Sauer seldom mapped anything himself, a seeming contradiction to Sauer's strong belief that maps and mapping were fundamental to geography (1970a:5–6, ch. 48 herein). There were, however, early exceptions where Sauer indeed either did his own mapping or supervised it (e.g., 1927a, 1936c). The 1920 dissertation/monograph is well documented in contrast to later studies by Sauer. In one of the few reviews, J. F. Unstead in five pages in the *Geographical Journal* in 1922 praised the book ("a notable contribution"), but argued that the discussion of settlement history was mostly inappropriate because it did not relate to the present (echoed later by Hartshorne, 1959:106). John Leighly (1976:338) many years after believed that *Ozark Highland* "is probably the best work of its kind and time written in the United States."

Sauer had an early interest in the teaching of geography in the schools, as indicated by his Chicago educational report (1911) and his teaching for a year at a teacher's college in Salem before going to Michigan. One of his first articles was "The Condition of Geography in the High School and Its Opportunity," published in the *Journal of the Michigan Schoolmaster's Club* in 1916 and reprinted in *The Journal of Geography* (1917a). This interest reappears later, especially in his school text *Man in Nature* (1939; see Lovell, 2003), the only textbook Sauer ever wrote aside from class syllabi.

In "The Condition of Geography," Sauer discusses the decline of geography, the lack of trained geography teachers, the failure of physiography, climatology's being boring because of a focus on technical aspects rather than processes, commercial geography's being excessively based on commodities rather than production in a regional context, and the need for courses on the "general principles" of geography. He points to the "Modern Geography" text by Salisbury, Barrows, and Tower in 1913 (actually *The Elements of Geography*, 1912) as the "new geography" which "attempts to give a general view of the relations of life, most particularly of man, to environment" (1917a:146).

This sounds similar to what today is one of the four primary subfields of geography, "nature and society," in the *Annals of the Association of American Geographers*.

Principles, Sauer indicates, should be packaged with regional studies, including fieldwork in the local region, or "home geography," which will particularly make the study of geography relevant and interesting for the school student. Such an approach to teaching geography will link "earth science and biology . . . to economic conditions . . . [and] can assure in time the general recognition of this course as a fundamental elementary science for the high school" (1917a:147–148).

The monograph on the Upper Illinois Valley is mostly on physical geography, with the remainder on settlement history and economy. Noteworthy is a section on "man as a factor in erosion" (1916a:140–143), possibly Sauer's first exposition on people as a factor in environmental change and anticipating his forthcoming work on soil classification and survey. As causes of accelerated erosion, he mentions deforestation, over grazing, and slope cultivation.

The same year Sauer published a little-known one-page note, "Man's Influence Upon the Earth," in the first volume of the *Geographical Review* (1916b, ch. 8 herein), a preview of work to come, culminating forty years later in the similarly titled *Man's Role in Changing the Face of the Earth* (1956a). He refers to modification of the earth's surface by mineral extraction, construction, agriculture, control of streams, drainage of lakes and swamps, accelerated runoff, groundwater depletion, removal of vegetation, fire, and disturbance of plant associations and animal life. He says that "man is an important ecological factor," probably his first use of the term "ecological." And, "Man has become the guide fossil of the present geologic period."

Sauer's first of several articles concerned with land surveys is "Proposal of an Agricultural Survey on a Geographic Basis," a Michigan Academy of Science report (1917b). At that time he was in his third year at Michigan, where there may have been some pressure toward applied work, as well as from the conservation movement itself. Related papers on land classification and mapping followed (1918b, 1919a, 1921a, 1924a).

In "Proposal" Sauer again states that the "the most urgent demand [of geography] is along lines of regional studies," and that these regions require "truely geographic surveys," heretofore lacking. Particularly needed are soil surveys, "not an end in itself, but as a means to a better understanding of the agriculture of a given district" by expanding the scope of the soil survey to "evolve into an agricultural survey" (1917b:79–80). The components of the proposed survey model, drawing on Sauer's own *Ozark Highlands* dissertation and influenced by soil scientist Eugene Hilgard, included location, topography, drainage, soil, slope, climate, and land use (farm type, crops, and management practices), utilizing photos and sketch maps. Examples are given of land use maps in Austria and Scotland, and classes of land use, actual and potential, are suggested for Michigan. "A map of this type unfolds a complete panorama of the rural conditions in a county. It will enable comparative studies in rural economy that are now impossible" (1917b:85). Sauer applied this land mapping approach to his *Geography of the Pennyroyal* (1927a), based on fieldwork in 1923.

The 1917 paper was a proposal for the establishment of a soil survey in Michigan, leading to the Michigan Land Economic Survey in 1922 (Martin, 2005:455–460), which mapped in detail the characteristics and utility of the cut-over pine lands of the Upper Peninsula. This influenced land-type and land-use mapping in various regional studies by geographers. Map overlays of multiple land characteristics, physical and human features, and potential use, were superimposed on base maps (later air photos) or were indicated by numerical or fractional code systems (see Finch's 1933 Monfort study). Today much the same is done more systematically, accurately, and rapidly with Geographic Information Science technology.

Sauer's work on land surveys included soils surveys, which led to an awareness of the neglect of soils by geographers. He addressed this in a little-known statement in the *Journal of Geography*: "Notes on the Geographic Significance of Soils: A Neglected Side of Geography" (1922, ch. 9 herein). He believed that the background of physical geography should include soils in course work, texts, and field research, but "one looks in vain." He differentiated soil science from soil geography, whose "guiding purpose is the study of the utilization of soils by man," including the identification of soil regions, "an important adjunct to the balanced study of the economy of the land." He also felt that "at the least every student of geography should have training in the origin, fertility, and management of soils." This is hardly the case today, just as it wasn't common when Sauer wrote these words. However for many geographers in the decades following 1922, a course on soils was recommended or required.

In the mid-1920s Sauer's writing turned to methodology, and in the 1930s his field research turned to Mexico. However, his applied interests in land use, soils, and erosion continued when he became a consultant to the new Soil Conservation Service (1936b, c, ch. 22 herein), and with his reports on land use to the Science Advisory Board (1934f; 1934g) (see Effland and Effland, 1992).

REFERENCES

Effland, Anne B. W., and William R. Effland. 1992. "Soil Geomorphology Studies in the U.S. Soil Survey Program." *Agricultural History* 66:189–212, esp. 190–195.

Finch, Vernor C. 1933. *Montfort: A Study in Landscape Types in Southwestern Wisconsin.* Geographical Society of Chicago, Bulletin 9, Chicago.

Hartshorne, Richard. 1959. *Perspective on the Nature of Geography.* Association of American Geographers, Monograph Series 1. Rand McNally, Chicago.

Leighly, John. 1976. "Carl Ortwin Sauer, 1889–1975." *Annals of the Association of American Geographers* 66:337–348.

Lovell, W. George. 2003. "'A First Book in Geography': Carl Sauer and the Creation of *Man in Nature.*" In *Culture, Land, and Legacy: Perspectives on Carl O. Sauer and Berkeley School Geography,* ed. Kent Mathewson and Martin S. Kenzer, 323–338. Geoscience and Man, vol. 37. Geoscience Publications, Department of Geography and Anthropology, Louisiana State University, Baton Rouge.

Martin, Geoffrey J. 2005. *All Possible Worlds: A History of Geographical Ideas,* 4th ed. Oxford Univ. Press, New York.

Salisbury, Rollin D, Harlan H. Barrows, and Walter S. Tower. 1912. *The Elements of Geography.* Henry Holt, New York.

Sauer, Carl O. For references, see the Sauer bibliography at the end of this volume.

Unstead, J. F. 1922. Review of *The Geography of the Ozark Highland of Missouri,* by Carl Sauer. *Geographical Journal* 59:55–59.

6

Exploration of the Kaiserin Augusta River in New Guinea, 1912–1913 (1915)

Carl O. Sauer

The basin of the large Kaiserin Augusta River, which roughly bisects Kaiser Wilhelm Land in the northeastern part of New Guinea[1] had been very little known until the German expedition of 1912–13 carried out its extensive explorations. The war has apparently deferred the publication in detail of the results of this important geographical work but it was summarized last year in a paper published by the Berlin Geographical Society [Behrmann, 1914], of which the following is an abstract:

The expedition for the exploration of the Kaiserin Augusta basin, equipped by the German Colonial Office, the Royal Museums, and the German Colonial Society, reached the Protectorate on February 8, 1912. The party was in [the] charge of Bezirksamtmann Stollé, and included Ledermann, botanist, Bürgers, zoologist, Roesicke, ethnographer, and Behrmann, geographer.

The main base was established near Malu, a village about 250 miles from the mouth of the river. Here the party was increased by four German officials, eleven Chinese boatmen, fifty native soldiers, and numerous carriers. The Kaiserin Augusta River is navigable by motor boats for a distance of 560 miles and its tributaries increase this mileage considerably. The main routes of exploration therefore were by water. In the nineteen months which the party as a whole spent in the field, all the tributaries as far as the Dutch boundary were explored to the extreme limit of their navigability and four overland exploring trips were made of an average length of three months.

The main stream is characterized by extensive meanders even in its upper course; cut-offs are common, but are destroyed rapidly by the rank growth of vegetation. The maximum variation of water-level was twenty-four feet, and the highest stage occurred at the end of the rainy season. The natural levees attain a height of ten to twelve feet and a width of 650 feet. Undercutting is active at low water stages, and is facilitated by seepage that softens the stream banks. Along the margin of the river, reeds and wild sugar-cane grow; farther back, at least on the upper course, is the forest, whose outer edge is usually a young growth of bread-fruit trees. On the lower course are vast grassy swamps, infested by mosquitoes. The native population is concentrated largely upon the banks of the river. Because of the strong shore-current, the Kaiserin Augusta has built no regular delta, but its sediments are shifted to the west and there have formed a lagoon three miles wide and twelve miles long. Below Malu, the country is still in an amphibious condition and for the most part impassable. At certain crevasses the river inundates regularly large areas, and in these backwaters fish abound. Here the natives build villages on piles, attracted by the easy supply of food.

At the mouths of the tributaries is a zone of grassy swamps. Above, there are luxuriant forests, which are however in many cases merely a screen in front of the sago swamps that intervene between river and upland. On the upper tributaries natives were found who had never heard of white men and were unacquainted with iron. All of the navigable affluents of the Kaiserin Augusta enter from the south, five above Malu and three below. The lowest, largest, and most interesting of these is the Töpfer River, which seems to receive a considerable part of its water from the Ramu River, which it approaches to within two miles. On its banks live advanced tribes, skilled in the making of pottery and engaging in commerce.

The density of population is less than had been expected. In the area visited by the main expedition there are not more than 27,000 people. The number of tribes is great, and between them are in many cases uninhabited buffer zones. The people are mostly small but no pygmies were found; the upper part of their bodies is unusually well developed. On the upper river the houses are characterized principally by their large size, accommodating about forty persons, and by being built upon piles thirteen to twenty feet long. The farther downstream, the easier are materials for building secured from the driftwood of the river. The artistic sense of the people expresses itself especially in the assembly-houses, which are well designed and elaborately decorated with carvings, paintings, and woven goods. In some sections the natives show considerable aptitude in the production of cultivated crops, and tobacco grown by them has been found to be of excellent grade.

A meteorological station was kept at Malu, and observations were taken at 6 A.M., 2 P.M., and 8 P.M., with the following averages (centigrade) respectively: 23.9°, 31.4°, and 26°. The annual mean was 27.1°C. The total precipitation was 114 inches for the year, which is supposed to have been abnormally wet. Rain fell with great regularity in the evening or at night, and was often accompanied by violent whirlwinds which uprooted trees and overturned houses.

From Malu to the Dutch border the lower courses of the tributaries are abnormally wide, and in many cases form lakes before they debouch into the main valley. The mountain mass is here sinking at a rate somewhat more rapid than that at which the rivers aggrade their flood-plains. Alluvial plains penetrate far into the interior of the western mountains and surround numerous outlying spurs. On the east, on the other hand, valley terraces record a recent elevation of the land.

The overland expeditions determined the hitherto virtually unknown character of the interior mountains. Progress was very slow because of the difficulties encountered, among which the daily fogs, the dense vegetation, the absence of anything edible in the forests, and the knife-like character of the ridges were chief. Most troublesome was a forest region, in which heavy washing had laid bare the roots of the trees to a depth of three to five feet, and the roots were covered with dripping wet moss. To secure a view it was often necessary to make a clearing. On one expedition the crest of the central range was reached at an elevation of 5,640 feet. This was south of the Hunstein Range, where the central range does not exceed 8,200 feet. At the eastern end of the area a high range, called the Schrader Range, leads off into the foreland. One of its summits, 6,890 feet high, was ascended; from this place a chain far in the interior was observed, which is estimated to rise to 11,700 feet. On the west, an expedition penetrated to the area previously explored by Leonhard Schultze. Rocky peaks, joined by knife-like ridges, are common. The bared rock surfaces are the result of landslides, caused by the falling of trees, whirlwinds, or cloudbursts. The mountains have sharply chiseled erosional forms.

The main mountain range extends into the former German colony from Dutch New Guinea, and takes a course far inland to the Bismarck Range. East of the international boundary it divides digitately into ranges which decline gradually and disappear beneath the alluvium of the Kaiserin Augusta basin.

The coast range consists, in so far as it has been examined, of coralline limestone, recently elevated. The structure of the island has been determined by block-faulting, one of the great lines of displacement being the depression occupied by the Kaiserin Augusta-Ramu-Markham valleys. The movements involved still express themselves in frequent earthquakes. The rivers are adjusted to the orogenic structure and have modified the relief of the mountains. The Kaiserin Augusta River flows out of the angle between the West Range and the Dutch mountains, and passes around the West Range in a broad arc. Tributaries occupy gaps between the other ranges.

The existence of mineral resources was not investigated. Judgment is also reserved as to whether plantations could succeed under present conditions.

NOTES

Reprinted with permission from the *Bulletin of the American Geographical Society* 47:342–345. Copyright 1915, American Geographical Society of New York.

1. Kaiser Wilhelm Land was acquired by Germany in 1885–86. The town and harbor of Friedrich Wilhelm, the seat of government, were occupied by the Australian naval forces at the end of September last [1914].

REFERENCE

Behrmann, Walter. 1914. "Geographische Ergebnisse der Kaiserin-Augustafluss-Expedition." *Zeitschrift der Gesellschaft für Erdkunde zu Berlin* 4:249–277.

7

Preface to *The Geography of the Ozark Highland of Missouri* (1920 [1915])

Carl O. Sauer

This volume is a study in regional geography, the most urgent field of geographic inquiry. Geography is among the youngest of the sciences. It is not ready, therefore, to announce many generalizations, but must concentrate on the systematic and comprehensive scrutiny of individual areas, inquiring into the conditions of the past as well as into those now existing. The collection of facts in this manner, and in this manner only, will lead to the establishment of the principles of geography. Such a study implies the attitude of the judge of conditions rather than of the advocate of theories. It is concerned with the impartial analysis of the conditions of life in a region, not with the enunciation of a theory for which evidence is to be adduced. It does not attempt to make out a case for the potency of any particular element of the environment, but contents itself with asking, What are the advantages and handicaps that are inherent in the region in question? The purposes of such a study are to furnish an adequate explanation of the conditions of life in a given area and to contribute proved statements which will aid in working out fundamental principles.

The preparation of regional monographs, numerously represented in European countries, has hardly commenced in America. A century ago the conditions and resources of various parts of our country engaged the attention of many observant writers. These accounts of early travelers constitute in fact the greater part of our geographic literature to this day. As facilities for observation increased, their number was reduced, until at present there is almost no contemporary geographic literature other than brief papers. If the curiosity which attaches to the unknown has disappeared, the need of correlated information about the parts of our country has increased as its parts have become settled and developed. This it is the province and the duty of the geographer to supply. The present paper considers a single geographic unit. The Ozark Highland of Missouri was selected because of its unusual wealth of geographic responses and because little is known concerning its conditions and possibilities. The size of the area, larger than Scotland and as large as Ireland, has precluded an exhaustive treatment of the subject. It is rather a reconnaissance, which, it is hoped, may lead to more detailed studies.

The topic is treated in three parts. The first is an outline of the environment, that is, a sketch of the region and a statement of the geographic factors. Only those things which are pertinent to an understanding of the conditions under which the people live are introduced. Rock formations are of significance in this connection in so far as they have determined topographic features, soils, and mineral resources, and in so far only. No attempt is made to sketch the physiographic history except as it contributes to the explanation of surface features, drainage conditions, and soils. The mineral resources need discussion only in so far as they have been a factor in the development of the region. Whatever is more than this may be of geologic, physiographic, or mineralogic interest, but is not pertinent to geography. The various factors of the environment differ in importance in different parts of the area. By evaluating them singly and collectively it is possible to establish contrasts between parts of the highland and thus to determine a number of smaller unit areas. Each of these subdivisions has internal unity of geographic conditions and is set off from its neighbors by important points of contrast. These natural subregions become the units of observation in the sections that follow, in which their past and present utilization is observed and compared.

The second part considers the influences of environment on the settlement and development of the different parts of the highland. Certain portions have had continuous advantages, as others have been permanently retarded in development. In some parts certain geographic opportunities resulted in a period of early growth, soon arrested, whereas other sections, later in securing a start, have forged to the front rapidly. Three racial groups have possessed a part of the area in turn, with curious contrasts in their fortunes under the same environing conditions. This historical portion develops its argument by the fullest possible use of source materials. Wherever possible, statements from original sources are employed to bring out the thread of geographic influences that runs through the history of the region.

Finally, economic conditions are represented as they exist today, together with their explanation in so far as they are not merely the continuation of institutions, the beginnings of which were traced in the historical section. In conclusion, a forecast is offered of the lines along which the future of the region will be worked out.

The study here submitted is the outgrowth of long acquaintance with the area and of deep affection for it. It is, in fact, a study in home geography, a study of the old home with its many and vivid associations. Later residence outside of Missouri has supplied a more objective viewpoint without destroying the old familiarity. Systematic field work in the fall of 1914 and summer of 1915 has supplemented the earlier acquaintance. To consider the region as an outsider has been impossible and will always be. With the increasing distance interposed by time and space there yet remain forever green the scenes of early years. The old white church, astride its rocky point, overtopped by cedars that grow on the warm rock ledges, forever looks forth upon the fairest valley. The lower slopes are abloom with red clover, or golden with wheat. Wide fields of blue-green corn border the shaded stream, where the bass lurk in transparent pools. In the distance forests of oak mantle the hillsides, up which, past spacious farmhouses, the country roads wind. The people who move upon the scene of this account are homefolks one and all. Some have succeeded better than others, some give greater promise than others, but they are all well worth knowing, and in all cases an understanding of their various problems of making a living goes far to explain their contrasted conditions. In this spirit the study is undertaken.

The first draft of the manuscript was presented before the Seminar in Geography at the University of Chicago in 1915 and there subjected to much helpful discussion. The several parts have profited by intensive reading and criticism at the hands of Professors W. S. Tower, H. H. Barrows, and J. Paul Goode. It is difficult for me to express in any adequate way the great debt I owe to my old teacher and friendly counselor, Professor R. D Salisbury, in the carrying out of this work. From its first planning to its publication his aid has been freely given in many ways. Grateful acknowledgments are due also to the Geographic Society of Chicago for making possible the publication of this volume, a study in a field in which avenues of publication have not yet been established.

NOTE

Reprinted from *The Geography of the Ozark Highland of Missouri*, vii–ix. Copyright 1920, Geographic Society of Chicago, Bulletin No. 7, University of Chicago.

8

Man's Influence Upon the Earth (1916)

Carl O. Sauer

In the study of the reciprocal relations of man and earth the emphasis is placed rightly on the physical environment as affecting the activities of man. The effects of man on the earth, although universally recognized, have received but occasional systematic attention. The geologic texts of Lyell, Dana, and Chamberlin and Salisbury make brief mention of man as a physiographic and geologic agent. Recent suggestive articles on this topic have been written by A. Woeikof (1901), C. P. Lucas (1914), and, most recently and most exhaustively, by Ernst Fischer (1915).

Fischer points out the fact that man is at present one of the leading agents in the direct modification of the earth's surface. He estimates the mass of material moved annually in the production of ores, mineral fuels, and salt at one cubic kilometer and would possibly double this figure for all mineral industry. To this amount are to be added the vast quantities of earth and rock moved in building and agricultural operations. Even more important is man's release or stimulation of physiographic forces, there being in general a great disproportion between the forces released and the energy expended by man in disturbing their equilibrium. The control of streams, the reduction of lake and swamp areas, the increased run-off, the decreased cover of vegetation, and, more especially, the rapid exploitation of the great ground-water reservoir represent, in the opinion of the author, the most extensive interference by man with the physiographic processes. The influence of man, as that of the other surficial agencies, is dominantly degradational, with occasional and minor aggradational activities.

Man is an important ecological factor, in other words, in modifying profoundly the paleontological record of the future. The adjustment of plant associations is being disturbed by the introduction of new species, by the clearing and cultivation of land, by fire, and other means resulting from the voluntary or involuntary meddling of man. Similar revolutionary changes are wrought in the animal world, especially among the larger animals and among numerous parasitic forms.

By his distribution through all climatic regions and his power to employ great physical forces, man has become the guide fossil of the present geologic period. A consciousness of this geologic role of man is necessary for the continued well-being of future generations. Conservation therefore becomes a problem of geology, as well as geography.

NOTE

Reprinted with permission from *The Geographical Review* 1:462. Copyright 1916, American Geographical Society of New York.

REFERENCES

Fischer, Ernst. 1915. "Der Mensch als geologischer Faktor." *Zeitschrift der Deutschen Geol. Gesell: A. Abhandl.* 67:106–148.
Lucas, C. P. 1914. "Man as a Geographical Agency." *Geographical Journal* 44:479–492.
Woeikof, Alexander I. 1901. "De l'influence de l'homme sur la terre." *Annales de Géographie* 10:97–114, 193–215.

9

Notes on the Geographic Significance of Soils: A Neglected Side of Geography (1922)

Carl O. Sauer

The science of soils has been growing vigorously year upon year with a very fine amount of achievement to its credit. It has long since achieved recognition as a distinct branch of knowledge. In some curious manner this subject has received but slight attention from the geographer. Our competence to evaluate soil conditions is in general slight, and does not extend much beyond a few poorly understood generalizations of soil fertility and geologic origin. It is possible that soils are to be ranked quite as high in the environment as is surface. In generalizations covering large areas they are of much inferior value to climate, but usually the smaller the region the more significant does an understanding of soil conditions become. And the future of geography as a field of research lies immediately in the more intensive working of restricted areas.

The matter of soils cannot be dismissed by the notion that soils are fertile or infertile any more than climate can be disposed of as being beneficial or hindering. There is little or no critical value attached to most of the meager statements about soils that appear in the literature of geography. How much attention is paid in our geographic instruction to geological formations as sources of soil materials or to physiographic processes as soil forming agents? The study of physiography as a group of dynamic processes modifying the earth's crust is really a phase of dynamical geology. As the study of the origin and nature of land forms and of soils, together with the study of weather and of climate, it becomes the physical background of geography. Such a course, however, is practically not in existence. This physical background is developed adequately only on the climatologic side. The study of land forms is still largely a hope and that of soils is essentially a missing quantity in physical geography today.

Physical geography needs redefinition from the geographic standpoint and soils should be considered in it not merely (1) as the products of geologic formations and (2) as the result of physiographic processes, but as well (3) as being biologically modified, (4) as being profoundly affected by climate, and (5) as being changed by the processes of time. These five things represent the genesis of a soil and are fundamental and similarly significant notions of soil science. With the idea of the differentiation of soils according to origin in mind, the geographer is further interested in this subject in two ways, the question of soil fertility and of the management of soils for specific production. Soil origin is the study of the environment that has resulted in the soil; soil fertility is the study of soil properties that affect its use; and soil management is part of the record of man's experience in gaining a livelihood from the land. Are not all three considerations so important that they must find emphatic recognition in the field of geography? One looks in vain in any textbook of geography or of physiography for an even partially adequate recognition of this factor. Only one school [Chicago] it would seem has considered a course in soil geography.

Soil geography is as appropriately a branch of geography as is climatology. The study of soils as such is a separate science from which the geographer should learn to borrow, but in which considered separately he is no more interested than he is in meteorology. Meteorology as such is properly considered a branch of physical science. It has come to be taught in connection with geography at a good many institutions because often the geographer has been more interested in it than has the physicist, but the observations and principles of meteorology are essentially not in the province of the geographer. Similarly, where no courses in soil science are given in agricultural or other divisions, it may be desirable that the geographer provide them for his own interests. But this is not soil geography. Soil geography as yet is unformulated, but its guiding purpose is the study of the utilization of soils by man. We should have a knowledge of the different types of utility of the various types of soils, of the problems and possibilities of their improvement based on economic rather than technical considerations; in short we should know the economy of each important soil. It is possible also to study soil regions in terms of use as we study climatic regions in terms of human activities and progress. Agricultural colleges in many cases do that sort of thing now as an appendix to soil work, but they do it incidentally and mostly from the standpoint of individual crops. A soil region embraces a group of soils, in which certain dominant soils usually determine the soil practices for the area. In the soil region the utility of the individual soil can be considered only in terms of what might be called the soil association. This group economy has as yet received no systematic investigation. A balanced soil geography will inquire into the whole matter of the economy of the soil region, the competition between farm crops, and between plow use and grazing and forest use of the land. It will register the experiences of past and present use as to destructive exploitation or sustained yields of more intensive production. It deserves recognition as an important adjunct to the balanced study of the economy of the land.

At the least every student of geography should have training in the origin, fertility, and management of soils. These matters should form part of any serious introductory course in geography, whether it be to freshmen in college or to the teachers who are preparing at normal schools. The professional student of geography should have a course in soil science comparable to his work in meteorology, physiography, and plant ecology. And finally, perhaps, it may be hoped that the geography student of advanced standing will have the opportunity of studying at competent hands the soil geography of the world, and that laterite, chernozem, and podsol will become as intelligible terms to him as are mesa, fiord, and doldrum.

He who would read on soils can do no better than to begin with *Soils* [1914] by E. W. Hilgard, which after fifteen years is still without need of major revision. Hilgard went into exile as State Geologist of Mississippi, as his colleagues who were hard rock geologists thought. On the Gulf Coast he gained the experience that made him the pioneer of American soil work. In this geologically apparently uninteresting area he found a virgin field of soil studies which have become of classical significance. With his widened experience he later became the builder of scientific research in agriculture through the monumental development of experiment station work under his direction at the University of California. Hilgard's *Soils* is a book of extraordinary experience and wisdom. Professor [F. H.] King's *The Soil* [1900] in the Rural Science series, is the other and earlier American classic, perhaps most highly interesting in its discussion of drainage and irrigation. To these two primers of soil science should be added [A. D.] Hall's *The Soil* [1910], summarizing the

British viewpoint in particular and based largely on the work of the world's most venerable experiment station, at Rothamsted, England. To this general list may be added a thoroughly modern handbook by [T. L.] Lyon, [E. O.] Fippin, and [H. O.] Buckman, *Soils: Their Properties and Management* [1915], the most symmetrical treatment of the general subject in English.

NOTE

Reprinted with permission from the *Journal of Geography* 21:187–190. Copyright 1922, National Council for Geographic Education.

III

TOWARD MATURITY

10

Introduction

Geoffrey J. Martin

> Barbed wire fences may be necessary in elementary curricula, but the pursuit of knowledge cannot afford to frustrate itself by building fences around narrow plots of learning.
>
> Carl O. Sauer, Correspondence, *Geographical Review*, 1932

By the 1920s the role of environmental determinism in American geography had been reduced. Terms, in descending order of rigor, lingered through the 1920s and 1930s: "controls," "influence," "geographic factor," "causation," "response," "adjustment." William Morris Davis, largely responsible for the inclusion of the relationship between environment and environed, had retired from his Harvard stronghold. He spent much of the 1920s lecturing in the West (accepting accommodation and other hospitalities from the sons of Harvard), where he continued his forte parsing the meanings of landscape both as hobby and profession, "reading God's thoughts," as Mark Jefferson was later to write. The Association of American Geographers, which Davis had founded in 1904, was under spreading sail; departments of geography were being established; and a body of courses had been developed for which a literature had been generated. Works of a very high order were being written by members of the profession, typified by H. L. Shantz and C. F. Marbut, *The Vegetation and Soils of Africa* (1923); I. Bowman, *Desert Trails of Atacama* (1924); J. K. Wright, *The Geographical Lore of the Time of the Crusades* (1925); E. Huntington, *The Pulse of Progress* (1926); M. Jefferson, *Peopling the Argentine Pampa* (1926); W. M. Davis, *The Coral Reef Problem* (1928); and E. Antevs, *The Last Glaciation* (1928).

Visitors (largely non-German) came to the United States from other countries, their books and articles preceding them. In the 1920s came more notably P. Teleki, the De Geers, E. Antevs, R. Blanchard, A. Penck (and assistant A. Haushofer), P. W. Bryan, J. Fairgrieve, A. G. Ogilvie, T. G. Taylor, and M. Aurousseau. The exception was Berkeley, where from 1925 to 1930 the German geographer Oskar Schmieder teamed with Carl Sauer; other German geographers were to follow (though not to the exclusion of briefer stays for non-German geographers).

Regional geography, having emerged with J. W. Powell most notably in the 1890s, became ever more dominant as the common denominator of what was considered geographic. A series of regional books was being written that embraced the continents of the world. Economic geography was commonly offered, and variants of political geography began their intrusion into the curriculum. Students absorbed these courses with interest and departments thrived.

It was in this milieu that Sauer's intellectual posture departed from the mainstream. Sensing regional geography as compilation, and not investigation, he sought a more inclusive study of landscape. He did this by adding the dimension of time to his inquiry in questing for understanding of man's occupance of the earth. His inquiry now became investigation, and investigation of smaller areas than was the practice of many of the regionalists. This was the early design of Sauer's cultural geography that sought the biography of human impress on the face of the earth. He then wrote his thought on the loom of his language, investing the remainder of his life in the cause.

In "The Survey Method in Geography and Its Objectives" Sauer urged that "distributional differences of life therefore represent the distinctive field of geography" (1924a:17), asserting that this posture secured geography from challenge by any other science. He dismissed "influences" as guidance and dwelled on the worth of study in the field. This enthusiasm remained with him for the rest of his days. Thirty years later he wrote, "Our greatest need is to recruit, train, and support competent field observers. This is the greatest weakness in American geography ... we are most woefully weak in the ability and desire to make original observations" (Sauer to F. Kniffen, L. Quam, and R. J. Russell, December 24, 1954). Invoking the natural region as place where the geographer might begin investigation, he pondered the notion that "no better definition of geography has been given us than this idea of the derivation of the cultural area from the natural area" (1924a:24).[1] Then it was that Sauer introduced the German terms *Naturlandschaft* and *Kulturlandschaft* (from Gustav Braun).

His work to this time had been invested in surveying smaller units of land, in studying their component parts, the interaction between those parts, and the meaning and impact of human intrusion. He studied the process of survey in works including publications of the General Land Survey, the Great Geological Surveys, other geological surveys, and his own studies, which included "Outline for Field Work in Geography" (with W. D. Jones, 1915c). By 1924, he used the term "natural region" in a sense different from, and devoid of reference to, A. J. Herbertson. It was then that Sauer offered courses in regional geography, and read the French geographers (especially Jean Bruhnes and Vidal de la Blache) on that and other subjects. In "The Survey Method," he defined the natural region: "In its simplest terms a geographical field study has the dual objective of representing the natural or original condition of the area and of showing the manner and degree of its utilization and modification by the people who live there. . . . Other expressions for the same idea as the term 'natural region' are physical or natural environment, equipment, and resources, and the common French term 'milieu'" (1924a:21).

Later, Sauer was to reconsider the concept of the natural region. He wrote to John Leighly (July 27, 1943): "I think we may consider that for the duration geographers will be considered as people who know something about regions. As you know, I don't disagree with this view at all, and have objected to the regionalists only because they have been too often short on knowledge and curiosity." And again, writing to Preston James (March 16, 1940): "We've taken refuge in a regionalism that is descriptive without being analytic, because we refuse to face the analysis of cultural processes . . . we've shyed away from the dynamics of cultural origins and cultural change."

In the same year as "The Survey Method" was published, Sauer and Leighly (then both resident in Berkeley) published in Ann Arbor *An Introduction to Geography: Elements* (1924b). It was an introductory text or syllabus (the authors referred to it as a "manual") that experienced numerous editions during the 1920s and 1930s. As a physical geography syllabus, the unifying element was climate rather than geomorphology, which dominated in other texts of the time. An antecedent to the Sauer-Leighly syllabus was *The Elements of Geography* by Chicago geographers R. D Salisbury, H. H. Barrows, and W. S. Tower (1912), with which Sauer was familiar. Later editions of Sauer and Leighly served as models for the long successful *Elements of Geography* textbook by V. C. Finch and G. T. Trewartha (1st edition 1936, with other editions to 1967) (Leighly, 1976:339; Martin, 1988:170; Mikesell, 2002).

The first chapter of *An Introduction . . .* (1927e edition, ch. 11 herein) was titled "The Field of Geography" and began with a history of the field dating from the time of Homer and Nordic ancestors. Originally written just before "Morphology of Landscape" (1925), it states that "geography is an orderly knowledge of the content of regions" (1927e:1). However, the authors also state that geography "is concerned with . . . the development of the cultural out of the natural landscape" (10). The regional viewpoint was developed and honed pursuant to Sauer's study of the Upper Illinois Valley in 1910 (published 1916a) and his 1915 dissertation on the Ozark Highland (published 1920b). It was rigorously pursued in his Michigan years (1915–1923) with the cut-over land studies, the establishment of the Mill Springs Field Station in Kentucky in 1920, a burgeoning regional geography on university campuses, his leadership role among the younger geographers of the Midwest, and his drive and enthusiasm that advanced regional thinking via survey.

All of this was set against Sauer's own very keen interest in the history of both science and more particularly geography. When he departed the University of Michigan in 1923 for the University of California, Berkeley, he took with him John Leighly, a person whose "intellectual curiosity is insatiable" (Sauer to Ruliff S. Holway, March 12, 1923). Leighly also showed a very keen interest in the history of science. Sauer's interest in this area of study led to Berkeley seminars concerning the history of geographical thought, and so placed him alongside C. T. Conger, J. Paul Goode, E. van Cleef, and E. C. Semple as one of those responsible for the earliest offerings of this genre in the United States. This interest and knowledge first appeared in "The Field of Geography" (1924b; 1927e, ch. 11 herein).

Then came "The Morphology of Landscape" (1925) and numerous other works prior to "Recent Developments in Cultural Geography" (1927c). Some three years only had passed between "The Survey" (1924a) and "Recent Developments" (1927c), but there emerges during that time a maturity of thought, research design, and literary expression that is both imposing and hard to explain. Possibly he felt freed from the anti-German posture of W. H. Hobbs as chair of Geology-Geography at Michigan (though Hobbs had been scrupulously fair in the matter of recommendation of Sauer to membership in the AAG in 1920 and to a professorship in his department). Possibly it owed to the fact that he now owned courses designed to his taste, that he could begin to work with AAG committees that interested him (in 1920 he had been appointed, incongruously, chair of "Geographic Illustrations" and a member of "Geographic Education"). Possibly there was a growth in confidence and experience from exchange with scholars of like mind, and success

Introduction

in developing his written expression. All could have contributed to a maturation hardly possible of definition. Later in life, Sauer indicated that "Morphology" (1925), "Recent Developments" (1927c), and "Geography: Cultural" (1931a, ch. 12 herein) "are best considered as successive orientations and have had utility as such; they belong to the history of geography" (Sauer to Richard Hartshorne, June 22, 1946).

Ill definition of the nature of geography still plagued the subject. A series of essays attempting definition had become familiar to American geographers: Davis, 1906; Tower, 1910; Roorbach, 1914; Fenneman, 1919; Dryer, 1920; Barrows, 1923; S. De Geer, 1923; Davis, 1924; and Bruhnes, 1925. The extended essay "Recent Developments in Cultural Geography" (1927c) and the brief "Geography: Cultural"[2] (1931a, ch. 12 herein) continued Sauer's published interest in the history of geography.[1] They also represented a summation and veiled quest for direction via evolution of published works in France, Germany, the British Isles, and the United States. This periplus, scribed around more recent work melding contributions into an order of his choice, explicated the late nineteenth- and early twentieth-century sweep of these conjoint geographies.

The choice of title for the "Recent Developments" essay remains something of a curiosity, for it was a comprehensive sweep across the components of geography and not particularly focused on cultural geography. Yet at this time, Sauer seems not to have defined cultural geography with the particularity that came later: e.g., "We are now sufficiently out of the rough in the field of physical geography so that we are becoming more concerned with cultural or human geography" (1927c:179). He considered that cultural geography derived substantially from the works of the French geographers La Blache, Bruhnes, Vallaux, Blanchard, and from the German geographers Hettner, Passarge, Schlüter, and Gradmann. He presented this at Berkeley as cultural geography, a tradition deriving from European geography wrought over several decades. The term "culture" first presented itself in the Berkeley course offering "Geography 2" in 1923–1924 (Macpherson, 1987:74); the course was taught by Sauer for some thirty years. This Berkeley geography stood alone for several decades, and was considered by some anomalous to American geography. In fact, American geography may have been anomalous to the larger body of established European geographical thought.

Early twentieth-century geography in the United States had, to a large extent, filtered through the experience and publications attaching to the Great Surveys and the work of geologists. W. M. Davis, who synthesized these works, contributed the cycle of erosion, proclaimed the environmental juxtaposition of environment and environed, and founded the Association of American Geographers, in which his posture would be exercised. And forms of determinism lingered in both courses and literature deep into the twentieth century. This geography, originating very largely from studies in American physiography, was a new geography for a new world.

Sauer's proximity to and his appreciation of European geographical thought never left him. In his earlier years, however, he could be outspoken in behalf of his appreciation of this geography, as in "Recent Developments." And in 1932 he wrote to Gladys Wrigley, editor of *The Geographical Review*, concerning reviews of three books that had been published in 1931 and 1932 in that journal. The books were written by Emmanuel de Martonne, Robert Gradmann, and J. Früh. Sauer accepted the review of de Martonne (*Europe Centrale*) by Griffith Taylor, but took marked exception to reviews of the work of Gradmann and Früh. "Gradmann's *Suddeutschland*, a beautifully symmetrical study full of mellow wisdom and incisive observation, fares ill." Sauer objected to the reviewer Eugene Van Cleef's proclaiming that landforms were the business of the geologist and complaining of a lack of attention "to the life responses to the physical conditions of the regions." Sauer stated that Gradmann had perhaps made an "unequaled series of intensive studies in culture geography" and has written "most lucidly, on objectives and methods in human geography." Sauer again urged the inclusion of the origin of landforms in such a study. His criticism of the review of the book by Früh (*Geographie der Schweiz*) again dwelled on reviewers' (Roderick Peattie and Van Cleef) concerns that questioned the place of geology and particularly the study of the origin of landforms in a geographic work. Sauer concluded this review of reviews with his comment on "building fences about narrow plots of learning" (1932b:528, ch. 13 herein; also see 1941b:4).

Editor Gladys Wrigley wrote to Sauer (March 15, 1932), "Now that you have risen up in your wrath do you not want to smite the Philistine in the Correspondence columns of the *Geographical Review*?" Sauer revised the statement in minimal fashion, and saw it published in the *Geographical Review* (1932b, ch. 13 herein).

Thereafter, Sauer reviewed little. His time was evermore invested in his correspondence, in which he thought out loud concerning geography and geographers and exchanged thought with others frequently beyond the pale of traditional geography, for he saw geography as a synoptic enterprise, drawing from a variety of other sciences and intellectual undertakings. Considerable amounts of correspondence that Sauer both generated and received in the 1920s is now unavailable—a great loss given that this was the seed-time of his mind. It is in the existing archival collection of correspondence that the man and his geographic point of view perhaps are to be best comprehended. By the early 1930s, his posture with regard to culture and diffusions had been designed; the remainder of his life was spent in its elaboration and appreciation.

NOTES

1. Although Sauer stated in a letter to Richard Hartshorne in 1946 (see Martin, 2003:39) that "Morphology" "was written in several weeks as a sort of habilitation," the basic ideas had already appeared in "The Survey Method" in 1924 and in the *Introduction to Geography* textbook, also published in 1924. And while Sauer supposedly repudiated "Morphology" according to Leighly (1963:6), the thesis continued to appear in later editions of the *Introduction to Geography*—"The most significant definition of geography . . . is the development of the cultural out of the natural landscape" (1932 edition, p. 7). Many of Sauer's subsequent writings incorporated landscape themes (Penn and Lukermann, 2003:252).

2. Volume editors' note: We originally intended to include "Recent Developments in Cultural Geography" in this section, but it turned out to be too lengthy. It is replaced herein by "Cultural Geography" (1931 [1962]), and is in part a summary of the 1927 article.

REFERENCES

Barrows, Harlan H. 1923. "Geography as Human Ecology." *Annals of the Association of American Geographers* 13:1–14.

Bruhnes, Jean. 1925. "Human Geography." In *The History and Prospects of the Social Sciences*, ed. H. E. Barnes, 55–105. Knopf, New York.

Davis, William M. 1906. "An Inductive Study of the Content of Geography." *Bulletin of the American Geographical Society* 38:67–84.

Davis, William M. 1924. "The Progress of Geography in the United States." *Annals of the Association of American Geographers* 14:159–215.

De Geer, Sten. 1923. "On the Definition, Method and Classification of Geography." *Geografiska Annaler* 5:1–37.

Dryer, Charles R. 1920. "Genetic Geography: The Development of the Geographic Sense and Concept." *Annals of the Association of American Geographers* 10:3–16.

Fenneman, Nevin M. 1919. "The Circumference of Geography." *Annals of the Association of American Geographers* 9:3–11. Also, *Geographical Review* 7 (1919): 168–175.

Finch, Vernor C., and Glenn T. Trewartha. 1936. *The Elements of Geography*. McGraw-Hill, New York.

Jones, Wellington D., and Carl O. Sauer. 1915. "Outline for Field Work in Geography." *Bulletin of the American Geographical Society* 47:520–526.

Leighly, John. 1963. "Introduction." In *Land and Life: A Selection from the Writings of Carl Ortwin Sauer*, ed. J. Leighly, 1–8. Univ. of California Press, Berkeley.

———. 1976. "Carl Ortwin Sauer, 1889–1975." *Annals of the Association of American Geographers* 66:337–348.

Macpherson, Anne. 1987. "Preparing for the National Stage: Carl Sauer's First Ten Years at Berkeley." In *Carl O. Sauer: A Tribute*, ed. Martin S. Kenzer, 69–88. Oregon State Univ. Press, Corvallis.

Martin, Geoffrey J. 1988. "Preston E. James, 1899–1986." *Annals of the Association of American Geographers* 78:164–175.

———. 2003. "From the Cycle of Erosion to 'The Morphology of Landscape': Or Some Thought Concerning Geography as It Was in the Early Years of Carl Sauer." In *Culture, Land, and Legacy: Perspectives on Carl O. Sauer and Berkeley School Geography*, ed. Kent Mathewson and Martin S. Kenzer, 19–53. Geoscience and Man, vol. 37. Geoscience Publications, Department of Geography and Anthropology, Louisiana State University, Baton Rouge.

Mikesell, Marvin W. 2002. "Textbooks that Moved Generations." *Progress in Human Geography* 26:401–404.

Penn, Mischa, and Fred Lukermann. 2003. "Chorology and Landscape: An Internalist Reading of 'The Morphology of Landscape.'" In *Culture, Land, and Legacy: Perspectives on Carl O. Sauer and Berkeley School Geography*, ed. Kent Mathewson and Martin S. Kenzer, 233–259. Geoscience and Man, vol. 37. Geoscience Publications, Department of Geography and Anthropology, Louisiana State University, Baton Rouge.

Roorbach, George B. 1914. "The Trend of Modern Geography—A Symposium." *Bulletin of the American Geographical Society* 46:801–816.

Sauer, Carl O. Correspondence (cited in text). Carl O. Sauer Papers, Bancroft Library, University of California, Berkeley.

———. For references, see the Sauer bibliography at the end of this volume.

Salisbury, Rollin D, Harlan H. Barrows, and Walter S. Tower. 1912. *The Elements of Geography*. Henry Holt, New York.

Tower, Walter S. 1910. "Scientific Geography: The Relation of Its Content to Its Subdivisions." *Bulletin of the American Geographical Society* 42:801–825.

II

The Field of Geography (1927)
Carl O. Sauer and John B. Leighly

THE CHOICE OF TITLE

We have chosen to say that this [textbook] is an "Introduction to Geography," not that it is an outline or something else having the intent to summarize the results of geographic study as a whole. At the outset it should be noted, for the proper understanding of this volume, that it is not a condensed review of what geography contains, but that it is a guide by means of which one may occupy one's self to advantage in the study of a field. Some text books are digests or recapitulations by means of which those who have but little leisure or minor interest in a subject may yet know its principal conclusions. Other texts are manuals, presenting in the main a proper set of tools and instructions, by means of which the student may learn to make for himself increasingly useful observations and judgments. To the latter class this book aspires to belong. According to the position hereinafter developed, geography is an orderly knowledge of the content of regions. To acquire such knowledge certain preparations are necessary which we have tried to supply.

Although the volume is designated a manual, it is not a collection of exercises or of questions, neither a catechism nor other mechanism of instruction. Props to teaching and study that are designed as class drills are not the authors' ideal for collegiate study. The attempt is made to include enough of content as to things that are first principles so that the coherence of the field is established and also so that the reader may use some selection in arriving at an adequate orientation to begin the serious study of regions.

HISTORICAL VIEW OF THE GEOGRAPHIC FIELD

[Geography is] an ancient field of interest. There is an insistence in some quarters that geography, at last as a subject appropriate to the college and university, is a new thing and different in principle from that which was called geography in the past. In general, however, we are dealing merely with a field of knowledge that has been under cultivation for a very long time and has changed from generation to generation with the changing intellectual interests of the time. Excepting philosophy, mathematics, and history, there is perhaps no recognized line of intellectual interest as ancient and none more persistent. The subject inevitably contains heritages from many ages and many cultures. We may begin therefore with a brief historic setting, since thus we are most likely to understand its scope, its relations to other fields of knowledge, and the problems that most demand our present attention. Our intellectual goods being determined as well by our cultural history as by logic and experiment, the historical approach is in reality the best means of discovering what a field of knowledge is.

The immemorial geographic themes are the explanation of terrestrial phenomena and a knowledge of the differences between countries using the word in the sense in which we speak of an "unsettled country" or a "distant country." In the former case, occurrences of sensible character, [such] as storms, tides, and earthquakes, received most attention; in the latter [are] the fortunes of man in contact and often in contest with the area that formed his home. In both these motifs primitive literature and especially mythology abound. The sagas of our Nordic ancestors and the Homeric epics for example are concerned in large part, however fancifully, with these themes. Homer indeed was regarded among the ancients as a geographical authority. Thus we have from the beginning an interest in what the Greeks called cosmology (theory of the universe), or Earth Science, and in chorology (knowledge of the nature of countries), or Regional Science.

The oldest school of Greek (Ionian) philosophers is known as the Cosmologic School. It sought to discover and explain the organization of the universe by a system of natural law. These philosophers were not so much concerned with what came later to be regarded as the principal province of philosophy, namely the process of thinking and of judging values such as questions of right and wrong, or of the beautiful and the ugly. They were trying to fit explanations to their observations of the external earth. They formulated remarkably shrewd theories regarding the planetary relations of the earth, about the movements of the wind and the ocean, concerning earthquakes and volcanoes, and many other phenomena of the physical world about them. From this distant time date concepts which have been carried down to the present, such as those of the continents, of the ocean, and of the torrid, temperate, and frigid climatic belts. Of these ancient concepts the only one which has maintained its full value to the present is that of the universal ocean. The continents and climatic belts which fitted the observational data of the Greeks sufficiently have since become impediments rather than aids to geographic thinking. The Pythagoreans are credited with developing the idea of a spherical earth, followed by observations on the celestial "sphere" that have made possible the modern map. Aristotle, in the fourth century B.C., wrote a treatise entitled *The Four Books of Meteorology*, really a general geography. It is especially interesting for its association of the weather as [a] causal element with a vast number of happenings on the earth and also for its climatology that divided the earth into habitable and uninhabitable regions, two habitable zones in intermediate latitudes being separated by an uninhabitable and probably impassable equatorial zone. It was not until nearly two thousand years later that this ingenious speculation was disproved or even seriously disbelieved.

Ancient Greece has also furnished us with the most trustworthy early accounts of regional geography. After the Homeric epics there developed the *historia*, in which the then known world was characterized as to its physical and cultural differentiation. The work of Herodotus in particular is invaluable for its observations on countries and their people, much of it based on what he saw during extensive and apparently painstaking travels. He was a master not only of narrative, which is an historical form of writing, but also of areal description, which is properly geographic.

The most famous geographer of antiquity was Strabo, who lived in the first century B.C. Like Herodotus he exalted observational knowledge derived from intimate travels. He therefore wrote largely about regions. As classical antiquity drew to its close, a considerable amount of observational knowledge of the earth had been accumulated, at the last through the empire-building of the Romans. Measurements of positions on the earth were undertaken increasingly. Roads and shore lines, mountains and rivers were charted, and maps were constructed according to plan. Eratosthenes (third century B.C.) is thought to have originated the word geography. He is to be remembered as having helped to lay the foundations of scientific map making by his measurements of an arc of the earth. Various projections to represent the earth's surface in maps were devised, proceeding from an appreciation of its sphere-like form. In particular, Claudius Ptolemy, who lived at Alexandria in the second century after Christ, prepared a geographic treatise that remained the model of map making and geographic description until well into the Modern Period, initiated by the great oceanic discoveries. The writers of the Classical Period are still impressive as to the quality of their learning and by their increasing realization of direct observation as the foundation of knowledge. We may say that a science of geography did exist at this time and that its final expression was primarily chorologic, that is, an effort at a precise and ordered knowledge of the earth's surface.

THE PERIOD OF GREAT EXPLORATIONS

After the long intellectual night that we know as the Dark Ages, the revival of learning set in vigorously in the fourteenth century with a renewed study of the classics. In this connection the old Greek, and even more numerously the less significant Latin geographies and natural philosophies were resurrected. In particular, the Arabs, then at the height of their cultural development, acted as conservators and intermediaries of classical learning. Their contacts with Sicily in particular served to bring critical revisions of the Ptolemaic geography to the attention of the European World at this time. Most European scholars at first copied freely and imitated poorly the classical tradition. Reverence for the ancients was at first more conspicuous than discriminating adaptation. To a world that had long been stifled in ecclesiastic formalism and authority, the observations and speculations of the ancient Greeks regarding our world constituted a revolutionary rediscovery. Thus the old knowledge that the earth was a globe became a doctrine fraught with some personal danger a millennium and a half after it had been the common property of every educated man. The enthusiasm with which the classical masters filled the scholars of the day entailed at first a blind respect for classical authority. The transfer of allegiance to Greek thought however was gradually followed by lessened docility and increasing inquiry, resulting in time in another period when the direct observation of nature was held in honor and a new advance of knowledge became possible.

Among the earliest printed books, geographic treatises of classical origin are to be found in number. Strabo was printed in many editions and much annotated. In particular, adaptations of Ptolemy were in great vogue, the name being used for many geographies that had less and less similarity to the prototype as the sixteenth century wore on. Finally it was no longer thought necessary to base a geographic work on the system of the great Alexandrine.

Toward the end of the Medieval Period the curiosity of man was directed again toward the exploration of unknown regions. To the European the world had shrunk until it embraced little more than the Mediterranean and North Sea countries. The crusades, contacts with the Arabs, [and] the oriental goods that came to the great Italian merchant cities were among the earlier expressions of the new period that was to break finally the snug confines of Medieval Europe. The first important expeditions were in the main overland to the courts of Tartar princes in eastern Europe and inner Asia. European missions of the thirteenth century brought back such strange accounts of the barbaric magnificence of Asiatic states that they were received as fantastic romances. The extensive and careful observations of Marco Polo on the great civilization of China, then almost at its height, received little credence and not much more attention. However merchants and missionaries were increasingly on the move to the East, the Arabs were trading diligently eastward to the East Indies, and occasional Chinese and Hindu seamen and merchants were seen in the Levant. The growing tide of commerce brought an infiltration of knowledge, shortly to express itself in great oceanic explorations.

In the first half of the fifteenth century, Prince Henry of Portugal launched a series of expeditions down the African coast, which resulted ultimately in the opening of the sea route to India by [Vasco] da Gama in 1498. Spanish explorations commenced with the brilliant success of Columbus, and England was but little behind in extending trade and explorations. The Low Countries and France followed suit in the sixteenth century. Oceanic explorations dominated the geographic field from the fifteenth into the middle of the seventeenth century. By the end of the eighteenth century all the seas and their coasts, except those severely ice-bound, were quite well known, and excepting Africa and inner Asia, the continental interiors had been reconnoitered. In the annals of geography the gallant company of explorers is forever memorable. Technically this delineation of the earth's features was made possible by improved methods of orientation, in particular by means of the magnetic needle, by more accurate determinations of latitude through crossstaff, astrolabe, and quadrant, by nautical tables that were the forerunners of our nautical almanacs, and perhaps above all by improved maps.

The fruits of discovery were expressed by a growing flood of maps. Maps were issued from many centers, steadily, containing in the latest discoveries the most important news of the time. The first collection of maps of wide circulation bore the significant name *Theatrum orbis terrarum* (the atlas of Ortelius, Antwerp, 1570). Terrestrial globes were to be found in every important library. Skillful globe makers lived in many of the principal seats of commerce and exploratory interest. More and more the growing extent of the known world made the classical models of map projection inadequate. These centuries were perhaps the busiest of all time in recasting the forms and symbolism of the map. In particular, the year 1569 is important. At this time appeared Gerhard Mercator's "New Map for the Use of Navigators" on the projection that has since then been the standard chart for navigation and still is the commonest and most useful of all world projections. The technical solutions of representing figures of the earth in maps were so numerous and so well done in this period that today we largely use the projections of past centuries.

Explorations, maps, and descriptive accounts of this period were all essentially chorologic. Interest in the content of regions was uppermost. The principal geographic books published were called cosmographies, concerned mostly with characterization of regions and their people. They kept a large reading public informed as to the latest status of knowledge with reference to the frontiers of the earth. Perhaps no other books of the time were as widely circulated. Certainly none were as much revised and reissued. Sebastian Münster wrote the first edition of the most famous of all cosmographies in 1544. In the century following, it went through scores of editions and numerous translations. Like Herodotus and Strabo he was eager to record what ever appeared to be the most dependable information about all parts of the earth. Today these books may appear somewhat ridiculous because of the credence placed in monstrosities and marvels, and we find many of the statements casual and bizarre mixtures of facts and superstition. It should be remembered however that they were for their time an adequate record of the progress of geographic knowledge.

THE BEGINNINGS OF CRITICAL SYNTHESIS

Information about the earth accumulated too rapidly for a time to be sifted and organized critically. Gradually, however, the data of exploration were compared and classified so as to yield orderly notions about the earth's surface. Varenius in particular is regarded as the principal founder of modern scientific geography. In 1650 he published his *Geographia Generalis,* with a division of the subject into general and special geography, defining the former as the "science which considers the earth as a whole and explains its properties," the latter as "instructing concerning the constitution of the individual regions." He revived therefore the Greek tradition of cosmology, the science of the earth as a whole, and of chorology, the science of regions, and joined them appropriately in an inclusive earth science of dual organization. He did not live to write the second part of his work but he laid the general outlines that geography followed for many years to come and which indeed it still follows in some quarters. On the Continent for instance the terms general and special geography are still largely employed precisely in the sense of Varenius, as marking a dualistic organization of the field. In the Varenian and usual European sense the present volume is general geography.

The seventeenth and eighteenth centuries witnessed great advances in the technique of making maps. Perhaps never again have maps been engraved as beautifully or designed as attractively as in this period. The issuing of great atlases became one of the most important branches of book publishing. Cartographic institutes of a quality now rarely encountered developed in number and flourished for generations in the greater commercial centers, first of all at Antwerp and Amsterdam in the Low Countries, but also at Augsburg and Nuremberg in Bavaria and in Paris and London. The symbolism of the map was elaborated to the extent that the maps became in plan for the most part essentially what they are today. Precision of survey was largely increased by geodetic measurements, and harbors and coast lines in particular were closely charted, especially in Europe and the East and West Indies. Topographic-ordnance maps of high quality were issued during the eighteenth century in a number of European countries.

RESTRICTION OF THE FIELD

The nineteenth century brought a rapid development of natural science as ordered observation as against the older natural philosophy which was most largely speculative. The increased insistence on systematic observation found expression also in geography. [Alexander von] Humboldt (1769–1859) in particular gave to geography a scientific basis by insisting that distribution of phenomena constituted the specifically geographic task. Thus he directed attention to the distribution of temperature values for different latitudes and altitudes and suggested conclusions to be drawn therefrom. Plant distributions were of so much interest to him that he really established the geography of plants. His *Travels to the Equinoctial Regions of America,* and especially his volumes on New Spain, invest regional studies with scientific qualities and are the first regional geographies in the modern sense.

The period of special disciplines was ushered in with the nineteenth century. The field of geography as the whole of earth science was too vast to be held. The new science of geology had found new methods and startling results in the study of rocks and their fossil remains. Geology became the science of earth history, based on the "testimony of the rocks," as Hugh Miller said. Geography relinquished to geology therefore the study of the earth's interior and of earth structure. It became restricted in this process to the study of the earth's surface, a definition that we can still accept as describing geography in the broadest sense.

During the nineteenth century the major attention of geographers was given over to the relief features. Patterns of elevations and depressions were observable that suggested possibilities of systematic explanation. In the end the origin of relief forms has come to be largely understood. At the same time, geologists, especially in Great Britain, discovered that the record of the rocks could be interpreted only by observing the processes that are now shaping the surface of the earth, for the sediments are reproduced in the sedimentary rock bodies and the contacts of rock masses may be ancient land surfaces. In consequence, both geologists and geographers worked freely in the field studying the modifications of the earth's surface by external processes. A special branch of earth science developed, called physiography. In this border field some studied surface features primarily in order to understand better the configuration of the land, others to interpret better the rock characteristics, still others to carry the history of terrestrial development through the record of the bedded rocks down to the present surfaces. For a time therefore most geography was physiography and its connections with the vigorously developing science of geology were so close that the ancient chorologic aims were much obscured by making the study of terrestrial processes the major end.

Another group of students, also in the field of general geography, examined the causal relations existing between man and his natural environment.

As the physiographers had isolated individual processes and their effects, students of human geography turned to the development of environmental causality in human fortunes. There are obviously some parts of the earth in which men have generally fared badly and others which are intrinsically suited to great developments. It was thought that one might therefore discover a general set of natural laws explaining the rise and decline of civilizations from physical conditions. Buckle's *History of Civilization* is perhaps the most famous statement of human events as determined by adequate and discernible natural law. The doctrine of environmental influence has been stated as follows: "A given environment, as a certain type of climate, will have a specific and constant effect on peoples." Under such a view a sufficient knowledge of the environment will make it possible to deduce the fortunes of the people concerned. Actually the hopes of such students usually are that they may measure the importance of the physical environment in human conditions. This study is sometimes called anthropogeography.

There has existed recently therefore a strong tendency for the development of two divergent branches of general geography, which lost sight more and more of each other and also lost interest in the study of regions, except in so far as regional studies provide demonstrations of universal physiographic or anthropogeographic principles. This dualism has been especially strong in America, with the human geographers latterly getting in general the upper hand of the physiographers. In contrast to the forking of general geography into two subjects, the choice here made is that of the traditional view, lately strongly reasserted in Europe, that GEOGRAPHY IS THE SCIENCE OF AREA.[1]

THE CHOICE OF POSITION

The history of geography gives us several choices: (1) We may maintain the plan of Varenius and concern ourselves with general earth science (General Geography) and with comparative regional studies (Special Geography), considering both to be of similar importance and interdependent. (2) We may consider the end to be universally applicable laws of human behavior derived from environmental stimuli and inhibitions. (3) We may define the subject simply in terms of a body of materials to be taken under observation, this body of materials to be an important, popularly given category of phenomena, in this case the features of the earth's surface.

The first choice is no longer generally attempted, because knowledge has developed into so many fields that a number of specific earth sciences are generally recognized as a necessary division of labor. Under the third position we readopt the chorologic view of Strabo and Münster for modern ends. We then consider General Geography, or general earth science, as a group of special disciplines which the geographer employs only in order to arrive at an adequate knowledge of regions. We thus reassert the ancient theme that geography is the study of the differentiation of the earth into regions.

THE RELATION OF GEOGRAPHY TO GEOLOGY

Historically, geology is part of the older General Geography, the original division meaning simply that the younger science developed a proficiency in the study of rocks, or of the earth's crust and interior, whereas geography became limited to the study of the earth's surface. This is still essentially the division of labor, and the two fields are still closely related. The contrasts and relationship between the two fields may perhaps be further illustrated as follows:

The study of minerals and rocks (Mineralogy and Petrography)	
The study of fossil remains (Paleontology)	History of the Earth (Historical Geology)
The study of the dynamics of the earth's interior (Geophysics in the narrow sense)	
The study of the processes of erosion and sedimentation (Physiography and Geomorphology in part)	

Geology employs these four "general" fields of earth science as methods of observation in order to arrive at genetic conclusions, which are in the final analysis predominantly historical. The symbol which geology uses most characteristically is the structure section, which is a chronologic profile of a succession of rock strata and structures.

An analogous organization of the field of geography may be thus expressed:

The determination of the relative position of points on the earth's surface (Geodesy and in part, Cartography) ⎫ The study of climates (Meteorology and Climatology) ⎬ The study of the forms of land and sea (Geomorphology or Physiography in part) ⎬ The distributional study of life forms (Biogeography) ⎭	Comparative Knowledge of Regions (Chorology)

The characteristic symbol of geography is the topographic map in the literal sense, that is the representation of places in their relations to each other. The term topography is often used as equivalent to relief; actually it means "place-description." The geographic place may be a relief or drainage feature, but it may as well be a city or a farming district. We could perhaps substitute the term "land-mark" for "place." The assemblage of the significant features of the land is the broadest task of the geographic map. In itself the map is therefore merely a synthesis of the natural and cultural phenomena of areas, a distinctive means of areal description. It contains the essential data of geographical study, the task of which is to determine the distributional relationships of these physical and cultural land-marks.

A simple observational division of labor may therefore distinguish for all practical purposes the two earth sciences of geology and geography. The former is based on the competence to read the record of the rocks. The latter finds its story written in the earth's surface.

GEOGRAPHY AS A SOCIAL SCIENCE

It is certain that physical nature exerts manifold influences over man. We feel well under certain climatic conditions, ill at ease under others. Climate has physiologic influences on life, so do altitude and drinking water. The direct effect of external nature on man is however most properly the field of study of the physiologist, not that of the geographer. For the rest, the environment sets limits within which man is constrained to choose his actions. Opportunities and handicaps are placed before him by his environment, but the choice of action remains still pretty largely with man. [Georg] Gerland has said: "The effects of terrestrial forces are not absolute and certain in the case of man as they are in the inorganic world . . . since the will of man interposes itself as an unknown quantity, X, between natural conditions and effects."

It is likely to be therefore a grave error to derive a condition of society from the place which it inhabits. Different peoples and different times have found different uses of the same or similar natural environments. The important thing therefore is to find out the forms of utilization that man has developed in given areas. We are concerned not with a particular causal relation, such as environmental influence, but with the meaning of the things that make up an area. Geography is a social science in so far as it concerns itself with the expression of man in area. It is a mistaking of the entire history of geography to consider it as a study of the effect of nature on man. Geography has never been a science of man, but of the land of the earth's surface. It is limited therefore on the social side to the attempt to secure valid judgments regarding areal expressions of culture.

THE CHOROLOGIC POSITION

The task of geography is conceived therefore to be to understand in its fullest meaning the areal scene. It is concerned with such study by a definition of great antiquity, representing a persistent interest of knowledge. Since Greek days geography has supplied the knowledge of countries, and not of areas from which we imagine the works of man as eliminated but of areas as they are and this means for most of the world areas of human culture. If one detaches the idea of physical area from that of inhabited area with its human impress, the traditional notion of geography is lost. The question then is, to what extent this knowledge can be organized, compared, and checked so as to have scientific validity.

The concept of area may be expressed in another fashion by saying that geography is concerned with the features of the landscape, the visible features of the earth's surface or generalizations that are derived from individual, visible features, always proceeding however from direct observation.

In a scientific study of the landscape we note connections between its constituent features and trace their origin, in both ways arriving at general group concepts. A catalogue of rivers or of towns is not science nor is it even a detailed description of what may be seen on a river or in a town. But if we note that there are repeating patterns in the form of streams, which we can relate to a cause, and that features of a stream are in definite relation to other features of the area, which we can explain as to their connection, the beginning of scientific analysis has been made. Similarly if we note that the settlements of man can be grouped according to types of common origin or function we have begun scientific generalization. Through such descriptive study of the landscape, geography follows the general method of science. In particular if we can sum up the features of a landscape and establish significant connections and contrasts with other areas so that areal differences and similarities can be understood and demonstrated we have organized our knowledge on an objective, comparative basis, which is scientific generalization.

The design of the landscape includes (1) the features of the natural area, so-called, and (2) the forms superimposed on the physical area by the activities of man, or the cultural landscape. Man is the latest agent, and the most definitely recognizable one, in the fashioning of the earth's surface. Since we are specifically interested in the utilization of area by man, this division is most convenient. The study of geography normally must begin with physical features, but coasts are marked by ports and mountains have flung over them in some measure the trails and scenes of labor of man. For our purposes therefore the most significant definition of geography is that it is concerned with THE DEVELOPMENT OF LANDSCAPE, IN PARTICULAR WITH THE DEVELOPMENT OF THE CULTURAL OUT OF THE NATURAL LANDSCAPE. These facts may be presented most precisely in the map, which is the representation of the content and arrangement of the landscape. This statement is true whatever the scale of the map, and involves with change of scale simply change in the amount of generalization.

The areas in which man does not live are not therefore excluded from the study of geography. They lack the last, and perhaps most generally interesting element in the differentiation of the earth's surface, but being part of the earth's surface they belong in the geographic field. We may however assign our interest in such fashion that we emphasize in particular the study of those areas in which man lives. Thus we pay attention mostly to the land masses and relatively little to the oceans, except in so far as their surface and floor contribute to the understanding of the land bodies. That we consider the earth primarily as the differentiated home of human groups is hardly to be denied as a leading motif in geography.

A somewhat broader expression than "surface of the earth" is "biosphere," the zone to which life is restricted. With reference to life in general, only a film of slight thickness enveloping the earth is of direct significance. It includes the lower part of the atmosphere, the outerpart of the solid earth or lithosphere, and the waters that cover the solid earth as oceans and lakes and which also penetrate into and fill the interstices within the rocks underground, known collectively as the hydrosphere. The surface of the solid earth, as the surface of contact between the atmosphere and the litho- and hydrospheres, is the habitat of most life, from which only minor vertical distances separate even the most dispersed life forms. The geographic field is this contact zone, in terms of its differentiation.

THE DATA OF GEOGRAPHY ARE THE FORMS OF LANDSCAPE

All geography is based essentially on observations of the differences and resemblances between areas. We attempt so [in order] to distinguish and classify the features of the earth's surface as to make them intelligible as to their connection and order. Each area is made up of many physical features that determine its character. Actually these features are all material, tangible, visible, but we have come to distinguish between physical features in the narrower sense of those not determined by human occupation and those that are caused by man, or cultural features. Since the latter have been derived out of or superimposed on the former, we must first learn to understand the physical facts of the earth's surface.

In this volume we therefore attempt to select and organize the physical facts of the earth's surface so as to prepare for the comparative study of regions. [We] view these [features] as inhabitants of the earth; [they] are the materials out of which we build our material culture. They are present in different areas in varying combinations. Each area is limited in some measure in the activity of its inhabitants by its physical equipment. If the use of area by man is not wisely determined or if the propagation of population continues to the limit of available resources, the group comes up unpleasantly against its natural limits. The biologist uses the term symbiosis to describe the balanced or harmonious living of a group within its habitat. We are much concerned with this idea in geography. We may be interested in a plain because of its physical origin, say as a body of glacial outwash, but we are also likely to ask ourselves questions regarding its utility, for its origin or generic character also implies certain attractions for agriculture or forestry, certain characteristics that need to be taken into account in the making of lines of communication. In short, in physical type there is implied a condition or use value that differentiates it as a site from other plains and land forms of other kinds. Since we study the physical area not only as to its origin but also as to its use significance, it does not follow however that the area alone determines its use nor that one may deduce culture from physical area, but only that one must know well both physical equipment and the people in order to know that differentiation of areas.

PHYSICAL FORMS OF LANDSCAPE

We may analyze the physical area into the following elements of form:

(1) Space relationships. Every area is characterized by the fact that it occupies space or has two-dimensional form and measurable distance from other areas. These general qualities of location are the concern of geodesy, always the foundation of geography, and their representation is the first and minimum essential of every map.

(2) Climates. The term is used in the plural; the first geographers recognized seven climates. It is not simply the sum of average weather conditions, but of atmospheric conditions as characterizing or distinguishing one area from another.

(3) Forms of the land. In the broadest sense, the geographic forms of the solid earth: (a) relief forms, or the configuration of the surface; (b) soil forms, or the differentiation of the earthy material which supports vegetation; (c) drainage forms, the water bodies which are part of the land masses rather than of the great seas; and (d) mineral resources. The last are concentrations of mineral constituents which man extracts and which are therefore directly and deliberately removed by his activity. They need not be taken into account at certain stages of human development, and they may be destroyed and thus cease to be significant.

(4) The oceans, for practical purposes especially their coasts, coastal currents, and other qualities of significance to the land. Remoteness of interest, not logical considerations, are in general making the study of oceans a separate field.

(5) We must consider as part of the natural landscape the vegetation, unaffected by human interference. In this case we are concerned with the asso-

ciation of plants as covering the surface, such as forest or steppe. The distribution of plants *en masse* is always one of the most important considerations in the study of areas.

Out of these form groups the physical landscape is composed. We wish to understand the association of these forms and therefore must undertake to analyze the landscape into its component forms. Only by performing such anatomical studies on the face of the earth can we understand the qualities of its regions. We thus come to a knowledge of the genesis and of the interrelation of the forms. This is the purpose of the present volume, to give a basis for the analysis of regions so that it may be possible to compare regions. Plains and mountains, deserts and rain forests, are to be studied as one may study plants, by noting the constitution of each constituent part, the manner in which the parts are put together, or their structure, and the effects of the parts on each other, or their functional association. The student may familiarize himself further with this analogy by examining the morphologic method in the biological sciences and by regarding this course as an introduction to the morphology of landscape.

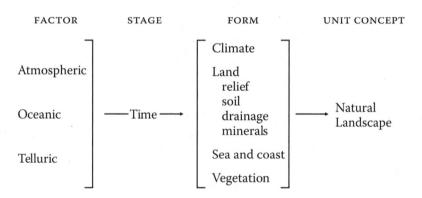

Diagrammatic Outline of Scope of Course

The physical forms of the earth are conditioned by the processes based on the contact between land, sea, and air, principally the action of the atmospheric forces on the solid materials of the earth. The surface of the earth is changed by forces acting at its surface, but also by those acting from within. Since the forms are the result of varying combinations of processes, rarely by an individual process on an undifferentiated material, the possibilities of classification must be derived primarily from the forms, which are the elements of all landscape. The forms moreover change with the time that has been involved in their development. In the forms themselves therefore is to be sought the principal grouping of the material, but these forms are referred back to their genetic bases. In the end we come again to the unit concept of the actual form association, the natural landscape.[2]

NOTES

Reprinted from *An Introduction to Geography: 1. Elements*, 4th ed., 1–13, 1927. Copyright 1926, Carl Ortwin Sauer and John Barger Leighly. Permission from Edwards Brothers, Ann Arbor. We were unable to locate a copy of the first edition (1924).

1. Volume editors' note: In the 1932 edition this was changed to "the science of areal differentiation."

2. Volume editors' note: The subsequent chapters are entitled "Space Relationships," "The Climatic Elements," "Climates or Climatic Regions" (including vegetation, landforms, drainage, and soils), and "The Sea and Its Coasts." Thus, the differentiation of the physical landscape here is based primarily on climate.

12

Cultural Geography (1931)

Carl O. Sauer

In the past century the geographers have been dislodged from their earlier carefree encyclopedic state, in which they discriminated only in terms of personal interest and made camp wherever the prospect pleased. The scientific tendencies of the time have brought external criticisms and internal compulsions and a large methodologic literature which marks the process of intrenchment within a recognizable realm. The earlier volumes of the *Geographische Jahrbücher* (Gotha, 1866–), especially the articles by Hermann Wagner, are much concerned with questions of objective and method. The most comprehensive epistemology is that of Alfred Hettner (1927). In these discussions essential unity has not been attained, and to this day there are irreconcilable camps. Therefore the question as to what geography is must continue to be asked, because the answer determines the premises under which the data have been assembled.

Geography is approached in various ways and to various ends. On the one hand, there is an attempt to find the limitation of study in a particular causal relationship between man and nature; on the other, the effort is to define the material of observation. This cleavage has attained increasing dimensions year by year and threatens perhaps to form a gulf across which no community of interests may be maintained. The situation dates from the beginning of modern geography but has grown acute only in the present century. The one group asserts directly its major interest in man: that is, in the relationship of man to his environment, usually in the sense of adaptation of man to physical environment. The other group, if geographers may be divided into simple classifications, directs its attention to those elements of material culture that give character to area. For purposes of convenience the first position may be called that of human geography, the second that of cultural geography. The terms are in use in this manner, although not exclusively so.

Carl Ritter, holder of the first academic chair in geography, especially emphasized human activity as physically conditioned. The thesis of the environment that molds civilization is of course very old but received special attention from the rationalism of the eighteenth century and found able spokesmen in Herder, Montesquieu, and later in Buckle. Ritter's position was vigorously attacked by Froebel and Peschel as impressionistic and unscientific. Even around the middle of the last century there existed a polemic literature concerning the physical environment as the field of geographic study.

Friedrich Ratzel (1882–1891) in his *Anthropogeographie* outlined the framework in which human geography in the narrower sense has moved since that time, a set of categories of the environment—ranging from abstract concepts of position and space to climate and seacoasts—and their influence on man. By this one work he became the great apostle of environmentalism, and his followers have largely overlooked his later cultural studies, in which he concerned himself with movements of population, conditions of human settlement, and the diffusion of culture by major routes of communication. The effect of Ratzel's environmental categories was not great in his own country; in France it was tempered by Vidal de la Blache's (1922) acute substitution of *possibilisme* for the original determinism; but in England and the United States the study of the physical environment as the goal of geography became well nigh the mark of recognition of the geographer. Apparently Ratzel did not regard his *Anthropogeographie* as anything more than a stimulus and an introduction to a human geography that was to be based on a study of culture. Whereas anthropologists have made large use of his analysis of the diffusion of culture, western geographers think of him only as an environmentalist. In the United States the *Annals of the Association of American Geographers* (published since 1911) show the rapid spread of human geography. So far the high point of this invasion has been marked by H. H. Barrows' presidential address before that body in 1923, a frank plea to constitute the subject solely on the basis of environmental adjustment. So prevalent has this view become in English-speaking lands, and so different is the objective of the continental body of geographers that the work done by the one group is largely ignored by the other.

The rejection of the environmentalist position in geography is based not on any denial of the importance of studies in environment but simply on the following methodological grounds: (1) no field of science is expressed by a particular causal relation; (2) the environmentalist inquiry lacks a class of data as field of study, there being no selection of phenomena but only one of relations, and a science that has no category of objects for study can lead, in Hettner's words, only a "parasitic existence"; (3) nor is it saved by a method that it can claim as its own; (4) special pleading is most difficult to avoid by reason of the fact that success lies most apparently, or at least most easily, in the demonstration of an environmental adjustment. Theoretically the last objection is least serious; practically it has been most so, as is illustrated by a flood of easy rationalizations that certain institutions are the result of certain environmental conditions. In this regard those students who have least troubled themselves with knowledge have reaped the greatest apparent success. The polemic against the position of geography as the study of environmental relations has received its sharpest definition by Schlüter (1906), Michotte (1921), and Febvre and Bataillon (1922).

The other school continues the major tradition of the subject. It therefore does not claim that it represents a new science but rather that it attempts the cultivation of an old field in terms acceptable to its age. It is not anthropocentric; rather it has shown at times excessive tendencies in the other direction. Cultural geography is only a chapter in the larger geography and always the last chapter. The line of succession passes from Alexander von Humboldt through Oskar Peschel and Ferdinand von Richthofen to the present continental geographers. It proceeds from a description of the features of the earth's surface by an analysis of their genesis to a comparative classification of regions. Since the day of Richthofen it has been customary also to use the term "chorology," the science of regions. During the latter half of the past century the work carried on was overwhelmingly physical, or geomorphologic, not because most geographers thought that the study of genesis of physical land forms exhausted the field, but because it was necessary to develop first a discipline to which the physical differentiation of the earth's surface would yield. Geographers are now in possession of a method by which the origin and grouping of physical areas can be determined and in which successive steps in their development are identified. Processes have been identified, measures of the intensity and duration of their activity have been determined, and the grouping of land forms into assemblages which constitute unit areas that can be genetically compared is well advanced.

The latest agent to modify the earth's surface is man. Man must be regarded directly as a geomorphologic agent, for he has increasingly altered the conditions of denudation and aggradation of the earth's surface; and many an error has crept into physical geography because it was not sufficiently recognized that the major processes of physical sculpturing of the earth cannot be safely inferred from the processes that one sees at work today under human occupation. Indeed, a class of facts which Bruhnes (1910) labeled "facts of destructive occupation," such as soil erosion, are most literally expressions of human geomorphosis. The entire question of narrowing subsistence limits which confronts man in many parts of the world, apart from the question of the greater number of human beings among whom the subsistence may need to be divided, is directly one of man as an agent of surficial modification. Even the most physically minded geographer is driven therefore to this extent into the examination of human activity.

There has, however, never been a serious attempt to eliminate the works of man from geographic study. The Germans have long had a phrase, "the transformation of the natural landscape into the cultural landscape"; this provides a satisfactory working program, by which the assemblage of cultural forms in the area comes in for the same attention as that of the physical forms. In the proper sense all geography is physical geography under this view, not because of an environmental conditioning of the works of man, but because man, himself not directly the object of geographic investigation, has given physical expression to the area by habitations, workshops, markets, fields, [and] lines of communication. Cultural geography is therefore concerned with those works of man that are inscribed into the earth's surface and give to it characteristic expression. The culture area is then an assemblage of such forms as have interdependence and is functionally differentiated from other areas. Camille Vallaux (1925) finds the object of inquiry to be the transformation of natural regions and substitution therefore of entirely new or profoundly modified regions. He considers the new landscapes which human labor creates as deforming more or less the natural landscapes and regards the degree of their deformation as the veritable measure of the power of human societies. In this sense then he finds the physical area expressed through two sorts of modalities, those that limit and those that aid the efforts of the group. A persistent curiosity as to the significance of the environment is unaffected here by any compulsion to dress up the importance of the environment. The facts of the culture area are to be explained by whatever causes have contributed thereto, and no form of causation has preference over any other.

Such a method of approach is entirely congenial to the geographer. He has been accustomed to regard the genesis of the physical area and he extends similar observations to the cultural area, which has a somewhat simpler and more exact form than the culture area of the anthropologist. The geographic culture area is taken to consist only of the expressions of man's tenure of the land, the culture assemblage which records the full measure of man's utilization of the surface—or, one may agree with Schlüter, the visible, areally extensive and expressive features of man's presence. These the geographer maps as to distribution, groups as to genetic association, traces as to origin, and synthesizes into a comparative system of culture areas. The experience in geomorphologic study provides the necessary technique of observation and a basis for evaluating the modalities stated by Vallaux. A geography such as this is still an observational science utilizing skill in field observation and cartographic representation, and geographic therefore in methods as well as objective.

The development of cultural geography has of necessity proceeded from the reconstruction of successive cultures in an area, beginning with the earliest and proceeding to the present. The most serious work to date has concerned itself not with present culture areas but with earlier cultures, since these are the foundation of the present and provide in combination the only basis for a dynamic view of the culture area. If cultural geography, sired by geomorphology, has one fixed attribute it is the developmental orientation of the subject. Such a slogan as "Geography is the history of the present" has no meaning. An additional method is therefore of necessity introduced, the specifically historical method, by which available historical data are used, often directly in the field, in the reconstruction of former conditions of settlement, land utilization, and communication, whether these records be written, archaeologic, or philologic. The name *Siedlungskunde* has been given by the Germans to such historical studies, and they have been furthered especially by Robert Gradmann, editor of *Forschungen zur deutschen Landes- und Volkskunde,* and Otto Schlüter. A compact view of attainments and problems is given by the former in "Arbeitsweise der Siedlungsgeographie" (Gradmann, 1928). August Meitzen (1895) gave great impetus to field studies by asserting the extraordinary persistence of field forms (*Flurformen*) and village plans as culture relics. Although many of his conclusions have fallen, the inertia of property lines has proved a most valuable aid in determining inherited conditions. Whereas much has been attained in the reconstruction of rural culture areas, the anatomy and phylogeny of the town as a geographic structure are less well advanced to date. They are at present being pioneered by numerous studies, most particularly in France and Sweden. Important generalizations have not yet appeared, but a technique of analysis is emerging.

A logically integrated development is also under way in economic geography [with participation] in the culture geography program (Pfeifer, 1928). Localization of production and industry is no longer the major aim as in the familiar economic geography, which taught distributions of commercial products and analyzed them. This now becomes a device in synthesis, not an objective in itself. The economic geography that is in the making is nothing else than culture geography carried down to date, for the culture area is essentially economic and its structure is determined by historic growth as well as by the resources of the physical area. The title of pioneer belongs to Eduard Hahn (1892, 1909, 1914), who broke the purely speculative culture stages of gathering, nomadism, agriculture, and industry and formulated a set of economic form associations, of which the system of hoe culture has become best known. Also he disproved a general succession in culture stages and demonstrated the lateness of nomadism as a culture form.

Cultural geography then implies a program which is unified with the general objective of geography: that is, an understanding of the areal differentiation of the earth. It rests largely on direct field observations based on the technique of morphologic analysis first developed in physical geography. Its method is developmental, specifically historical in so far as the material permits, and it therefore seeks to determine the successions of culture that have taken place in an area. Hence it welds historical geography and economic geography into one subject, the latter concerned with the present-day culture area that proceeds out of earlier ones. It asserts no social philosophy such as environmentalist geography does but finds its principal methodic problems in the structure of area. Its immediate objectives are given in the explanatory description of the data of areal occupation which it accumulates. The major problems of cultural geography will lie in discovering the composition and meaning of the geographic aggregate that we as yet recognize somewhat vaguely as the culture area, in finding out more about what are normal stages of succession in its development, in concerning itself with climatic and decadent phases, and thereby in gaining more precise knowledge of the relation of culture and of the resources that are at the disposal of culture.

NOTE

Reprinted from *Readings in Cultural Geography*, ed. Philip L. Wagner and Marvin W. Mikesell, 30–34. Copyright 1962, Univ. of Chicago Press, Chicago. Editing and references are mainly by Wagner and Mikesell. Originally published as "Geography: Cultural" in the *Encyclopedia of the Social Sciences* 6:621–623. Copyright 1931, Macmillan, New York.

REFERENCES

Barrows, Harlan H. 1923. "Geography as Human Ecology." *Annals of the Association of American Geographers* 13:1–14.

Bruhnes, Jean. 1910. *La géographie humaine*. Felix Alcan, Paris. 3rd ed., 1925. Trans. T. C. LeCompt as *Human Geography*, Rand McNally, Chicago and New York, 1920.

Febvre, Lucien, and Lionel Bataillon. 1922. *La terre et l'évolution humaine: Introduction géographique à l'histoire*. La Renaissance du Livre, Paris. Trans. E. G. Montford and S. H. Paxton as *A Geographical Introduction to History*, Knopf, New York, 1925.

Gradmann, Robert. 1928. "Die Arbeitsweise der Siedlungsgeographie in ihrer Anwendung auf das Frankenland." *Zeitschrift für Bayerische Landesgeschichte* 1:316–357.

Hahn, Eduard. 1892. "Die Wirtschaftsformen der Erde." *Petermanns Mitteilungen* 38:8–12 and map.

———. 1909. *Die Entstehung der Pfugkulture*. Carl Winter, Heidelberg.

———. 1914. *Von der Hacke zum Pflug*. Quelle and Meyer, Leipzig.

Hettner, Alfred. 1927. *Die Geographie: Ihre Geschichte, ihr Wesen und ihre Methoden*. F. Hirt, Breslau.

Meitzen, August. 1895. *Siedlung und Agrarwesen der Westgermanen und Ostgermanen, der Kelten, Römer, Finnen und Slawen*. 3 vols. and atlas. W. Hertz, Berlin.

Michotte, P. L. 1921. "L'Orientation nouvelle en géographie." *Bulletin de la Société Royale Belge de Géographie* 45:5–43.

Pfeifer, Gottfried. 1928. "Uber raumwirtschaftliche Begriffe und Vorstellungen und ihre bisherige Anwendung in der Geographie und Wirtschaftswissenschaft." *Geographische Zeitschrift* 34:321–340, 411–425.

Ratzel, Friedrich. 1882–1891. *Anthropogeographie*. 2 vols. J. Englehorn, Stuttgart. 2nd ed., 1899–1912.

Schlüter, Otto. 1906. *Die Ziele der Geographie des Menschen.* R. Oldenbourg, Munich.
Vallaux, Camille. 1925. *Les sciences géographiques.* Felix Alcan, Paris.
Vidal de la Blache, Paul. 1922. *Principes de géographie humaine, publiés d'après les manuscrits de l'Auteur par Emmanuel de Martonne.* Armand Colin, Paris. Trans. Millicent T. Bingham as *Principles of Human Geography.* Henry Holt, New York, 1926.

13

Correspondence [on Physical Geography in Regional Works] (1932)

Carl O. Sauer

It happens that the last two issues of the *Geographical Review* brought reviews of three unusually significant regional geographies and that these reviews illustrate how uncertain may be the fate of an author at the hands of his critic.

Griffith Taylor (1931) very evidently enjoyed the feast of facts and ideas spread before him by De Martonne's *Europe Centrale* [1930], citing them in number and with gusto. In particular the reviewer commends the author for the adequacy of the morphologic statement, saying: "The reviewer has long felt that the geography of the future must be written by scientists who not only thoroughly understand the geology and morphology of a country but are willing to devote considerable space in their books to this aspect of the subject."

That [Jacob] Früh, who has tramped [the] Alps and their foreland through a lifetime of devoted study, should at the last be presenting his long-awaited synthesis of the origin of the Swiss scene, *Geographie der Schweiz* [1930], moves his reviewers not at all (Peattie and Van Cleef, 1932). They pass over the whole matter of land forms with "the question as to the extent to which geology, or, in general, the origin of land forms, may (!) be incorporated in a geographic treatise." As to the content of the volume, three climatologic observations find brief notice in the review.

[Robert] Gradmann's *Süddeutschland* [1931], a beautifully symmetrical study full of mellow wisdom and incisive observation, fares ill (Van Cleef, 1931). Not a single one of the many thoughtful syntheses, original conclusions, or significant queries with which this monograph is packed gets a word from the reviewer, who, proceeding from the position that land forms are the business of the geologist, complains of the lack of attention directed "to the life responses to the physical conditions of the regions." What Gradmann does say about man, as for instance about urban centers, "takes on an almost encyclopedic character with a minimum of geographic interpretation." I wish to submit that the review is unfair in giving the impression that the study is primarily physical and for the rest largely encyclopedic. The first volume contains 20 pages on land forms out of a total of 215, and 141 are specifically and expressly devoted to human geography! Even a "human geographer" could hardly dispose of the land forms in less space than that, although he might substitute non-explanatory circumscription for the genetic accounts which Gradmann supplies. The second, regional, volume contains physical and human considerations in more intimate association. The section on the Neckarland, which Van Cleef selects as an illustration of the geomorphologic preoccupation of the author, has an equal number of pages on land forms and on human geography, and this ratio is typical of the whole volume.

The disregard of this large mass of "human" material, it appears, is to be explained because Gradmann just does not happen to be the reviewer's kind of human geographer. But the front-rank position which Gradmann has won in this very field is evident in the *Forschungen zur Deutschen Landes- und Volkskunde,* which he has made an unsurpassed and perhaps unequaled series of intensive studies in culture geography. And he has written largely, and many of us think most lucidly, on objectives and methods in human geography. Gradmann sees "life responses" as only a part of the problem, which is the harmonious analysis of geographic forms of culture, as expressed in their historical succession. The "environmentalist" may find slim picking in *South Germany,* but to the culture geographer it is a marvelously rich field, with its remarkable persistence of historic monuments. Stuttgart as an industrial and commercial city has very slender roots in the environment; as a study in urban evolution from a medieval background it becomes intelligible. It is none the less significant geographically to Gradmann because it is only in slight measure a "response to physical conditions." His "encyclopedism" is in fact a well-thought-out methodical basis which prevents the impoverishment of the subject to a few conclusions based on a single set of "influences."

Geographers who do not care to do physical geography have the right to limit themselves, although it is a bit queer that the environmentalists should care so little about the physical qualities of regions. But whence do they derive the authority to correct or even to read out of the party those who concern themselves also with the origin of land forms? To the majority of the world's geographers and to educated people generally, the great geographers are men like W. M. Davis, Richthofen, Penck, and De Martonne whose work has lain exclusively or in major part in physical geography. A field is occupied by preemption and cultivation, and on these terms there is no other field to which geographers have so valid a claim. The important question is whether the work is significant, well done, and opens up larger horizons. I am at a loss to understand this pernicious anemia that has seized upon some geographers, expressed in a weary shaking of the head at sight of a fellow student happily productive and in the monotonous, sepulchral query: "But is it geography?" Imagine geologists saying about their work: "The work is good, but is it geology?," or economists forever saying to themselves: "We must not look into this matter, for it might turn out to be history or political science!" Barbed wire fences may be necessary in elementary curricula, but the pursuit of knowledge cannot afford to frustrate itself by building fences about narrow plots of learning.

NOTE

Reprinted with permission from *The Geographical Review* 22:527–528. Copyright 1932, American Geographical Society of New York.

REFERENCES (ADDED BY THE EDITORS)

Peattie, Roderick, and Eugene Van Cleef. 1932. Review of *Geographie der Schweiz*, by Jacob Früh. *Geographical Review* 22:175–176.
Taylor, Griffith. 1931. Review of *Europe Centrale*, by Emmanuel de Martonne. *Geographical Review* 21:688–690.
Van Cleef, Eugene. 1931. Review of *Süddeutschland*, by Robert Gradmann. *Geographical Review* 21:691–692.

IV

ECONOMY/ECONOMICS

14

Introduction

Kent Mathewson, Martin S. Kenzer, and Geoffrey J. Martin

> For better or for worse this discipline [economics] has become primarily a study of wealth, one might almost say the science of money. With all of our modern disciplinary growth, there remains, tremendously neglected, the study of human economies....
>
> Carl O. Sauer, "Regional Reality in Economy," 1984 (1936)

PREFACE (MATHEWSON)

Almost always prescient, and often prophetic, by the 1930s Carl Sauer had well-formed and critical perspectives on the ancient Greek root word *oikos* (home, household, dwelling) and its main modern derivatives—economy, economics, and ecology. Throughout his career he offered occasional comments on these domains both in their practical applications and academic articulations. He was never moved like polymath Kenneth Boulding or heterodox economists such as Herman Daly to propose a fusion of economics and ecology into a single evolutionary social science; rather, he was content to point to the increasing depredations of nature's economies by the fictions and forces of the human-directed ones.

Sauer, however, never fully embraced ecological science during his academic career, or "ecology's" manifestations in popular culture during the last decade or so of his life. He voiced a skepticism regarding American ecology's roots in Davisian notions of stages and cycles, as well as the implications of its teleological assumptions of tendencies toward equilibria. His opinion of most economic theory and practice went beyond skepticism. He discounted it, along with much of social science modeled on similar philosophical and methodological bases. This is not to say that Sauer felt economic studies were unimportant. To the contrary, he considered them a crucial concern. "Economy" for Sauer, in contrast, "lies largely outside the field of economics" (1984:41, ch. 16 herein), and consists of the older tradition of *economia* or description of "the ways of making a living" (1976b:74, ch. 49 herein). His approach was both regional and cultural-historical.

"GEOGRAPHY AS REGIONAL ECONOMICS" (MARTIN)

Early in the 1920s, economic geography (then sometimes referred to in course titles and texts as commercial geography, industrial geography, or commodity geography) was rapidly becoming a significant part of the disciplinal skein. J. Russell Smith's *Industrial and Commercial Geography* (1913) and *North America* (1925) were dominating the field, though other of his numerous books and articles were also much read, as were those of other writers, including V. C. Finch and O. E. Baker. On the occasion of the sixteenth meeting of the Association of American Geographers in Chicago, in 1920, Sauer presented "Geography as Regional Economics" (Abstract, 1921b, ch. 15 herein). In reply to a questioning letter concerning this paper from J. Russell Smith, he wrote on May 19, 1921:

> With regard to my paper on "Geography as Regional Economics," I presented this for the sake of discussion at the Chicago meeting [AAG]. . . . My main argument was that Geography is suffering from a confusion of purposes and I made the plea for a concentration of effort on something that lies central to the subject, has major significance, and may supply a definite focus. I also objected to the special pleading that is bound to come out of an interpretation of geography as the study of geographic influences. I proposed the study of areas in terms of their economic performance, with due emphasis on their opportunities, handicaps, and stage of development, but without any partiality to the consideration of physical factors. We can develop a discipline for this type of work that will rid us of the odium of trying to make a case for one set of influence.

This desire to develop "focus" (1921b, ch. 15 herein) was an informal quest of definition devoid of entangling "influences." Expression came in several of Sauer's early writings (1924a, 1924b, 1925, 1927c, 1931a). "This paper . . . seemed to provide a new direction for American geography in the twenties" (James and Martin, 1978:72).

COMMERCIAL GEOGRAPHY AND LAND USE (MATHEWSON)

As Geoffrey Martin (ch. 10 herein) points out, Sauer looked back to an older practice in economic geography, that of "commercial geography," where the focus was on the movement of commodities and the modes of producing them. Later (1976b:73, ch. 49 herein) he lamented that one of the little recognized casualties of World War I were the "bright young geographers . . . invited to Washington" to serve on the shipping boards. After several years plotting statistical data on maps, they became "economic geographers; that is they became geographers of statistics." Sauer's own field studies in the early 1920s in Michigan put him squarely into the economic arena. He was charged with coming up with land use plans for areas that had suffered destructive exploitation during the lumbering boom. He called for regional approaches to sound land use and planning.

His move to Berkeley in 1923 brought an end to this phase of his direct concern with economic issues and planning. However, as the Great Depression deepened throughout the decade of the 1930s, Sauer once again was called to address questions of land use and regional economics. His service included a stint with the Soil Conservation Service surveying soil-depleted lands in the American South. After this, his contact with economists and his comments on economics and economies were tempered by his increasing commitments to seeing things in their cultural and historical dimensions, as in his address to economists in 1936.

"REGIONAL REALITY IN ECONOMY" (KENZER)

In October of 1936 Sauer was invited to present a paper at the fifteenth annual conference of the Pacific Coast Economic Association to be held in Eugene, Oregon, later that year.[1] The invitation, by Blair Stewart, called for Sauer to be the dinner speaker, and he was given *carte blanche* regarding the topic of the talk. Stewart informed Sauer that the Pacific Sociological Society was also to meet in Eugene at that time and that a joint session or even a joint meeting could be arranged, leaving the decision entirely up to Sauer. The idea of speaking to a combined group of social scientists intrigued Sauer, and he accepted straight away. He later wrote to Stewart and suggested "Regional Reality in Economy" as a tentative title for his address.

About two and one-half weeks prior to the Eugene conference, Sauer wrote back to Stewart and graciously backed out of the dinner speech, but offered to go ahead and fulfill his agreement to Stewart "if it would inconvenience you or . . . [Reed] College in any way to suspend it." Sauer's annual commitment to the Guggenheim Foundation weighed heavily on his mind, and he also felt badly about having to leave his family in Berkeley, having earlier promised them a "holiday trip" to Oregon. Prior to bowing out of the program, however, Sauer had prepared a draft of the talk.

"Regional Reality in Economy," now over seventy years old, was published posthumously (1984, ch. 16 herein). And yet, many of the issues it addressed are still pertinent today, as well as suggestive of problems Sauer would raise in subsequent years. For those familiar with his wide variety of interests and his seemingly capricious positions on methodological matters, this article is also significant for providing a partial bridge between the ideas expressed in "The Morphology of Landscape" (1925) and those in his "Foreword to Historical Geography" (1941b). Moreover, "Regional Reality in Economy" helps target the types of questions that Sauer was pursuing in the mid-1930s. Michael Williams has explored Sauer's forays into methodological matters, opining that Sauer "did not get involved again [after 1932b] in any methodological debate until he wrote 'Foreword to Historical Geography' . . ." (Williams, 1983:8). While that may be true, the 1936 paper reveals that Sauer was very much concerned with methodology during this period and, had he attended that Eugene meeting and presented and published "Regional Reality in Economy," it seems likely that he would have again been caught within the jaws of methodological controversy. Indeed, a careful examination of the Sauer Papers indicates the subject of methodology haunted Sauer his entire life (Entrikin, 1982:35).

The interested reader may wish to compare this 1936 essay with "Folkways of Social Science" (1952b), which, although written sixteen years later, was the one piece Sauer penned specifically for an audience of social scientists. He may have been a geographer by profession, but few academic geographers had such a deep concern for the state of social science as Sauer. "Regional Reality in Economy" is his personal statement about the condition of social science in the mid-1930s. It was written by a scholar who sincerely believed that university departments were little more than administrative conveniences; it reflects what Sauer regarded as both the potential foundation of a true social science and the inherent weaknesses in trying to create a science of human action. It is curious that Sauer apparently made no attempt to publish this important statement. Possibly he intended to make major revisions but never got around to it.

History and geography are seen by Sauer in this paper as the foundation subjects of research in the social sciences, excepting perhaps psychology. Contemporary social science is criticized as being heavily burdened with social values and social philosophy, lacking the analytical rigor of the other sciences. Economic man is declared dead; economics has become the science of money. Instead, anthropology appears to hold more promise as a model for the social sciences; methodologically, it is the most advanced field. Anthropology owes a great deal to Friedrich Ratzel's concept of diffusion. By focusing on culture traits, and by utilizing a historicogeographic method, social scientists can identify the economic region and the "intellectual climate." There are no stages of culture; there are only inventions and culture hearths. These hearths, Sauer believed, are the key to all questions of culture dynamics. Sauer ends by telling the economists that a "simple but tremendous [and 'little-developed'] theme is that of destructive economies" (1984:49, ch. 16 herein).

REGIONAL DEVELOPMENT (MATHEWSON)

The title of Sauer's talk to the West Coast social scientists befitted the times. During the 1930s regionalism was not only the reigning paradigm in geography, but also the touchstone in much of social science as well as the humanities (Steiner, 1983). The same could be said for many domains of both intellectual and popular expression outside academia as well, whether architecture, cinema, literature, music, painting, photography, theater, or practical endeavors such as urban planning or rural restructuring. Iconic figures of American arts such as Frank Lloyd Wright, Frank Capra, John Steinbeck, Aaron Copeland, Thomas Hart Benton, Dorothea Lange, and George Gershwin were largely formed in the regionalist climate, and in turn helped give the movement and sensibility definition and momentum. Visionary planners and critics such as Lewis Mumford and Arthur E. Morgan (initial director of the Tennessee Valley Authority) were advocates of practical action in tune with these times. Nor was the regionalist aesthetic and overall mentality just a product of American experience and expression—it had global manifestations. Sauer was not only in touch with the pulse of the movement, but also with some of its key figures, such as Mumford and Lange. Although Sauer at times expressed approval for the ideas and works of regionalists, such as the southern sociologists Howard Odum and his associates at the University of North Carolina at Chapel Hill, he did not believe that geographers would find their *métier* in merely regional studies (1941:3). As for the geographers that did think that geography had finally found its calling in chorology, in retrospect they had little impact on, or visibility within, the larger enterprise of regionalism, whether inside or outside of the academy. Sauer might have predicted as much.

Outside of geography, regionalism's unraveling or denouement was as abrupt as it was broad. All the arenas of cultural production mentioned above experienced radical shifts in theory, technique, and execution immediately following World War II. Modernism's penchant for abstraction, geometrics, and universality reached new heights. Regionalism's celebration of the concrete, local standards of measurement, whether of quantities or qualities, and particularism in general, increasingly seemed to be artifacts of a world that had been turned asunder if not eclipsed altogether. Characteristically, geography as a discipline was a bit behind the times. While the spatial analytical movement's challenge to regionalism in geography began in the 1950s, it was not until the mid-1960s that enough beachheads were established, practitioners trained, and publications produced that the latest "new" geography could claim to be the true heir to the realm. In that Sauer's own stock in chorology as the preferred mode for geography's future advances had long since been cashed in, he was not moved to explicitly rise to its defense in the face of its new critics and competitors. However, in what might be seen as tacit support, at least complicity in light of what threatened to supercede it, Sauer (and his student and colleague James Parsons) directed a number of field projects during the 1950s and 1960s that had an overarching regional rationale to them.

The theater of operations for these place-based, process-oriented research projects was the Circum-Caribbean realm, including a number of its islands and portions of its mainland perimeter. These were studies sponsored by the Geography Branch of the U.S. Office of Naval Research (ONR). As a prelude to this move into the Caribbean and its rimlands by Sauer and his students, Sauer ended his 1950 summer field season in northern and central Mexico with a side trip to Yucatán and then on to Cuba for a week of reconnaissance (West, 1979:150). His objectives in Cuba are unclear, but it probably was in preparation for the initial proposal for an open-ended program of Caribbean field research that Sauer made the following year to the newly formed ONR. The proposal was approved in late 1951 and the program began the following summer. The projected objectives included study of coastal morphology (beaches, mangroves, coral reefs, *ria* [estuaries], harbors), coastal plain morphology (marine and alluvial terraces), karst landscape formation, climate, vegetation change (especially savanna formation), the *conuco* planting system, the *hato* (ranching) system, and corporate plantations. Thus, the main human geographic foci were directed toward landscape change and agricultural systems, which certainly involved economic questions and issues, but were not mainstream economic geographic concerns even at the outset of the 1950s. The funding was renewed on an annual basis for the next seventeen years (1952–1969). Some thirty-five individuals, mostly Berkeley geography students, were funded by this Caribbean program. The combined output yielded some fifty publications, including journal articles and monographs and thirty-six mimeographed ONR reports (including five doctoral dissertations and seven masters' theses).

The initial trip in the summer of 1952 was to the Dominican Republic. Sauer led it, accompanied by James Parsons and four graduate students. Investigations included grass succession on hillside pastures (Carl Johannessen), marine and alluvial terraces (Ward Barrett), and general reconnaissance of western Haiti (John Street). Sauer himself spent time at the Jewish refugee settlement of Sosua on the north coast, looking at their dairying operation, but chose not to work this up into a publication, which had he done probably would have been a typical contribution to the Latin Americanist economic geography literature of the era. Sauer went on from Hispaniola to visit other field sites, including those of his earlier student Robert Bowman in Puerto Rico, Gordon Merrill, who was working on an ONR-funded historical study of St. Kitt's and Nevis, and Charles Alexander, who was doing an ONR-funded study of Margarita Island off the coast of Venezuela. Along the way Sauer stopped in Antigua, Trinidad, and Jamaica, most likely to scout other promising sites for the kind of physical (mostly) and human geographic research that his ONR program called for.

After the first field season, Sauer turned over most of the field direction to James Parsons, who also assumed much of the responsibility for Latin Americanist teaching and student advising at Berkeley after Sauer retired in 1957. Parsons, very much a product of the Berkeley program and who later became perhaps the single best exemplar of the "Berkeley school" both in spirit and practice, was originally hired by Sauer to cover the economic geography courses taught by Jan Broek after Broek moved to Minnesota in 1948. Given this, one might have assumed that under Parsons's direction the ONR studies after Sauer's initial involvement would have trended more toward economic topics and approaches. But this was not the case. They largely continued on the trajectory that Sauer had established at the outset—focusing on physical or cultural-historical processes rather than economic analyses or chorological syntheses. The official ONR rational for these studies, however, can be viewed as exercises in regional geography, albeit in the post–World War II Cold War environment they fell under the rubric of "area studies."

Just as an abstract modernism swept through both the precincts of cultural production and the academic social sciences, a de-natured or at least a de-geographically contextualized social/political/economic planning movement took hold under the banner of "economic development" for "underdeveloped" nations. While Sauer did not reject the call to study places, peoples, and their products, to point out problems, and to propose solutions, he also insisted on the importance of cultural-historical and environmental perspectives in posing and probing economic questions. In this regard, he had a foot in the past (specifically the 1930s regionalist perspective on planning) and a foot in the future (with the later demands for environmentally appropriate approaches to economic development). For example, his focus on the Caribbean found him diagnosing economic problems and making recommendations for their remedy. Not surprisingly, Sauer's prescription was to look to the past and rehabilitate traditional systems of production, rather than surrender completely to modernity's mandates, including its economic "logic" and, perforce and more importantly, its ecological "illogic."

"ECONOMIC PROSPECTS OF THE CARIBBEAN" (MATHEWSON)

If we take a closer look at Sauer's one venture into making economic prescriptions for the purpose of regional development since his early days in the Midwest, we see flickers of this former land-use analyst, but in the main it is still the culture historian that speaks. The piece in question is his chapter "Economic Prospects of the Caribbean" (1954c, ch. 17 herein) in A. Curtis Wilgus's edited collection *The Caribbean: Its Economy*. Sauer divides the short piece into eight sections, making one or more points in each, and raises questions and issues that most economists would not be likely to consider.

His first point is "advantage of position." He argues that the "American Mediterranean" is one of the chief crossroads of the world, "second perhaps only to the Near East." Thus, the region is destined to play an increasingly important role in both hemispheric commerce and global affairs. His second point reinforces the first: the "energy resources" have shifted from colonial wind power (for both transport and processing products) to hydrocarbons. The Circum-Gulf-Caribbean region will remain one of the key producer and transport regions for the foreseeable future. However, his third point, "harvest of the sea," highlights the finite nature of "natural" resources. The destruction of the sea mammal and sea turtle populations foretells the fate of many other useful sea species. Point four concerns "cropping tropical forests." Here, modern chemistry has sidelined once important sources of medicinal and industrial products for world markets, but local use still validates them. Deforestation has eliminated much of the forest cover in the region, but reforestation, especially in the rimlands, may foster a viable commerce in wood products. Sauer calls attention to the wisdom of local folk in their use of natural sources of building materials, fuel, and other necessities rather than relying on monetary market exchange transactions. Digging deeper into the store of local wisdom, Sauer's centerpiece, point five, is "the *conuco* as an economic system." At the time the *conuco* system was a little-studied, or even recognized local adaptation to Caribbean island conditions. A form of intensive multi-cropping (especially root crops) using mounds, it combined Amerind, West African, and other regional elements to provide a highly productive base for local peasant culture and economy. He closes this section with the admonition that "a major obligation of agricultural science and economy will be to learn the merits of the native systems."

Sauer's sixth point is reserved for unexpected praise for the plantation. Beyond its objectionable entanglements with slavery, Sauer suggests that unless one wishes to "decry" all industrialization, it is perhaps the most efficient and efficacious mode of agro-industry that the region could expect. He points out a number of improvements, however, that would make it an even more appropriate enterprise for these lands and their peoples. His seventh item is "tropical pastures." He sees a bright future, just as when in colonial times livestock were a key economic concern, for livestock raising in the Caribbean, especially in the rimland areas. Here one would have to say that Sauer's legendary power of prophetic foresight was limited. He doesn't seem to have anticipated the ecological consequences of widespread cattle production in these tropical environments. The eighth section is devoted to "sum of prospect." Again, he comes back to the redemptive potential of the *conuco* system and its inherent if overlooked logic and wisdom, the necessary enterprises of plantation and pasture commerce, and the need to learn from, as well as to transfer knowledge to, this region. Less sanguinely, Sauer forecast a dim future for the islands that could not "stabilize their populations," but he also averred that he didn't know how it could be done. He held out little promise for either emigration or for tourists' dollars as offering a permanent fix for these demographic and economic quandaries.

This Caribbean paper, then, is as close as Sauer comes in his later years to playing the role of economist, whether by prescription or simply as prognosticator. He doesn't seem to be uncomfortable in doing so, but one has to wonder how his ideas and examples were received by economists and regional development experts. Given that this was something of an outlier in his publication list, one might surmise that he was not invited to follow it up with recommendations or ruminations on other regions. It does, however, provide a window on Sauer's thinking about economic development issues somewhat in advance of the "development decade" of the long 1960s. In addition, it provides us with some insights into his brief encounter with Caribbean peoples and landscapes. Save for his voyage to South America in the 1940s (1982) and a brief visit to the Cayman Islands in 1968 en route to Costa Rica, this interlude at the beginning of the 1950s was Sauer's only on-the-ground-and-sea exposure to the Caribbean region. Yet, his book *The Early Spanish Main* (1966a) is viewed as one of the classics of Caribbean history, and especially environmental history—a field that at the time had not yet formally announced itself. In addition, it may be Sauer's best-known book to scholars and general readers outside the field of geography. It is therefore somewhat ironic that among Sauer's gifts to posterity—and as he said himself, "the treasure a scholar lays up on earth is largely the printed page"[2]—it is his printed pages on the Caribbean that may turn out to be among the crowning jewels. Also ironic perhaps is that economy rather than ecology constitutes the focus of this fairly obscure paper on the Caribbean, but nonetheless an important statement in the absence of others.[3] But then Sauer did not see economy apart from ecology, and from his cultural-historical perspective the Caribbean could hardly be viewed apart from North, South, and the rest of Middle America.

NOTES

1. This commentary utilizes and revises the "Commentary" and "Abstract" by Martin Kenzer which precede the version of Sauer's address that was published posthumously in the *Yearbook of the Association of Pacific Coast Geographers* (Kenzer, 1984:35–38. Revisions, reorganization, and editing have been undertaken by W. M. Denevan and M. S. Kenzer.

2. This quote comes from a letter from Sauer to Earl J. Hamilton, March 18, 1943.

3. At least two other publications by Sauer on the Caribbean should be mentioned. William Denevan transcribed a talk on "Indian Food Production in the Caribbean" that Sauer gave at the University of Wisconsin in 1965 and published it (1981b) after Sauer's death. "*Terra Firma:*

Orbis Novus," published in Germany in 1962 (ch. 30 herein), previewed *The Early Spanish Main* (1966a).

REFERENCES

Entrikin, J. Nicholas. 1982. "Sauer on Social Science: Howard Odum's Institute for Research in Social Science." *History of Geography Newsletter* 2:35–38.

James, Preston E., and Geoffrey J. Martin. 1978. *The Association of American Geographers: The First Seventy-Five Years, 1904–1979.* Washington, D.C.: Association of American Geographers.

Kenzer, Martin S. 1984. "Commentary" on "Regional Reality in Economy," by Carl O. Sauer. *Yearbook of the Association of Pacific Coast Geographers* 46:35–37.

Sauer, Carl O. Correspondence (cited in text). Carl O. Sauer Papers, Bancroft Library, University of California, Berkeley.

———. For references, see the Sauer bibliography at the end of this volume.

Smith, J. Russell. 1913. *Industrial and Commercial Geography.* Henry Holt, New York.

———. 1925. *North America.* Harcourt Brace, New York.

Steiner, Michael. 1983. "Regionalism in the Great Depression." *Geographical Review* 73:430–446.

West, Robert C. 1979. Carl Sauer's *Fieldwork in Latin America.* Dellplain Latin American Studies, No. 3. University Microfilms International, Ann Arbor.

Wilgus, A. Curtis, ed. 1954. *The Caribbean: Its Economy.* Univ. of Florida Press, Gainesville.

Williams, Michael. 1983. "'The Apple of My Eye': Carl Sauer and Historical Geography." *Journal of Historical Geography* 9:1–28.

15

Abstract: Geography as Regional Economics (1921)

Carl O. Sauer

There have been numerous discussions of the scope of geography, and especially there have been examinations of the periphery of the science. Much less attention has been given to the determination of particular objectives within the field of geography. Geography is suffering from a scattering of interests over too broad a field for the limited number of workers engaged in it.

The focussing of attention on certain phases of the field alone appears to give hope of establishing the science solidly. This involves consideration of the content, aims, and methods of such a specific type of inquiry. In this country historical geography has been treated in such a manner.

A voluntary limitation of research by a group of workers to the field of regional economic geography is probably the most urgent need of the science today. Regional economics has not been preempted by economic science and belongs most appropriately to geography. The essential problems are (1) the determination of bases of unity of the area, (2) the inquiry into advantages and handicaps inherent in the area, (3) the time element as affecting stage of development, and (4) the analysis of the entire economic complex of the region. It follows that any area, geographically defined, is an appropriate subject of inquiry, and that the inquiry must not be limited to the evaluation of so-called geographic factors.

In the method of research, work needs to be done in forming a scientific discipline for (1) agrogeographic research, as referring to rural conditions, especially the utilization of the land, (2) urban studies, and (3) movement of trade.

A logical as well as pragmatic sanction is at hand for such studies, and by means of them geography may knock successfully at the door of the business world and, as well, it may present itself as an advisor to governmental policy.

NOTE

Reprinted with permission from the *Annals of the Association of American Geographers* 11:130–131. Copyright 1921, Association of American Geographers.

16

Regional Reality in Economy (1984 [1936])

Carl O. Sauer
Edited by Martin S. Kenzer

At this reunion of some of the tribes of the confederation called the social sciences, a visiting member of another tribe, having partaken of the food of his hosts, rises to make [an] observation on the state of the confederation.

The report is heard near and far that the renown of the group is not growing as it should. It is said that more territory is preempted than is used well. It is also said that everywhere the shamans are growing in power and wealth, whereas the workers, those who fetch things and work goods, are poorly regarded. It is also said that among the young men, in consequence, there are far too many training to become shamans and too few who will carry on the crafts of their people. We know that the confederation was formed at different times by the adding of new tribes; the oldest tribes became those of History and Geography. Lately others have grown more powerful. We of the older people recognize the energy and statesmanship of you of the younger people. There is, however, some feeling that our ways, which are the older ways, are being somewhat overrun and forgotten by the newer ways, which the younger tribes have brought in. On this occasion some of the older traditions are raised to see whether they have maintained their appeal to the larger group.

In one sense we represent special disciplines concerning the cultivation of which we may be appropriately jealous as to specific qualifications of training. There are numerous divisions of our fields in which success is related to the specialization of skill. In another sense, however, there are approaches to social science, both in facts and methods, that belong or may belong to all, even though their use is more general in one field than another. It is in the latter manner that I wish to speak of geography, and also somewhat of history, as pervasive of the social sciences in general. We are therefore not concerned with the raising of higher partitions between disciplines but with the possibilities of extension of geographical and historical objectives as underpinnings throughout the social sciences. From these claims possibly psychology should be excluded in major part. It is another foundation subject, utilizing in the main, however, a method quite distinctive to itself among social sciences, though there are indications that its methodology can be extended into other social fields. Its congruence with the central social sciences—economics, political science, and sociology—is probably as great as that of history and geography, but is of markedly different character.

The so-called central fields just mentioned have probably exerted their major influence in terms of social philosophy. No one of us can disregard the continuing scrutiny of social philosophy if we wish also a more varied and more realistic inspection of social processes and relations. But, perhaps we are not at all times sufficiently aware of the difference between moral philosophy and social science. The continuity of this awareness appears to be greatest in the so-called peripheral fields of social science. A lengthening period of years in participation in doctoral examinations has given the present commentator the opinion that the preoccupation of economists, sociologists, and political scientists is heavily with social values, [but] that in the other fields it is with evidence and classification. This generalization is perhaps somewhat gross, but few will dispute that it is an approximation of a real distinction. Neither a psychologist nor an historian would be likely to say that we suffer from a plethora of facts but that we need more rigorous thinking to distinguish between right and wrong answers to social problems. Yet the economic and political theorist can and does take this position. At times we of the periphery look with amazement on the dialectic prowess and perhaps on the capacity of exultation or protest on the part of the theorist of society. We see also from time to time the rapidity with which these talents develop, so that even extreme youth may express itself in masterly fashion in the form of theory, whereas over our workshop is written:

> The life so short, the craft so long to learn [; or, The assay so hard, so sharp the conquering.[1]]

We would not diminish or dampen the clash of sharp wits, this thrust and parry of acute minds. We think, however, that we may assert the need for more emphasis on analytic studies which do not involve social values. We do not abdicate thereby our interest in a better world if we do say, "*Die Weltgeschichte ist das Weltgericht.*" We simply say that our main task is that of being critical spectators rather than to be field coaches. We do not wish to dissemble, however, our pride in cherishing most the quality of unabating curiosity. This curious world is unendingly fascinating. It is not primarily our business to make it conform to our reason. It is our business to try to understand it kinetically. In using this figure, we do not assert that the properties of man will ever be approachable as are the properties of matter.

Social science and social theory run side by side, sometimes touching, sometimes disparate. It is an intellectual obligation that social theory always be cultivated, even if no great social theory, other than a romantic one, is possible for our age or perhaps may be possible again. Perhaps the logic of a given situation normally is to be abstracted from that situation and not to be judged by general postulates? It is no matter. The philosophical critic serves his purpose nevertheless. But many of us do think that social processes and categorical imperatives have limited connection.

Leaving then the universal and normative we turn, at least for working hypotheses, to a pluralistic world. We assert nothing as to the beginning or end of being in this connection. Logos and logic are not matters of our concern. We can do without economic man. We do not need to define society. We can say that these are constructions of a given civilization that is dated and localized, the expressions of an "intellectual climate"—and climate is a quality of place and subject in course of time to change.

Instead of a valid theory of society we wish immediately to build up a body of organized knowledge of identifiable social groups. Economy[2] becomes a plural term and, apparently, lies largely outside of the present field of economics. That it should do so is not asserted. For better or for worse this discipline [economics] has become primarily a study of wealth, one might almost say the science of money. With all of our modern disciplinary growth, there remains, tremendously neglected, the study of human economies and what may be called again "intellectual climates." These are to each other as obverse and reverse. Planned and projected over and over again, especially by the founders of sociology, the comparative study of human groups has made excessively small progress.

An honorable exception should, I think, be made of the science of anthropology, which appears to me to be methodically the most advanced of the social sciences and the one of greatest intellectual return for the amount of effort expended. Indeed, I should go so far as to indicate the history of anthropology as a major importance to all students of man. Extraordinarily fecund in theory, its theories nevertheless are only working hypotheses that stand or fall by the increasing evidence. Divided into embroiled schools, each is yet continually enriched by the finds of the other, which are significant apart from their particular interpretation. The joy of inquiry, which is almost a generic trait of the anthropologist, is referable in large measure to the fact that he has learned successfully to begin with the specific. He is a very good craftsman because he knows his materials and their assembly. Had he begun with culture *per se*, he might still only be making speeches. Instead he begins with culture *traits*, which he places, dates, and labels. He assembles these into cultures and considers their genetic and functional relation. He has not faced reluctantly the work of description and classification, but is imbued with encyclopedic curiosity. He has learned much because of this driving force of curiosity. But curiosity is nourished by a successful methodology, by means of which the individual item becomes united into the generic, the trait related to other traits. Here again anthropology has profited by an approach which it did not originate, but which it has used more largely than any other discipline and has developed further, namely the analysis of geographic distribution.[3]

The development of this method is due primarily to Friedrich Ratzel's rich, varied, and inconsistent personality. Unfortunately, the major influence of Ratzel on social science has been through the first volume of his *Anthropogeographie*, which was an attempt to form a dynamics of society in terms of physical environment. As the result, to this day there are some geographers, and far more other social scientists, who think that geography must interpret the fortunes of man as derived from his milieu. Ratzel was guilty of excesses in this respect, though perhaps of fewer than have been charged against him. At the other extreme is the study of human history as though it happened physically *in vacuo*, probably a more common failing in actual practice than excessive environmentalism.

In the less-known second volume of his *Anthropogeographie* Ratzel concerned himself with the diversification of cultural phenomena over the earth. First, he said [that] there was to be regarded first of all the inhabited as against the uninhabited part of the earth. We may redefine this as the presence or absence of man and add that its complete expression is the spread of man from his earliest homes to his modern range. Next comes the variation in density of population, the gross contrasts in [the] massing of inhabitants and the specific differences such as between town and country. It may be noted incidentally that we do not have today a population map of the world for the present which begins to make use of the demographic data available and that hence, of course no similar representation of the world for past periods has been attempted. Ratzel emphasized the need of population cartography as a measure of economic potential. Compare Ratzel and our population experts: In addition to urging the study of a cartography of what we may call cultural *quanta* of various sorts, he also urged examining the distributional occurrence of *non-numerically significant culture qualities* or traits. It is by this latter type of study that diffusion of culture traits is mainly examined, and it is from Ratzel that the diffusionists mainly stem. Large insight has been gained into spread and even source of cultures by careful charting of the limits of occurrence of traits, by their varietal diversity within those limits, and by the specific nature of the limits themselves, including exclaves and enclaves.

The objective of such study is in principle historical. It supplies information concerning spread and retreat of culture that is otherwise missed and it has even served to correct documentary statements. Both archaeology and ethnology work largely with the most careful cartographic layout of their data. This method, like that of statistics, is an auxiliary rather than an absolute method toward understanding. It does arrange the data of a given trait as to spatial occurrence. From these data, it has been implied that certain inferences can be drawn directly as to the direction of cultural movement and hence as to dynamic centers of culture hearths. By superimposition of several such distributional studies, further insight is [gained] into their interrelations. The method has made possible many of the discoveries of anthropology and, especially, has encouraged the view that parallel invention is an exceptional occurrence in human history.

Outside of the "primitive" field this method of inspection has been less used. It does have certain notable exceptions, however. An examination of housetypes, village plans, and field systems of continental Europe, thus executed, has forced significant revisions of documentary history as to movement of nationality groups and economic institutions. Linguistic and dialectic atlases similarly have given unexpected indications of cultural connections and blocks. Other illustrations might be cited to show that the method of geographic distribution is not unused above the primitive level.

It should be underscored that this method is simply a method of inspection. By itself it is no method of classification. Comparative distributions, however, provide an insight into relationship that is preferable in many cases to statistical correlation because [comparative distributions] emphasize the spatial arrangement of associations. Natural history has long availed itself of this mode of gaining knowledge. In culture history it is capable of indefinite expansion. It is based simply on the concept of extension and of association of phenomena, and has nothing to do directly with influence of environment. Unfortunately, interest in the classification of culture traits is as yet not very great among social scientists and even less [is] any realization that these qualities were developed in some particular spot, [that they] have won their way over a particular territory, and [eventually] come to rest against certain physical or cultural blocks.

Yet these interests are fundamental to what I have called the economic region and the intellectual climate. Human culture is a question of multiple origins of individual skills, attainments, and attitudes, of their migrations and rejections, of their combinations and the stimuli to change [in turn] what they develop. We know no culture in the singular, yet we have hardly begun a comparative study of our cultures in the sense that the anthropologist has ordered the knowledge of primitive ones. The whole situation boils down to something like this. In the process of spreading over the earth and occupying it, man has developed a great variety of institutions expressing very different economies and combined into divergent societies. These divergences, parallels, borrowings, and reconvergences of human history invite comparative study. One of the promising means of approach is by breakdown into culture traits and the study of their spread or retreat: [Here] the map is invaluable as a device in analysis, both for quantitative recording and for charting limits that are significant without regard to the frequency of the phenomenon. [Next,] the culture traits are examined as to their connection. Here again accordances and discordances in distribution—that is, once [again] inspection by means of the map—are useful. The third step is the association of culture traits as complexes or areas.

Be it noted that this descriptive procedure is [both] geographical and historical in method. It asserts the reality and persistence of areal variation and tries to characterize this difference. It does not need to commit itself as to the bases of such differences, but it is concerned with discovering them. The differentiation of life is in part a matter of environmental adaptation, in part a question of cultural growth and diffusion. The origin and spread of ideas and skills is of course not to be thought of as taking place by any evolutionary sequence. There are no stages of culture; there are only inventions that make their way out into a wider world. In general, therefore, we are dealing with culture history, not with the history of personalities but with the history of institutions. (Perhaps we have here again the dual cleavage of the social science groups; the great man and the great idea on the one side, the impersonal movement of events on the other.) The spread of these things of culture is conditioned both by the receptiveness of the land itself and of the people in the land. History is an idiot's tale without careful regard to the environment at all times. The whole business of human differentiation of the world rests upon the never-ceasing interplay of time and place, of heredity and environment.

It is therefore no idle business that puts emphasis on the study of the association of human attainments in a human area, well characterized as to its physical qualities. These are attempts to get at the larger aggregates of human beings that have functional reality. Though in the modern world we have gotten far away from the supposedly neat distinction between tribe and tribe of primitive society, and modern commerce has in limited measure made one economic region of the world, there is still, and still will continue, a consciousness of likeness and unlikeness between groups. Distances still separate or bind people; race and tradition anew demonstrate their vitality; environment still prescribes limits of activity.

There is a curious resistance in our academic, though not in our intellectual, life to this sort of inquiry. For an active and successful concern with the American scene and its component areas, one must turn largely still to the field of literature. I submit that there is more of American regional economy and intellectual climate in the *Saturday Evening Post, American Mercury, Harper's Magazine,* and the *Atlantic,* than is in all the works of all the social scientists of the country. Not only this but I suspect largely that future students will pay very much more attention to these sources than to ourselves. A small part of the answer has been suggested: Art as well as science is involved. No weirder term could have been thought of for our field than to label it "social science." First of all it turns out to be dominated by philosophy rather than science. Next, the only well-developed and recognized method is the highly limited one of statistical correlation. Finally, a (large) part of the understanding aimed at is not within the reach of [the] scientific approach, but frankly is and must be an art.

Much of the difficulty rests, however, with our professional compartments. May I ask your indulgence in again referring to my friends, the anthropologists? If, instead of one such working association, the anthropologists had farmed out the study of primitive man to a half-dozen professional groups, each jealous to vindicate its difference from the next group, what would have been the resultant knowledge of primitive man? Yet we have followed predominantly the latter procedure. The program of this conference indicates a ray of hope in the other direction. If we say that we wish to characterize and consider life in the Far West as an important part of our work, we are making an admission that too few American social scientists have made. The congresses of Americanists and Orientalists have an enthusiastic and often combative attendance, which indicates that people do strike fire when they get down to areal reality in culture. Should not we within the social sciences make similar provisions for those of us who find the intellectual implications of the log cabin and the chilled-steel plow of bracken-choked cut-over lands, or the mail-order catalog of the Mennonite, more exciting by far than any correction of theory by Marshall or definition of rent? We should suffer no loss in respectability and would gain in liveliness by encouraging attention to the meaning of life in our part of the world.

Another form of resistance I am included to blame on the political map, which incidentally is almost the only map which the American knows. An honest-to-goodness map is a generalization of an actual scene. If nature and life are varied, the map becomes an intricate assemblage. If the land is a sparsely inhabited plain, it is represented by a paucity of symbols. But the political map is a dogmatic and mendacious sort of thing. It draws sharp and violent boundaries. It says that some parts of the world are red, some blue, and some yellow, and that if they are yellow they cannot be blue. It does such violence to the thinking of people in order to impress territorial sovereignty. Almost all boundaries other than political ones are anything but lines, yet the whole business of studying areal differentiation suffers because of the simplification of the concept due to the political map.

The regional differentiation in economy must in the end be determined by trait analysis and their synthesis into complexes. The sudden interest in national planning gave a sharp impetus to interest in economic regions that bear little resemblance to each other in plan or delineation, but which illustrate the hazard of synthesis without sufficiently broad and critical analysis. When our social scientists become map minded as they are number minded and learn to distinguish between maps and cartograms, we shall have a lively time with plotting and interpreting cultural data. Then we shall be well onto our way to a comparative culture morphology.

For the present the distinction between regional economy and intellectual climate may have utility. The picture is borrowed from physical geography where we distinguish between the physical forms of the earth's surface and the atmosphere that envelops them. It is not difficult to see the resemblance to the material and spiritual elements of culture of the anthropologist. Some prefer the word ecology to that of economy. Plant ecology at least provides numerous valuable methodologic suggestions to the study of human distributions and associations. Fundamental to an economy are the forms or habits of food, clothing, and shelter, and the means (skills and tools) of providing them; hence the manner of utilizing the productive possibilities of the area and their supplementing by trade must be resolved. This leads to the most important question as to whether the economy is self-sustaining or self-destructive, and also that of its sensitiveness or insensitiveness to world conditions. Manner and quality of utilization of the land and the distribution of population are the most specifically geographic elements of such study. Division of labor and return—i.e., the part of the individual in the group economy—are equally necessary elements of the analysis and equally necessitous of distributional study. In terms of agricultural lands, attention need only be called to the significance of knowing the distribution of size of operations, of tenancy, of seasonal labor.

Finally, we may ask how much descriptive analysis and synthesis will contribute significantly to understanding. (1) If we are to have a science of man we cannot begin nor remain with the general nor the unique, but must deal with recurrent phenomena. This we have called culture traits. These communicated and applied ideas, if I may be permitted to call them such, are the stuff of which cultures are made: They are the pattern of life of a group. Their identification and classification is a general obligation on all social science. Their weighing in interpretation can come only with experience. (2) We know enough of the development of human societies to say that these do not go through a general series of stages but that they have become differentiated in terms of a series of experiences, which may have happened first to a given group in its area or which may have been brought in from the outside. The spread of such experience is necessary to all understanding of cultural differentiation. The older term for this is diffusion of culture traits, the one in current vogue is acculturation. (3) Diffusion depends, for survival, on the applicability of the idea to the given physical environment. The world has become differentiated at the hands of man because ideas have sprung up in different places at different times; have had unequal facilities for spreading and unequal chances for survival. Culture survival and growth are alike dependent on history and environment; without the proper tracing out of both we are dealing with something unidimensional and nonsensical as against the four-dimensional realities of cultures. (4) Culture growth, in particular, demands attention to greatly differing receptiveness of areas to diffusion, depending in part on physical accessibility, in part on physical adaptiveness of area, but also on purely cultural resistance. We may learn to make useful distinctions between areas that are normally given to change and those that resist change. (5) An intriguing problem of culture area is the determination of culture hearths. From our limited knowledge of the world it appears that at any one time very few centers originate the vital ideas that diffuse strongly. There are such centers of culture origins which exist for a time and then fade out. There are others which have indefinitely long significance. These 'key areas,' or 'hearths,' or *Kräftezentren* are at the root of all questions of culture dynamics. (6) Of equal significance are the areas of culture degeneration. There is a little-developed pathology of culture areas. One simple but tremendous theme is that of destructive economies which destroy the subsistence base of the population. No small part of the history of the world and of economic planning for the future needs to be written in these terms.

These are some of the major themes that may be found in a study of culture content, growth, and succession, [all of] which are based on [an] historical-geographic descriptive approach. They are as broad as all social science and they invite widest participation. May I suggest that our position on the far western fringes of the Occidental economic world invites our efforts at determining its regional realities?

NOTES

Reprinted with permission from the *Yearbook of the Association of Pacific Coast Geographers* 46:38–49. Copyright 1984, Association of Pacific Coast Geographers.

Two manuscript copies of this proposed address to the Pacific Coast Economic Association in 1936 are preserved in the Carl O. Sauer Papers at the Bancroft Library, University of California, Berkeley. In editing, I have relied more heavily on the typescript copy in Carton 3, which contains numerous handwritten corrections by Sauer. Correspondence between Sauer and Blair Stewart regarding the invited presentation are also in Carton 3. The second is a mimeographed copy found in Carton 2. The original publication (Sauer, 1984) is by permission of the Bancroft Library, and I would like to thank Marie Bryne for her assistance in this regard. (M. S. Kenzer)

I have made a few changes in Sauer's written manuscript, including corrections of punctuation to convert an oral presentation into a written one. Words added for clarity are in brackets. Stylistic "errors" and awkwardness stand as written. I have done as much as possible to retain Sauer's language and meaning, and any mistakes or misrepresentations are attributable to me. The reader needs to keep in mind that this essay was originally prepared for oral delivery and that, in all likelihood, it was composed in a short period of time and would have been subject to changes by Sauer before, or even during, its actual presentation. Both are from Geoffrey Chaucer's *Parliament of Fowls*. (M. S. Kenzer)

1. In Sauer's typescript copy, the second phrase—"The assay so hard, so sharp the conquering"—is penciled in, but the first phrase exists as well: "The life so short, the craft so long to learn" was his initial choice. It is evident from the flow and context of the oral paper that only one of these phrases would have been read aloud by Sauer. However, since Sauer was undecided and left both versions on the typescript, I have opted to include both phrases in this paper as well. (M. S. Kenzer)

2. Volume editors' note: By "economy," Sauer means the way a cultural group makes a living from the land. He is not interested in "economic man."

3. Volume editors' note: The digression that follows is one of Sauer's earliest expressions of interest in the distribution and diffusion of cultural traits.

17

Economic Prospects of the Caribbean (1954)

Carl O. Sauer

The American Mediterranean is one of the chief crossroads of the world, second perhaps only to the Near East. This is an elemental and permanent fact of position of sea and land to which different times may give different value and expression, but which no political or economic pattern of the world can diminish in the long run.

Our own Middle Atlantic seaboard became important three centuries ago as the opposite and colonial shore of northwest Europe, and it has continued to depend mainly on the business crossing the North Atlantic. In the not-too-distant future, southward communication, across climatic belts through the West Indies and along them to South America, may take first place in the commerce of our eastern ports.

ADVANTAGE OF POSITION

Middle America, by mainland corridor and island stepping-stones, as well as through its sheltered seas, links the two Americas. Between its island guards and across the isthmuses of its mainland passes, must pass, in the future, the main traffic between Atlantic and Pacific. Florida, Panama, and Trinidad bound the strategic triangle of the New World; it is not by chance that these have become world centers of air lanes.

These large features of our globe are so salient that they must never be overlooked. They have operated in the geologic past in the dispersal of plants and animals. This history of man in the New World—aboriginal, colonial, and contemporary—is a series of solutions of the tactical positions and interior lines of this area as lying between the two continental Americas. Lately we have become aware of the superior transport position and strategic implications of the great and critical mineral resources lying within the Caribbean rim—petroleum, bauxite, and iron ore—as affecting the industrial future of the New World.

Man always economizes expenditure of energy, and so his assembly points grow up where the assembly costs are least; his busiest routes of trade involve the fewest ton-miles. The more commerce draws upon the ends of the earth for primary materials, the more advantage accrues to the most central locations. Middle America has such [a] distance-saving position, on which routes converge from all quadrants; in particular, the cheap routes of the sea, the fast ones of the air.

ENERGY RESOURCES

On the Lesser Antilles the ruined towers of windmills recall the days when the trade winds were used to perform the labor of grinding cane. At the same time, wherever cane was grown the woods were heavily depleted for boiling the cane juice. The modern fuel era began in the nineteenth century with imported bunker coal, and in the twentieth has shifted to fuel oil. The greatest proved oil districts of the New World are at the northern and southern rims of our Mediterranean, between the Mississippi and Panuco [Mexico] rivers, and along the southern shores of the Caribbean. Low-cost fuel oils are available on short hauls by pipe line and tanker—perhaps the cheapest and most flexible of all means of transport—to island and Central American shores, where nature has made generous provision of natural harbors, beyond the necessities of commerce. The processing of primary materials, the growth of service industries, and facilities of communication within the Caribbean area are not held back by transport cost of introduced fuels, nor will they be so long as these major oil fields last.

Cane bagasse is probably, at the moment, in first place as [the] energy source for the islands. Hydroelectric energy is available in largest amount from the rainy mountains of the Dominican Republic, where economical hydropower stations are now functioning.

HARVEST OF THE SEA

The harvest of the sea is nearly abandoned. The salt pans on which the North Atlantic fisheries once depended are gone, except for a little salt still made on Turks and Caicos Islands. The pearl fisheries along the semiarid coasts of the Caribbean mainland became devastated very early and have not been restored.

In aboriginal days the land provided man with carbohydrates, the sea with proteins. The manatí, or sea cow, weighing up to a ton, furnished superior meat to the Spaniards. It was declared to be fish by the Church and thus proper for meatless days. The water pastures of the stream mouths where once it fed in large numbers were gradually depopulated, and are today unutilized by man.

The great herbivorous green turtles were a mainstay of Indian populations, produced in such abundance all about the West Indies as to provide for many years the chief flesh food of the black slaves. Fleets of turtles were engaged the year round out of Jamaican harbors for the provisioning of the plantations. When the Jamaican waters were exhausted, these fishing vessels operated on the keys and shores of Cuba and the Caymans. A few Caymanian vessels continue the business to the present, no longer in home waters but along the wilder shores of Central America. The former common food of slaves has become a delicacy of luxury markets in the United States, the turtles being shipped to our urban markets by air freight.

The green turtles have been depleted by the practice of catching them while mating, by taking females when they come to shore to lay eggs, and by stealing the eggs. Protection might be had for beaches still visited by sea turtles. As the situation stands, they are rapidly going the way of the manatí. A conservation policy for the resources of the sea would not be difficult to draw up, but would be difficult of acceptance because of the many political jurisdictions and interests. None has been tried; as a result, the plankton-rich areas of the Caribbean and Sargasso seas are not being utilized beneficially by man.

A further difficulty is that most of the present inhabitants know nothing of fishing as a means of livelihood, excepting the colored folks of the Bahamas and the Cayman whites. Thus, even where sunlit shallow banks afford superior fishing grounds, as about most of Cuba, sea food is scarce and expensive. It is likely to cost more than beef, and it may be imported from the United States.

CROPPING THE TROPICAL FOREST

The islands have more people than the whole of Canada. Need of tilled land has caused the removal of most of the forest cover, except from the steeper mountain slopes. The Dominican Republic, with least pressure of population, has [the] most forest left. Neither on islands nor on Caribbean mainland is effective attention given to trees as a permanent resource, except perhaps by the United Fruit Company. The Caribbean pine of the lowlands, nearly related to our southern slash pine, has been almost wholly cut out in Cuba. It is now also being cut heavily on the mainland, from British Honduras to Nicaragua, mainly for export. The highland pines, being less accessible, have been less invaded, but logging roads are now being pushed into the mountains throughout the island of Haiti. Watershed protection is thereby being lost where it is greatly needed. The valuable cabinet woods have been logged out earlier. Tropical cedar (*Cedrela*), mahogany (*Swietenia*), and blond mahogany (*Tabebuia*) are being priced out of the market by their growing scarcity.

Both cabinet woods and pines have been cut with little thought given to what becomes of the land afterward. The pine lands in many cases are of low fertility and little suited to agriculture or even to pasture. Reforestation is warranted commercially in many localities, especially for cabinet woods. These are natives of the rain forests in which trees, to succeed, must grow tall, straight, and fast. The mahoganies and tropical cedar may make saw logs in as little as twenty to thirty years, and they establish themselves readily because they seed freely into clearings. Given simple management, the highly diverse rain forest may be simplified into stands dominated by the fine species. Queensland has introduced these Central American trees successfully into forest plantations. In their native home they are in [the] process of being destroyed, but might be increased as a valuable and permanent resource, especially on the Central American mainland.

The once-important industries of cutting dyewoods, collecting aromatic resins, gums, and drugs have given way to the laboratory products of the organic chemist. There may be a modest place in the future economy for some of these woody plants. A little cash income to supplement subsistence farming is a sharply increasing need of the growing and often unemployed populations. For many of the people it is not a question of how much they can earn, but whether they can earn anything. Possibly, with some aid in marketing, some of these products can compete here and there with those of the chemical and pharmaceutical factories. Perhaps there is a valid customer preference for some natural flavors, perfumes, and cosmetics, as against the products of coal mines and gas wells. Perhaps some plants synthesize certain compounds better than do the laboratories. Perhaps applied science is thinking too exclusively in terms of large enterprise.

The native economy depends in numerous ways on trees and shrubs and will continue to do so because they may be most useful and least expensive. (1) I neither think nor hope that the functionally admirable native house, the *bohío*, will give way generally to concrete-block construction and roofs of tin or aluminum. The palms that provide so characteristic an accent to Caribbean landscapes do so because they are good primary structural and household materials: the fan-leaved sabals for their excellent thatch, the corozos split into thin strips that serve as weatherboarding, as well as for their fruits used as mast, the chontaduras, termite-resistant, for house posts and frames. (2) *Postes vivos*, living fences, continue to be preferred because they are not subject to termite attack and wood rot and serve admirably and inexpensively in enclosure of field and garden, yielding also a variety of fruits and other household items. (3) Commercial fuels and the expensive stoves they require cannot replace firewood and charcoal where people live at minimal income levels. We of the North, living in easiest circumstances, need awareness that our neighbors to the South have made most sensible adaptations of resources to severely limited economic capacities. We need to use caution in urging the transfer of the pattern of our commercial culture into situations that must remain very different from ours.

THE *CONUCO* AS AN ECONOMIC SYSTEM

The native Indian crops were a varied lot of high-yielding roots or tubers, yuca or cassava, sweet potatoes, yautia or malanga, arrowroot, certain yams, peanuts, and, I think, plantains. These, with fruits such as of palms, pineapple, and mamey, gave starches and sugars in plenty, as well as greens. Corn, beans, and squashes were of secondary importance. Plant proteins were few and little required since sea food was abundantly taken. The diet was ample, balanced, and varied. When Indians gave way to negro slaves, the latter took over for themselves, rather than for their masters, the cultivation of the Indian crops, and added thereto such African things as the greater yam, the pigeon pea or guandul, okra, and the keeping of fowls. In the Spanish and French colonies of the Caribbean, these Indian and African foodstuffs and ways of preparing them passed gradually into the kitchens of their masters and became the creole cooking of the present. In the English settlements they remained mainly the provisions of the colored folk. The island of Haiti, both on the Spanish and French side, still grows about every plant described by Oviedo when he first informed the Europeans of the crops of the Indies. In the Dominican Republic these provide abundance; on Haiti they make possible survival. Cuba, largely resettled only in late years, knows and uses the fewest root crops.

The food potential of the traditional *conuco* planting, or provision ground, is hardly appreciated by ourselves, be we agricultural scientists, economists, or planners, because its tradition as well as content are so different from what we know and practice. Yields are much higher than from grains, production is continuous the year around, storage is hardly needed, [and] individual kinds are not grown separately in fields but are assembled together in one planted ground to which our habits of order would apply neither the name of field nor garden. And so we are likely to miss the merits of this system.

The proper *conuco* is, in fact, an imitation by man of tropical nature, a many-storied cultural vegetation producing at all levels, from tubers underground through understory of pigeon peas and coffee, a second story of cacao and bananas, to a canopy of fruit trees and palms. Such an assemblage makes full use of light, moisture, and soil—its messy appearance to our eyes meaning really that all the niches are properly filled. A proper planting of this sort is about as protective of the soil as is the wild vegetation. The *conuco* system can make intensive use of steep slopes and thereby may encounter erosion hazards that should not be blamed on the system itself, as commonly they have been. Nor do I see chances of success for pulling down by decree the ever-crowding populations from their hillsides.

Its [*conuco*] commercial disadvantages are that most of the things produced are difficult to store and ship, that it is best suited to the small producer, and that it resists mechanization. However, no field agriculture can match it in overall productivity and continuous production. It is the best way of subsistence agriculture and of support for increasing numbers of people. Cacao and coffee fit well into *conuco* culture, can be satisfactorily marketed by small producers, and add cash income to self-sufficiency in food. The Dominican Republic offers especially good illustrations of peasant subsistence at a good level, with cash items from cacao, coffee, peanuts, and tobacco. A major obligation of agricultural science and economy will be to learn the merits of the native systems and aid them by discovering, developing, and distributing superior kinds and races of plants.

A PERMANENT PLACE FOR THE PLANTATION

A good word is next in place for the plantation system which came as naturally to the West Indies as did the industrial revolution to the English Midlands. It has dominated our area for three centuries and, despite its critics, seems to have life in it for quite a time ahead. It is the classical model of the factory farm that is now making such inroads on the family farm system of present-day United States. The plantation has had a very bad press ever since people developed a conscience about slavery, but it has a well-reformed character and hardly deserves to be the whipping boy it often is for politicians, unless one is consistent in decrying all industrialization.

Sugar was the earliest and still is the first product of the plantation. The sugar cane is not only the most effective producer of sugar, but its toll on soil fertility is relatively low (especially as to nitrogen and phosphorus), the energy demand in tillage is low, and this giant grass forms a superior ground cover against surface runoff.

That there is at present overproduction of sugar is true only because of political restrictions. Sugar entered world markets as a luxury and was immediately levied upon for the support of treasuries. Window and salt taxes have disappeared, but sugar is still in nearly all countries subjected to a variety of special taxes, direct and indirect. No other commodity has been entangled so long and so deeply in mercantilistic controls and manipulations.

There is no way back from the large sugar factory, or *central*, to small processing units except at increased cost of the product. Optimum size will be determined by cost sheets, and these register the disadvantage of small factories and plantations. Much the same is true of banana, pineapple, and tropical fibers for world markets. Size is self-regulating in industrial competition, and the present direction is toward larger units. Bigness is no more bad in sugar or banana than it is in shoes or automobiles. If the Caribbean is to prosper in access to world markets it has to produce attractive goods at attractive prices. No one has discovered anything for the area equal to its plantation products, for which it has real natural advantage.

I think there is nothing seriously wrong with the cane-sugar business except a heavy incrustation of political interventions, external and internal. The gradual reduction in late years of the role of American capital in the over-all industry is tactically to the good, since it diminishes the national sensitivities to foreign domination. Except for the inflow of American capital, ultimately from the savings of very many individuals and not from some mythical money colossus, the sugar industry and the internal improvements of Cuba, Puerto Rico, and the Dominican Republic would largely not have been achieved. Scores of millions of dollars in such investments have been written off, again mainly at the expense of citizens of the United States, in the forced reorganizations during depression years, mainly in the 1930s. The industry is now operating on valuations greatly below actual cost and even more so below replacement costs. The investment in and valuation of *centrales,* railroads, ports, power plants, and housing, and the necessary accumulation of working capital are far in excess of that in lands and crops. For one of the largest corporations, for example, producing mainly from administration-grown cane, the land represents about one-third of the physical property and one-sixth of the total investment at current book value. The record of American management is good as to wages, housing, sanitation, and health services. We ought to stop being apologetic about the role we have held as fair, and even generous, partners in the economic life of the Caribbean. It is in the common interest that this partnership be maintained and developed into the future, with mutual growth of understanding and good feeling.

The sugar industry is admirable technically in its extraction of raw sugar, but the utilization of by-products has made less headway. I am inclined to think that in part it is not technical but psychologic blocks that are in the way, the big gains or losses that a minor price change on the sugar market brings, the exhausting tension of the grinding season, the priority making the sugar *zafra* has over everything else. A *central* will carry to the mill hundreds of thousands of tons, even a million or more, of cane within a few months, a tenth to an eighth of which will become raw sugar. The residual molasses is a fourth or somewhat less of the sugar tonnage. The remainder of crushed stalk, or *bagasse,* is the fuel by which the *central* operates, which also may provide electric power to be distributed to towns round about. This admittedly may not be the most economical conversion of the great mass of residual organic matter, but it is always available and is tax free, which imported fuel oils may not be. The potash-rich ash in some cases is returned to the cane fields.

The disposal of the great flood of blackstrap molasses is a chronic dilemma, which has grown worse now that industrial alcohol is produced more cheaply from natural gas and petroleum. The industry has shown little initiative in helping the use of molasses as stock feed. Indeed, the Cuban common distributing agency has priced molasses lower to purchasers for industrial alcohol than it has to buyers for feed. Its cheapness and high feed value for cattle, hogs, and poultry has long been well known. The difficulty has been in handling the sticky stuff, especially to and at the farm. It may be that this problem is at last being solved and that a large, dependable, and profitable market is opening overseas.

The economy of the sugar islands is under seasonal strain by reason of the long dead season, when there is little employment either in cane fields or *central*. However good the wages during employment, they are likely to have been spent before the beginning of the next *zafra*. Obviously this may suggest a possibility, still virtually unexplored, for off-season employment in light industries, such as small factories making work clothes, shoes, and the like, for which we have successful examples in many small towns of the Middle West. The efforts both by companies and government to get the sugar workers to produce their own food rather than to buy everything at the stores are not new and are continuing. It is not that the workers cannot have land to plant but that they are really not farmers or peasants, and are accustomed only to work for wages. In the Cuban sugar areas, for example, native provisions and fruits are woefully wanting, and California rice and fruit juices, as well as Midwestern flour and meat, are staples in rural stores. This uneconomic situation rests on social habit rather than on denial of opportunity.

TROPICAL PASTURES

The Caribbean lands have immediate, and in part long-range opportunities in livestock. We must not forget that the New World was stocked with cattle and hogs via Hispaniola, where the Spanish colonists had found that all forms of livestock thrived exceedingly well. In the old plantation days, cattle were important on most estates, and cane tops and leaves were fully used as feed. The nineteenth century saw the introduction of a series of valuable African forage grasses that naturalized readily. Colombia has well shown that the tropics may have superior advantages for growing livestock. Lately, a number of sugar companies, in particular in the Dominican Republic, have become meat and dairy producers in a large way, feeding cane waste and rotating cane fields with pasture. Molasses is at hand to be added to local feed. Nitrogenous feed is still short, but the American tropics are perhaps the foremost area in the world in diversity of nitrogen-rich leguminous plants. We have hardly begun to look into their potentialities because we are accustomed to the clovers of the high latitudes.

SUM OF PROSPECT

There should be moderately good days ahead for the Caribbean lands, or for some of them. I can see little prospect for the sadly crowded islands, unless they stabilize their populations, nor do I know how that may be done. Emigration never is more than [a] momentary easing of population pressure. The world as a whole is no longer receptive to immigration. The wistful hope for tourist dollars cannot be realized by every spot that lies in beauty upon the Carib Sea.

By sum of position, climate, and soil, these are about the world's best lands for sugar, bananas, pineapples, cacao, and other tropical fruits. Such should remain the major source of cash income, properly divided between worker and the always necessary venture capital, for such operations require large enterprise. Nowhere is a climate of international and internal good will and respect more needed and the voice of the demagogue more mischievous. Sugar, at the moment once again beaten to earth, will rise again, for the political follies of the past will not all be repeated, and new consumer demands will come.

There are important possibilities in *conuco* planting, given some helping hand by science. This system makes the most intensive and balanced use of the soil, and it gives work to most hands. Also, it is well, perhaps best suited to certain commercial products such as cacao, coffee, and peanuts. In parts, a permanently valuable tropical-forest industry can be developed. In others, the raising of cattle and hogs has superior attractions.

A good ecologic balance of culture and nature is attainable without upsetting either by prefabricated action programs, with or without doctrinal blueprints. We neighbors of the North do not need to think that we can, or even should, supply the know-how. There is a lot of experience and ability below our borders, from which we may learn and to which we may perhaps join ourselves as associates.

NOTE

Reprinted from *The Caribbean: Its Economy,* ed. A. Curtis Wilgus, 15–27. Copyright 1954, Univ. of Florida Press, Gainesville.

V

CULTIVATED PLANTS

18

Introduction

Daniel W. Gade

> As artificer of cultural change, man has become increasingly powerful in modifying the plant and animal world surrounding him.
>
> Carl O. Sauer, *Agricultural Origins and Dispersals*, 1952a

A major theme of Carl Sauer's work was the material basis of human life, especially the domesticated plants and animals that emerged in different parts of the world over the past ten millennia. Sauer's study of these biological organisms brought together the dimensions of time, space, ecology, and culture. He provided a grand view of their diversity and their status as artifacts of human manipulation and diffusion. Sauer had little interest in crops as commodities; economic geography was for someone else to pursue. Sauer's thinking on the culture history of agriculture formulated for American geography the debate about process and pattern in the world's crop and livestock inventory (Veblen, 2003). Geographers since Sauer, many of them part of his intellectual genealogy, have continued this line of investigation through their field observations around the world and reconstructions of the past.

Sauer's approach to crops can be expressed by the untranslatable German word *Anschauung*. It refers to the mental processes by which we grasp, visually through observation and spontaneously by intuition, a thing in its wholeness. That way of thinking made Sauer's work on plants notable, not some privileged knowledge unknown to others about any one of them. Johann Goethe in the eighteenth century invented *Anschauung* as a philosophical perspective that focused on a holistic vision of organic life. Sometimes now called Goethean science, it offers an alternative to positivist discourse (Seamon and Zajonc, 1998).

Observations of peasant fields, dooryards, and markets in northwestern Mexico in the early 1930s crystallized Sauer's critical thinking about crops. At some point he recognized that a cultivated plant had both a natural history and a cultural history, which required asking a range of new questions that went beyond what he had learned in his graduate training. In the Missouri of his youth, farmers valued maize as a direct or indirect source of income; in Mexico, maize was more than that. Diverse in its types and uses, *Zea mays* was not only a primary source of sustenance for Mexicans, but it also held a deep symbolic and spiritual meaning for the people who grew it. Sauer saw this one plant also as a key to understanding the Mexican past. The associations that Sauer made in the field prompted him to ask when and where maize and other crops arose, where they spread to, and how and why they became the basis of Mesoamerican civilization.

Sauer's early awareness of crop histories came from reading Alphonse de Candolle (1806–1893), who used historical, linguistic, and botanical evidence; Elizabeth Schiemann (1881–1972), a politically courageous scientist who studied the origin of crops; and especially Nikolai Vavilov (1887–1943). Vavilov traveled the world observing and collecting crops and their wild relatives for his plant institute in Leningrad. He made a great breakthrough—one based more on intuition than data—when he proposed a limited number of crop centers of origin. In 1936, when Vavilov came through Berkeley on his way to Latin America, he met with Sauer. Little is known about that meeting, but it seems to have energized Sauer's interest in crops, in cultivated plants in general, and in Mexican crops in particular. Vavilov had gone to Mexico in 1930 after presenting a paper at Cornell University, so he could speak with Sauer knowledgeably about what he had seen and collected south of the border. In addition to his extraordinary fund of knowledge, the Russian was a man of charisma, surely one of the reasons his ideas influenced so many people who knew him (Harlan, 1995:50). For more than two decades, Sauer concentrated much of his research and writing efforts at Berkeley on understanding cultivated plants. He also had a subsidiary interest in domesticated animals, which meant that he could conceptualize agriculture more broadly than those trained in the plant sciences. Sauer instinctively appreciated that Vavilov's notion of centers of origin was tailor-made for a geographer trying to make spatial sense out of diversity, distributions, and diffusions. Sauer's major contribution to this was to put a time perspective on these centers despite the scarcity of an archaeological record.

The first of Sauer's writings on this subject, "American Agricultural Origins" (1936a), was published as an essay in the *festschrift* to A. L. Kroeber, his colleague and friend in the anthropology department. That text, which applied the Vavilov idea to Mesoamerica, now seems naive, but one should remember how little was known about crop origins in the 1930s before the carbon-14 method, invented in 1946, was used to date pre-historic plant remains. Vavilov was trained as a geneticist, but he seems to have also accepted the title of plant geographer (Roden, 1988). His idea that centers of diversity were also centers of origin was too broad to be true in all cases, and he later had to refine that linkage. Harlan (1975:52–59) questioned the validity of most of Vavilov's centers and later went so far as to assert that not a single modern study had agreed with Vavilov's theory (Harlan, 1995:54). Brücher (1971) even challenged the whole idea of gene centers as a figment of Vavilov's imagination, though such an extreme assertion had an ideological and personal basis (Gade, 2006). In spite of this dissent and the flaws that other researchers have recognized in Vavilov's concept of centers, it has continued to offer a productive way to consider the geography of agricultural origins.

Reinforcement of Sauer's crop plant interests came in the early 1940s. In 1942 he and his son, Jonathan D. Sauer, who later became a specialist on crop origins, went to western South America for the Rockefeller Foundation (1982). The primary mission of the trip was to identify in their home settings budding scholars who might then be encouraged to study in the United States. This mission to capture the hearts and minds of influential Latin Americans was part of the war effort. Sauer's secondary objective was to visit peasant fields and markets and talk with knowledgeable local people about their farming. He structured his itinerary to permit him to look at blue-egged chickens on the Chilean island of Chiloé and to seek primitive maize in Colombia that might suggest its place of origin and its antiquity. Little was then known about maize in the New World. About this same time, Sauer and Edgar Anderson began a mutually beneficial exchange and friendship. Anderson, a brilliant botanist of wide-ranging interests, had become interested in maize as a plant of great diversity and puzzling origins. With the help of a Guggenheim Fellowship from the same board that Sauer sat on, Anderson came to Berkeley in 1942 specifically to work with Sauer. Although they never coauthored, each imparted to the other much to enhance their respective knowledge bases.

When Julian Steward, editor of the planned *Handbook of South American Indians*, asked Sauer to contribute a chapter on the geography of South America, Sauer agreed, provided that he could write an additional chapter on the New World crop inventory. The latter provided the opportunity to combine what he had learned in western South America in 1942 with what he knew about Middle America from previous trips there. Although originally submitted in 1943, the manuscript for this article did not appear until 1950, when volume 6, delayed by the war and by financial problems, was finally published. Before publication, he updated the story of cotton, which suggested an amazing case of pre-Columbian dispersal. This work provided the first detailed survey of crops in the New World since the Russian work of the 1920s. Many American scholars and students became aware for the first time of the rich diversity of New World cultivated plants and their histories.

Sauer recognized the possibilities of putting the beginnings of agriculture into a world context. Using teaching as a way to fill in his gaps, he began in 1944 to offer seminars in plant and animal domestication. The grand opportunity to formally synthesize this wide-ranging knowledge came when he was asked to deliver a series of lectures in 1950 at the American Geographical Society in New York. In 1952 the Society published them as *Agricultural Origins and Dispersals.* Its world scope provided a coherent and panoptic perspective of the far-reaching consequences of human intervention into the biotic realm. Its organization into centers, which Sauer called "hearths," largely followed Vavilov. However, the Southeast Asian hearth was Sauer's own idea. Unlike Vavilov, Sauer was also concerned with how and why wild plants were domesticated and with the movement of crops through space and time. Several key arguments of the book were speculative, yet its luminous presentation inspired a sense that Sauer's intuitions were right. His understated and sometimes epigrammatic writing style has played no small part in the acceptance of his ideas. His emphasis on diffusion, which as a process has exceeded independent invention over the face of the earth, converged flow, movement, and change. Since the 1960s Sauer's ideas have had to share the stage for this subject with other reconstructions, including those that focused on environmental change, co-evolution (which downplays voluntaristic human participation in the domestication process until a late stage), and population pressure. However imperfect it may now seem, Sauer's thinking on agricultural origins and dispersals remains one of the grand syntheses of knowledge of the earth over the last century.

Sauer orally delivered versions of three papers, two reprinted here, at conferences in San José (Costa Rica) in 1959, Vienna in 1960, and London in 1964. This helps to explain why these articles seem as if they were written to be heard rather than read and also why referencing is notably sparse.

"AGE AND AREA IN NEW WORLD CULTIVATED PLANTS" (1959C, CH. 19 HEREIN)

The idea that plants with larger distributions were older than those with more restricted spatial patterns came from an hypothesis that Joseph Hooker enunciated in the mid-nineteenth century. Taxonomists got excited about it around 1916, and Willis (1922) developed the notion the farthest when he published a small book with "Age and Area" in the title. Under scrutiny, however, it became apparent that the spatial spread of an organism cannot be predicted by the length of its evolutionary history. The efficiency of a plant's dispersal mechanisms is at least as important as the time it has to disperse. Notwithstanding its abandonment by taxonomic plant geographers, Sauer applied this expression to the origin and spread of cultivated plants, but without examining its conceptual foundation. With crops, dispersability depends, with some exceptions, on humans or possibly their domesticated animals carrying them from place to place. Short distance movement, as when one farmer gives a propagule to a neighbor, has always occurred; long-distance transfer of plants to distant places with satisfactory growing conditions was also possible, but accounting for such movement poses conundrums and elicits controversy.

Sauer proposed that most New World crop plants originated in tropical latitudes and that wet-and-dry climates created conditions encouraging edible roots or tubers. These ideas now seem self-evident and have elicited little controversy. Less accepted is Sauer's notion of the Andean highlands of Colombia as the hinge area for crops from north and south. Seed crops he viewed as having come mostly from Mesoamerica and root/tuber crops from South America. This distinction as a cultural-historical tradition has some merit, yet the Andes have yielded several seed crops that strain the generalization. Sauer understood the ecological zonation of crops in the Andes as a process in which one plant is displaced by a more cold-tolerant one with increases in elevation above sea level. It followed Sauer's line of reasoning that the idea of agriculture in South America began in warm settings and spread to other kinds of environments. This contradicts the idea that humans acquired the concept of planting many times in pre-history. At issue here is a fundamental cleavage between those who see humans as uninventive and those who do not. The points Sauer made in the "Age and Area" essay were mostly made in his earlier publications, but are focused on the Americas.

Re-reading Sauer on crops makes it apparent how important archaeology is for verifying his statements. Until dating is available, his assertions are still only interesting and plausible ideas. At the time when Piperno and Pearsall (1998) reviewed New World crop origins, the chronologies of older vs. more recent domestications were still often not clear. Before the 1960s Sauer could speculate freely about New World domestication because the archaeological knowledge of it was still minimal. It is interesting to note that for two seed plants once cultivated in Chile, *Bromus mango* and *Madia sativa,* Sauer drastically changed his thinking. In his 1950 work on plants, he suggested that they were relicts of a period before maize (1950a:495; Gade, 1999:195), but in this 1959 paper he argued that they had been weeds in crop fields which were selected and cultivated in zones where potatoes, quinoa, and maize did not do well. Since nothing in the subsequent half-century of archaeology has clarified their chronological place, the validity of neither possibility can be assessed. Without the grunt work of digging, sifting, and identification, Sauer's speculations remain arm-chair formulations.

"CULTURAL FACTORS IN PLANT DOMESTICATION IN THE NEW WORLD" (1965)

"Cultural Factors in Plant Domestication in the New World" appeared in the Dutch agricultural journal *Euphytica* through the good offices of J. G. Hawkes, who organized the Tenth International Botanical Congress in 1964 in London. Hawkes, a *Solanum* systematicist with much field experience collecting wild and weedy potatoes in Latin America, was also interested in the culture history and geography of the potato as a crop. At the London congress, Sauer largely reiterated what he had written more than a decade earlier about agricultural origins.

Although the article contained no new ideas, it reached a different and more science-oriented readership than did his 1952 *Agricultural Origins and Dispersals* book. One pet idea repeated in this short piece was Sauer's essentialist belief in the sharp distinction between root crop agriculture and seed farming. The former, centered in lowland South America, came first, whereas in Mesoamerica seed farming was dominant. To illustrate his point about two different planting concepts, Sauer used the example of native farmers on volcanic slopes in Central Mexico. They collected and ate the wild potatoes found there, but because the Mexican farmers were in the seed agriculture tradition, they did not take the next step and select these tubers for cultivation as South American farmers had done. This distinction was not as sharp as Sauer contended, for several crops were domesticated for their seeds in South America. Yet in the three decades since his death, the model has not been archaeologically overturned.

In the Dutch journal, Sauer also alluded to the motives for fruit domestication. He mentioned mamey (*Mammea americana*) and pineapple (*Ananas comosus*) as plausible Caribbean domesticates that were selected for their fermentable juice rather than for their edible fruit. The notion that alcohol was more valued than food has also been used to explain the domestication of barley. Certainly, substances that put people in an altered state of consciousness have been sought almost everywhere. Sauer's remarks referring to the Caribbean indicate his preoccupation at the time with *The Early Spanish Main* book (1966a), on which he was then laboring.

"MAIZE INTO EUROPE" (1962C, CH. 20 HEREIN)

Of Sauer's three conference articles on plant domestication and diffusion, "Maize into Europe" was the most original but also the most contentious and tendentious. It sought to shed light on when and where *Zea mays* first reached Europe. Sauer used Renaissance herbals for evidence to time the arrival of maize in different parts of the continent; in fact, Finan (1948) had already done that a decade earlier. But Sauer had his own unorthodox take on maize as an artifact of great movement. He combined the information in herbals with his analysis of names used for maize in order to derive pathways of diffusion. Identity confusion has complicated this quest, for Europeans initially called maize by the names of other grains that it resembled. For example, in Portuguese, maize is called *milho*, which in the sixteenth century referred also to millet. In Italy, maize is still called *granturco* ("Turkish grain"), a sixteenth-century name that to Sauer strongly suggested an entry to Europe from an easterly direction. As evidence for that, Sauer mentioned Giovanni Ramusio's unequivocal iconographic representation of maize from the *Veneto* published in 1554. However, this date was not early enough to persuade skeptics that maize had come to the Italian peninsula from the east, not the west. More significantly, it is possible that Sauer did not know that Ramusio had been in Seville in the 1520s, saw maize being cultivated there, and that a Venetian diplomat, Andrea Navagero, had carried seeds of red flint maize back to Venice from Spain (Messedaglia, 1927:178). Nor did Sauer seem to know that the herbal of Mattioli, which included maize in its 1565 edition, had emphatically denied the Oriental origin of this plant (Dioscorides and Mattioli, 1565). Mattioli called it *"formento indiano,"* a qualifier which in this case did not refer to the India of Asia, but rather to the Indies of the New World.

Moreover, Sauer failed to mention that in Eastern Europe most of the names for maize (*kukuruz* in Croation, *kukorica* in Hungarian, *kykypy3a* in Russian) follow a derivation that suggests no such diffusion from the east. For that matter, neither do the Rumanian *porumbul* or the Turkish *misir* names for maize hint at anything Oriental. More than four decades after Sauer wrote this article, there is still no good evidence that maize was in the Balkans before the seventeenth century. In his attempt to use language to understand maize diffusion, Sauer failed to thoroughly explore the many names for maize in the dialects of Italy. He was probably unaware of Luigi Messedaglia's (1927) monograph on maize in the Italian peninsula, as the book was not in any of the University of California, Berkeley, libraries. If he had consulted it, he would have learned that the "Turkish" allusions in Italian maize names were few compared to other names. Sauer provides the name *"formento"* as the current word for maize in Venice, Lombardy, and Emilia. This word, if unqualified, is actually an antiquated dialect word for wheat, not maize. Today the common term for wheat in Italian is *"frumento."* A related word, *"frumentone,"* has sometimes referred to maize. A rich dialectology must be carefully sorted out if one is to make useful commentary about plant diffusion.

Most readers of this piece would not have been aware of Sauer's motives in proposing (a) an entry of maize into Europe via the eastern portal, and (b) a pre-Columbian diffusion. The article never explained that he wanted to promote the idea that maize in Europe had come via India, not the Americas. The controversial proposal of pre-Columbian maize in India was fostered by an article prepared by Anderson and Stoner (1949) in which was described in some detail the primitive character of maize Stoner had brought back from Assam and that Anderson analyzed. Their conclusion made two points: that this primitive maize must have come to South Asia a very long time ago, that is, before Columbus, and second, that maize may have, in fact, originated in Asia. Maize scientists met both assertions with disbelief. They were confounded that Anderson, recognized as a brilliant, innovative scientist who between 1940 and 1968 authored or coauthored forty-one papers on maize and its close relatives, would have come to those unseemly conclusions in the pages of a respected scientific journal. In retrospect, this brouhaha can be seen as one of the several anomalies in Anderson's scientific career that confused his contemporaries (Stebbins, 1978).

Sauer's reluctance to mention the India connection suggests that his position on maize diffusion was unsettled at that time. Everything he had learned in the field had strongly pointed to its origin in Latin America. Sauer was also aware of opposition from maize scientists, especially from Paul Mangelsdorf et al. (1954), who published a paper unequivocally asserting a Mesoamerican origin for maize. Sauer realized that, given the evidence that had accrued, proposing an Asian Indian origin for maize was untenable, but he felt on surer ground in speculating about its pre-Columbian diffusion. The eastern connection seemed plausible to him, especially in light of Anderson's analysis of the primitiveness of the popcorn and green corn found in the remote Assam region of northeast India. Sauer was convinced of Anderson's scientific genius in understanding maize as a plant. Although Sauer was not able to develop a coherent case for how and when maize might have spread from the New World to India, he nevertheless held on to the possibility of such an extraordinary transfer after most interested parties had rejected the hypothesis. Sauer understood Peter Martyr to have been a reliable eyewitness to the presence of maize in Spain and Italy before 1492. A year before Sauer died, Mangelsdorf (1974:201–206) published a book in which he offered his objections to the pre-Columbian diffusion of maize outside the Western Hemisphere.

That assessment has not deterred Carl L. Johannessen, one of Sauer's surviving Ph.D. students, who since the 1980s has accumulated evidence for a pre-Columbian movement to India of a complex of crop plants. Johannessen (2003) interpreted stone-carving designs on ostensibly ancient Hindu temples as representing maize. Linguistic parallels were constructed to reinforce the connection. Possibilities of this interpretation tantalize, but the skeptics await further evidence. So far no archaeological maize of commensurable age has been found in India. As for Sauer's major thesis in his article, that Columbus was not the first person to have brought maize to Europe, the evidence is for me not persuasive, but neither is the notion outrageous. Clarification may eventually emerge, but in the last analysis, diffusion controversies are mostly about mind sets: open vs. closed. Questions of pre-Columbian diffusion require a fundamental willingness to accept ambiguity, uncertainty, and contingency in the development of knowledge (Gade, 2003/2004). As I reconstruct the best possibility from my background knowledge, maize cultivation in the Old World began as an irrigated garden crop in the Guadalquivir Valley near Sevilla between 1500 and 1506. It never became important there, however, unlike its later and more successful diffusion to the northwest Iberian Peninsula.

Sauer broadened this same essay with other diffusionary visions by attempting to corroborate M. D. W. Jeffreys's (1957) idea that maize in Africa was pre-Columbian and that the Portuguese took it to Portugal from Africa, not from South America. Jeffreys's reconstruction, based on reviewing mainly linguistic evidence, claimed that nothing supported the notion that Columbus was the first to bring maize to Europe, nor that the plant came to be disseminated to other parts of the Old World by way of Spain or Portugal. When Miracle (1966:87–100) laid out the evidence for a Portuguese introduction of maize to West and South Africa, it clarified for many the most plausible possibility. More recently, McCann (2005) reviewed the early maize evidence and came to a similar conclusion about the Portuguese introduction of maize from the New World to West Africa; for eastern Africa, he constructed a diffusion of the plant via Italy. In my opinion, Sauer placed too much faith in the linguistic acrobatics of Jeffreys, which asserted that maize went from Africa to Brazil and from Africa to Portugal rather than from Brazil to Africa and from Brazil to Portugal.

The three relatively short pieces of writings examined here convey several aspects of Carl Sauer's focus on and interest in cultivated plants. He did not try to trump the botanists, for he was aware of the taxonomic complexity of the plants with which he was dealing. He had a competent enough knowledge of Mendelian genetics to appreciate how and why wild plants evolved into crops. He took into account the agronomic aspects of cultivated plants when they shed light on distributions, though he himself never did any careful analysis of soils or climate as they related to agriculture. The economic geography of crops as manifested in land use or as important commodities for feeding the world or generating international trade scarcely occupied his thinking at a time when many geographers devoted their attention to just such matters.

Conversely, concern about the erosion of crop diversity, though scantily understood at the time, was since the 1930s (1936a) a subject of disquietude for Sauer, and one which today makes him seem like a visionary. His first allusion to this as a problem was in a paper he published in the *Journal of Farm Economics* (1938b:770–771). He also expressed it more concretely several years later when he wrote, "We should not permit by ignorance, the progressive loss of the cultural heritage of plant and animal domestication to continue" (1942a:65). That issue was part of a broad interest in process rather than pattern. Diffusion and domestication were two time-dependent themes that especially took hold of Sauer's thinking, and the reprinted articles that follow focus on one or both of them. Historicism was his strongest scholarly motivation, and it is what gave meaning to the other dimensions of space, culture, and resources.

These articles which reconstructed the past have not stood up well to the test of time. The paper on "Age and Area" did not plumb the concept its title seemed to promise. Both it and his *Euphytica* article, "Cultural Factors in Plant Domestication in the New World," largely rehashed parts of his *Agricultural Origins and Dispersals* book that appeared years earlier. A fresher text was his paper on maize in the Old World, but here Sauer was in over his head. Unwilling to think metaphorically about the names of introduced plants because he had a larger objective, Sauer misinterpreted the historical record for Italy and Africa. Messedaglia (1927) had warned about the "dangerous homonyms" ("*omonimi pericolosi*") that are pitfalls to understanding crop histories.

With the use of molecular techniques[1] and genome sequencing, the study of crop origins and dispersals has entered a new and rapidly moving phase. Startling new discoveries in geographical place of origin or crop species and the identity of crop ancestors can be expected (Motley, Zerega, and Cross, 2005). But there will always be aspects of humans and crops that evolutionary biologists cannot or do not wish to answer. Nor do new advances diminish Sauer's contribution to the subject of crops as human artifacts. That Sauer was wrong about certain suppositions he made does not invalidate his larger thinking on crops as human artifacts in space and time (Harris, 2002). Many works, especially by geographers, betray Sauer as their source of inspiration even as they have moved far beyond what he himself was thinking (Blumler and Byrne, 1991). His convergent thinking that tied together the historical (his greatest competency) with the archaeological, botanical, and linguistic is probably his main intellectual legacy in the culture history of agriculture. Second, Carl Sauer's intellectual charisma induced several generations of geographers, anthropologists, botanists, and others to explore this subject. He was a grand thinker and formulator who inspired others to get involved in these issues of culture history. In continuing his convergent thinking, they have pushed away the false boundaries of disciplines to follow their curiosity wherever it leads.

NOTE

1. Large-scale molecular analysis of genetic diversity offers a line of evidence to confirm dispersal in time and space. For example, this method has led to an assertion among French geneticists that Northern Flint maize was introduced into France in the sixteenth century by Jacques Cartier and/or Giovanni Verrazzano (Rebourg et al., 2003). The study of crop plant diffusion by use of molecular techniques has the potential of revoluntionizing many suppositions about the origins and early movement of crop plants (Jones and Brown, 2000; Zeder et al., 2006).

REFERENCES

Anderson, Edgar, and Charles R. Stoner. 1949. "Maize among the Hill Peoples of Assam." *Annals of the Missouri Botanical Garden* 36:355–404.

Blumler, Mark A., and Roger Byrne. 1991. "The Ecological Genetics of Domestication and the Origins of Agriculture." *Current Anthropology* 13:23–54.

Brücher, Heinz. 1971. "Zur Widerlegung von Vavilovs geographisch-botanischer Differrentialmethode." *Erdkunde* 25:20–36.

Dioscorides, Pedanio, and Pietro Andrea Mattioli. 1565. *Medici commentarii: In dex libros.* Ex Officina Valgrisiana, Venice.

Finan, John J. 1948. "Maize in the Great Herbals." *Annals of the Missouri Botanical Garden* 35:149–191.

Gade, Daniel W. 1999. *Nature and Culture in the Andes.* Univ. of Wisconsin Press, Madison.

———. 2003/2004. "Diffusion as a Theme in Cultural-Historical Geography." *Pre-Columbiana: A Journal of Long-Distance Contacts* 3:19–39.

———. 2006. "Converging Ethnobiology and Ethnobiography: Cultivated Plants, Heinz Brücher, and Nazi Ideology." *Journal of Ethnobiology* 26:82–106.

Harlan, Jack R. 1975. *Crops and Man.* American Society of Agronomy/Crop Science Society of America, Madison, Wisc.

———. 1995. *The Living Fields: Our Agricultural Heritage.* Cambridge Univ. Press, New York.

Harris, David R. 2002. "'The Farthest Reaches of Human Time': Retrospect on Carl Sauer as Prehistorian." *Geographical Review* 92:526–544.

Jeffreys, M. D. W. 1957. "The Origin of the Portuguese Word *Zaburro* as Their Name for Maize." *Bulletin de l'Institut Français de l'Afrique Noire, Series B. Sciences Humaines* 19 (1–2): 111–136.

Johannessen, Carl L. 2003. "Early Maize in Europe? A Case for 'Multiple-Working Hypotheses.'" In *Culture, Land, and Legacy: Perspectives on Carl O. Sauer and Berkeley School Geography,* ed. Kent Mathewson and Martin S. Kenzer, 299–314. Geoscience and Man, vol. 37. Geoscience Publications, Department of Geography and Anthropology, Louisiana State University, Baton Rouge.

Jones, Martin, and Terry Brown. 2000. "Agricultural Origins: The Evidence of Modern and Ancient DNA." *The Holocene* 10:769–776.

Mangelsdorf, Paul C. 1974. *Corn: Its Origin, Evolution, and Improvement.* Belknap Press of Harvard Univ. Press, Cambridge, Mass.

Mangelsdorf, Paul C., Richard S. MacNeish, and Gordon R. Willey. 1954. "Origins of Agriculture in Middle America." In *Handbook of Middle American Indians,* ed. Robert C. West, 1:427–445. Univ. of Texas Press, Austin.

McCann, James C. 2005. *Maize and Grace: Africa's Encounter with a New World Crop.* Harvard Univ. Press, Cambridge, Mass.

Messedaglia, Luigi. 1927. *Il mais e la vita rurale italiana: Saggio di storia agraria.* Federazione Italiana dei Consorzi Agrari, Piacenza.

Miracle, Marvin. 1966. *Maize in Tropical Africa.* Univ. of Wisconsin Press, Madison.

Motley, Timothy J., Nyree Zerega, and Hugh Cross. 2005. *Darwin's Harvest: New Approaches to the Origins, Evolution, and Conservation of Crops.* Columbia Univ. Press, New York.

Piperno, Dolores R., and Deborah M. Pearsall. 1998. *The Origins of Agriculture in the Lowland Neotropics.* Academic Press, San Diego.

Rebourg, C., M. Chastanet, B. Gouesnard, C. Welcker, P. Dubrueil, and A. Charcosset. 2003. "Maize Introduction into Europe: The History Reviewed in the Light of Molecular Data." *Theoretical and Applied Genetics* 106:895–903.

Roden, I. Y. 1988. "N. I. Vavilov: Geographer and Explorer." *Soviet Geography* 29:658–665.

Sauer, Carl O. For references, see the Sauer bibliography at the end of this volume.

Seamon, David, and Arthur Zajonc, eds. 1998. *Goethe's Way of Science.* State Univ. of New York Press, Albany.

Stebbins, G. L. 1978. "Edgar Anderson, November 9, 1897–June 18, 1969." *Biographical Memoirs of the National Academy of Sciences* 49:2–23.

Veblen, Thomas T. 2003. "Carl O. Sauer and Geographical Biogeography." In *Culture, Land, and Legacy: Perspectives on Carl O. Sauer and Berkeley School Geography,* ed. Kent Mathewson and Martin S. Kenzer, 173–192. Geoscience and Man, vol. 37. Geoscience Publications, Department of Geography and Anthropology, Louisiana State University, Baton Rouge.

Willis, John C. 1922. *Age and Area: A Study in Geographical Distribution and Origin of Species.* Cambridge Univ. Press, Cambridge.

Zeder, Melinda A., Daniel G. Bradley, Eve Emshwiller, and Bruce D. Smith, eds. 2006. *Documenting Domestication: New Genetic and Archaeological Paradigms.* Univ. of California Press, Berkeley.

19

Age and Area of American Cultivated Plants (1959)

Carl O. Sauer

THEIR TROPICAL ORIGIN

The plants and animals domesticated in the New World originated almost in their entirety within tropical latitudes, some in tropical lowlands, a number in temperate altitudes, and others in cold highlands, but well within the limits of the two tropic circles. What was added beyond in higher latitudes was minor, marginal in location and utility, and apparently late; it included no animals; none of the plants became major crops, nor were any of them adopted far from their place of origin.

North of Mexico there were added to cultivation several plums, and two grapes, fruits sufficiently ameliorated by Indian planting and selection to have yielded cultivated forms. The tuberous sunflower is native to the Mississippi Valley and was grown in Indian gardens on the East Coast; once planted it needs hardly any care and multiplies readily year after year. The plums and grapes are at home in the open, deciduous eastern woodlands; they received attention in historical time especially at the hands of Southeastern tribes such as the Muskhogean ones, who were pretty good farmers and lived in permanent "towns" with "gardens." To an ancient seed complex brought out of Mexico by people unknown, there were added here to cultivation some fruits and a tuber as minor extensions of an introduced complex. It cannot be claimed that the lands to the north of the tropics lacked plants or animals suitable for cultivation and domestication. We have in the United States a flora rich in edible seeds, fruits, and nuts; we have divers tubers and bulbs that were dug for food; a number of native plant species are at present the object of successful selection. No animal was domesticated in the north, though kinds attractive to man and amenable to breeding by him are probably more numerous than to the south. The southern Mexico turkey became the domestic animal, not any of the northern species. The methods and the plants of agriculture came from the south long ago; the north added a few minor elaborations.

Extratropical South America was less deeply penetrated by agriculture than was North America, its principal plants again having been carried in from tropical latitudes. The farthest extension was in Chile, where two plants were added to and altered by cultivation, a grass (*Bromus mango*) and a tarweed for its oily seeds (*Madia sativa*). Both colonize freely on fallow or poorly cultivated fields and may be considered as having entered tilled land as weeds towards the boreal limits of potatoes, beans, quinoa, and maize. In seasons in which these staples did poorly, mango and madia provided an alternate harvest. First tolerated and increased under cultivation of the standard crops, they came to be adopted as crop insurance and became to some extent changed from their wild state.

The development of New World agriculture took place in our low latitudes, where man fashioned to his uses a great diversity of plants, further largely diversified by breeding. These constitute as varied and genial a solution of producing food and other prime materials as was achieved anywhere in the world. In time some of these cultigens were carried into far parts of North and South America, extending the aboriginal limits of agriculture nearly to the climatic limits of modern farming. Still later they contributed greatly, both in New and Old Worlds, [to the] foods and fiber plants that made possible the modern world.

The primordial area of domestication was greatly less than that of the tropics. All wet lands and also the year round rainy climates are out of consideration. In contrast to the Old World we lack hydrophytic cultigens; the New World crop plants are not suited to land that becomes water logged or is more than briefly flooded. Our native cultivators hardly knew anything of artificial land drainage. Where agricultural settlement entered poorly drained areas it sought out spots of adequate natural drainage, such as natural levees, or it improved soil aeration by heaping earth into mounds, as in flood plain cultivation of manioc [yuca] and maize. The Mexican *chinampas* may be cited as a special device to aerate the soil. The sites eligible to primitive cultivation were selected for good drainage and loose soil, river banks and levees, small valleys of sufficient gradients, piedmont slopes, ridge crests, and even steep hill and mountain flanks of sufficient depth of soil. Relief as such was no deterrent to primitive tillage, but lack of relief was, the more so the greater the rainfall. The more rain and flooding the less were lowland plains of valley or coast amenable to agricultural occupation. Uplands in rain forest were colonized in the course of time where the soil was sufficient and friable, but the plants cultivated are not natives of the rain forest. At the other extreme, permanently arid lands became accessible to cultivators only after skills in spreading water were known, nor are the plants grown there native to arid climates.

The growth habits show to some degree a common climatic background. They fit into the rainfall regimen of the outer tropics (sometimes called savanna climates), with rainy season at the time of overhead sun, dry season at the low positions of the sun. On the equatorial side such seasonal contrast may merge into all-year round moistness, on the polar side into permanent aridity. The plants in question perhaps can be assigned, each to its niche, somewhere within the range of savanna climates or to their mesothermal equivalent at higher elevations. The storage of starch in fleshy underground stems is an adaptation by perennial plants to alternation of dry and wet season. Such make use of the rainy season for growth, including the setting of tubers and the like, and, their growth accomplished, are ready for the following dry season, indifferent to its length and severity. The annuals mature after the rains stop. Each, we may say, has built into itself its own adaptation to a particular rhythm and amplitude of wet and dry season. Arrowroot (*Maranta*) and allouia (*Calathea*) tolerate a lot of dryness; yuca (*Manihot*) has markedly strong drought resistance. On the other hand the sweet potato (*Ipomoea batatas*) needs abundant moisture for several months and the basic maize is strikingly mesophytic, as shown by its shallow root system, large growth, and free transpiration from large leaf surfaces.

THE AGRICULTURAL HEARTH NORTH OF THE EQUATORIAL AREA

The southern side of the American tropics is not advantageously located to have witnessed the beginnings of agriculture. Ethnologically and archeologically it is cultural hinterland, primitive, or late of penetration by advanced cultures. Nor does it have the appropriate flora ancestral to our domesticated plants, nor yet the properly attractive physical geography. The indicated location for a major agricultural hearth is on the mainland northern side of our tropics, about the Caribbean, southward into Ecuador, and northward into Mexico.

Here we meet with (1) an extraordinarily diversified terrain, due largely to the virgation of the Andean ranges and basins against the Caribbean Sea and to numerous volcanic cones and flows, (2) soils derived from diverse parent materials, including deep, friable, and fertile soils of volcanic origin, (3) marked differences in climate within short distances, due to relief and exposure, (4) great diversity of higher plant forms, including the near relatives of almost all the cultivated and domesticated kinds, (5) streams and lakes in number, rich in aquatic and riparian animal life. (6) The high physical and biotic diversity provided a diversity of habitants for man, each of restricted extent, favoring cultural provincialism but not to the extent of being adversely isolative. (7) In the larger geographic context the area was the great corridor and crossroads of the New World. (8) In this connection we should also note that the land leads northward by easy approach to the Mexican highlands and southward into the Andes. By its geologic structure Colombia provides good agricultural experiment stations from tropical coasts to high *páramos*. Also the structural basins of Magdalena and Cauca are ramps leading gradually upward and southward to the Andean *altiplano*. These interior passageways, rather than the external flanks of the Andes with their barrier zones of cloud and rain forest and of desert southward, were the avenues by which agriculture, if it spread from lower lands, could most readily establish itself in the cold interior of the Andes.

The hearth indicated provided also, by means of fishing and hunting, aquatic and riparian, the possibility of living in sedentary communities before agriculture was known. Such precondition I hold necessary. The initiators of domestication required a comfortable and dependable margin above mere survival, permanent homes, and a living in communities in which they could share observations and have the leisure to begin the long range experimentation that led to domestication. The business of plant growing and selection did not proceed from "prelogical minds" by hocus-pocus or chance. It required ease, continuity, and peace. It was carried out by acutely observing individuals, primitive systematists and geneticists we may assert, who taught others to identify and select, by lore and skill handed from generation to generation. The plants fashioned by man are artifacts of skilled craftsmen; plant breeders anywhere are still few and exceptional individuals. I have difficulty in visualizing the spontaneous and independent origins of agricultural living and arts by reaching an unelucidated "stage" or "level" of cultural advance, or by assuming that people turned to producing food because they were getting hungrier. Distressed folk were least likely to have the capital reserves for investment in deferred returns. Such progress I should look for as originating in a most favored area, with a society amenable to new ways and recognizing original talent in its individuals. Were such congenial physical and cultural situations presented as well anywhere else in the New World?

VEGETATIVE PLANTING

To the English colonists of North America Indian agriculture meant Indian corn, Indian beans, and squash for which they borrowed the Indian name; to the Portuguese in South America mandioca (yuca) was the Indian staple. Seed agriculture in the north; roots grown from cuttings in the south. Where the two systems meet anthropologists identify the Mesoamerican as against Circum-Caribbean culture. Did each system arise independently, or is one derivative from the other? Which the elder? Why the difference? What effects where they have been in contact? These are questions which we must ask and to which we may bring pieces of evidence, if not complete answers. If seeking understanding of events and processes, using data of whatever kind [that] appear related on reasonable examination, is speculation, it is only by such circumstantial evidences that the events and their order may be reconstructed.

The vegetative reproduction that dominated South American agriculture has a common distinctive pattern: (1) Reproduction is asexual, that is the new plant is grown from a piece of another plant, not from its seed. A tuber or piece thereof, a division of the root stock, a shoot, or a cutting of stem is planted, to grow into a complete plant. In its beginnings this is the most primitive means of propagation: Digging tubers gave an incomplete harvest; what was missed set a new crop in the disturbed ground, to be dug in another year. Such digging plots tended to become perennially productive. Another instance is the reproduction of waste parts of plants that were dropped on village refuse heaps. Transition from collecting to cultivation was thus facilitated. For each plant that became the object of continued cultivation a standard or conventional method of plant division seems to have become established. The plants thus taken under cultivation are perennials; their use by man rarely was for their seeds.

(2) With the change from digging to planting, perhaps from digging stick to planting stick, the way was opened for deliberate and individual selection.

The prepared planting ground, or *conuco,* became the spot into which pieces of preferred individual plants were set. Where the planter recognized that the progeny was like the parent he began to select and reject, and amelioration of the planting stock was under way. The multiplication of the desired plant was further advanced by the discovery, as in yuca and batata [sweet potato], that numerous stem cuttings could be taken to increase the progeny of a single parent. The whole road of domestication here rests on individual selection. After the initial selections from wild populations, further variation was provided by occasional (root) bud sports and by attractively variant accidental seed progeny that was saved for vegetative reproduction. Peasant cultivators in the Island of Haiti where yuca often still produces viable seeds, are well aware that the seed progeny is highly variable; such are now usually destroyed, since these are less likely to produce desirable plants. In early days however such variants provided an important means of improvement. By individual plant selection of the more desirable forms, long continued in various environments, a large diversity of races was developed.

(3) Because the planter was always concerned with vegetative reproduction the plants that were thus fashioned in many cases lost the ability to reproduce themselves by seed. The forms of such sterility are various and in part still unstudied. Sterile plants became wholly dependent on the care of man for their survival. Where Europeans rudely overran native cultures many such forms were lost by flight or death of the natives, by the abandonment of *conucos,* or were rooted out by the introduced pigs. A generation after the beginning of Spanish occupation of Haiti, Oviedo noted with regret that a number of the best kinds of batatas, for which he wrote down the native names, had become extinct.

(4) The attention to food production was mainly directed to underground parts, fleshy roots and tubers, to be harvested by digging. Unlike the Old World tropics (*Alocasia,* sugar cane, *Musa ensete*), no plants were developed for the food value of their stems. A few plants became vegetative cultigens for the fruits they yielded, with various resultant degrees and forms of sterility. Such are the peach palm or pejibae (*Bactris utilis*), the better races of which are reproduced by cuttings, the pepino (*Solanum muricatum*), resembling in quality both cucumber and melon, and the pineapple (*Ananas*), which may have had early attention as a fiber plant.

The case of the peanut (*Arachis*), called groundnut by the English, is peculiar. It is grown for its seeds and is planted from seed, but since it buries its pods in the ground and thus simulates a root crop and is harvested by digging or pulling the whole plant out of the ground we have a functional resemblance to root culture. Does this help to explain why out of a host of herbaceous and suffrutescent species of Leguminosae in South America it stands nearly alone as [a] product of cultivation?

(5) The plants vegetatively domesticated contribute starch and sugar to the diet, but very little protein, fats, or oils. It seems obvious that there was no interest in, or need of a balanced vegetable diet. The plants were developed to increase the supply of carbohydrates and they did so effectively. The lack of attention to seed production means that no need was felt to increase the availability of proteins and fats. In contrast to the Old World tropics the great cultural possibilities of palms in the New World remained largely unstudied by Europeans; one of the few exceptions, pejibae, was selected for increase of the starchy flesh surrounding the seed and toward the elimination of endosperm.

The meaning of such one-sided agriculture is clear: protein and fats were provided from animal sources. The inferred hearth area, the mainland adjacent to the Caribbean, was richly thus stocked, especially along both fresh and salt water. Fish, shell-fish, and turtles abounded. Flocks of migrant water fowl from the north come here in winter. The sea cow or manati, feeding in the lower stream courses, was once the great game animal. Various hystricomorphs, partly of aquatic habits, provided excellent meat in quantity. Tapirs, and in part peccary and deer, fed along the side of streams. Animal food, and thereby protein and fat, was in surplus supply; the need was for more carbohydrates and planting was limited in attention to supplying such need. This specialized agriculture points emphatically to an origin with fishing and hunting people living along streams and lakes, in permanent communities.

(6) Unlike seed agriculture, hardly any special procedures are required for harvest or storage. The roots remain in the ground and are dug as needed; usually they keep better in the ground than in storage. Provision is day by day; the calendar may have a planting time when the rains begin but it lacks a season of harvest and attendant ceremonial.

(7) The processing of food is simple; it merely needs to be cooked. Milling is unnecessary. The roots may be roasted at the edge of the fire or they can be wrapped in mud or in leaves and baked. Steam cooking in covered pits over a bed of coals is an ancient and excellent means of preparing a meal of roots, flesh, and fish. The *barbacoa* is characteristic of this area; such a grill over a low fire serves as well for cooking roots as for cooking and drying meat or fish. That pottery vessels were preceded by gourds is recorded in the pre-

ceramic agricultural sites of coastal Peru, one of the few areas where [early] agriculture was not associated with the making of pottery. The Caribbean lands are the home of the domesticated calabash tree (*Crescentia cujete*), selected vegetatively to a variety of sizes and forms of calabashes, light and durable containers. Columbus noted them as used for bailing boats. They are still thus prized, as well as to carry water, for fermenting drinks, and to some extent for cooking.

In marginal areas a few crops were less simply prepared. In the Andes leaching and the complicated *chuño* process were added and in eastern South America the elaborate procedures of preparing bitter manioc.

THE DIVERSITY OF VEGETATIVELY PRODUCED STARCH FOODS

Since they were concerned with getting more carbohydrate food the primitive fishermen collectors of the tropics may well have exploited every edible root. Many of these took root on village refuse heaps, and the best were transplanted into *conucos* when such were begun. But why have so many been kept in cultivation, changed so greatly from their original form, and diversified into so large a number or races? It is reasonable to think that at the dawn of agriculture many things were planted, but why have so many been continued to the present merely to supply the same kind of starch food? That has been and is really their one use. Some are used a little for greens. None serve for fiber or other domestic or personal ends.

The consumer takes in calories and that is about the size of it; he may be expected to discriminate against the less nourishing kinds and the less productive ones and in time thus to have dropped their cultivation. It is hardly reasonable that the same people should have kept up the breeding of a lot of different plants serving quite the same purpose and requiring the same attention at the same time. With one satisfactory or promising starch source available the long and tedious effort to ameliorate another wild plant to serve the same needs though inferior in yield would not have been continued, yet the characteristic situation today is that several such, economically more and less rewarding, are cultivated in the same locality and in the same ground.

Aboriginal cultivation by vegetative reproduction dominated one great, continuous New World area and only that area. It looks very much as though the idea was spread by contagion from one community to another culturally receptive one. (That this was not a matter of presence or absence of suitable plants will be considered later.) Useful plants were dug or grew on refuse heaps in many places over the world without giving rise to agriculture and plant breeding. The decisive first step, followed by the next ones necessary to create an enduring agricultural system, perhaps was taken once and in one restricted area. We shall hardly expect to locate place, time, or plants of earliest domestication since we cannot say that higher age and larger area agree, nor even that the diversification of a cultigen is an expression of its age.

The most apparent reason for the diversity of cultivated starch plants in tropical America is that they fitted into different climatic conditions. Breeding in time has blurred such environmental advantage or disadvantage, though much less on the whole than in seed cultigens. Much needs to be learned about the climatic range of each species and its component forms; but we can recognize that arrowroot (*Maranta*) and allouia (*Calathea*) are *tierra caliente* plants indifferent to drought, whereas yampee (*Dioscorea trifida*) and pejibae (*Bactris*), also of the hot lowlands, are exacting of moisture. The marked climatic contrasts within short distances support the hypothesis of substitutive domestications. Thus may the agricultural way of life have moved upslope from *tierra caliente,* if that was its earlier home, through the *tierra templada,* and into the *tierra fría,* finding in each amenable and rewarding plants for cultivation. Racacha (*Arracacia*) and Ilacón (*Polymnia*) I know only from temperate lands in northwestern South America. Oca (*Oxalis*), ulluco (*Ullucus*), and ysaño (*Tropaeolum*) are restricted to Andean highlands. *Canna edulis* and yautia or malanga (*Xanthosoma*) are grown here and there in both hot and temperate country.

Locally racacha, yautia, and pejibae may still be staples. Mainly the plants named are being cultivated less and less. Originally they may have enjoyed environmental superiority in their home locality, but they are held in cultivation to the present mainly by the persistence of cultural tradition; other plants have been developed that grow as well and yield more, on the same sort of land, with the same kind of labor used on the same calendar. Their persistence suggests that they were grown before the economically superior yuca, potatoes, and sweet potatoes were available.

In the course of time the ascendancy of yuca (manioc) in the low country and of potato in the highlands became so marked that Clark Wissler could with some justice divide South America into two agricultural regions identified by these plants. Manioc types were developed that did well in *tierra templada* and others that succeeded under rain forest cultivation. The range of variation of this cultigen is not as yet well studied. Sweet forms extend throughout the range of cultivation of manioc; the bitter ones (with their special techniques of processing) are Atlantic and did not enter Central America or western South America.

What happened to the potato at the hands of man is one of the better known chapters of domestication. Wild potatoes, with edible tubers, range in a very large number of species from Colorado to southern Chile and Brazil. Within this range they are unrecorded only in *tierra caliente* and deserts, but are numerous in temperate as well as cold climes. Only a few have entered into forming the great complex of cultivated potatoes, diploid and polyploid. One hearth is in Colombia and Ecuador, the other in southern Peru and adjacent Bolivia, with cultivated forms occurring from the lower parts of *tierra templada* to the cold limits of agriculture. (Knowledge of their phylogeny has been advanced especially by the work of J. G. Hawkes in the Imperial Bureau of Plant Breeding and Genetics, Cambridge.) Most of the cultivated potatoes set fertile seeds; accidental hybrid offspring, in part involving one wild parent, has maintained variation at a high level. Selection in cultivation has been by tubers so that the possibilities of multiplying and increasing clones are almost unlimited. Cultivated potatoes were spread throughout the Andes from Venezuela to Chile to the upper climatic limits of agriculture. They were also taken from the Bolivian highlands to extratropical lowland Chile as far as the polar limit of aboriginal agriculture in Chile.

THE LIMITS OF VEGETATIVE PLANTING

In South America the system of vegetative reproduction was extended to the limits of agriculture, both latitude and altitude. With certain exceptions, such as the arid west coast, it continued to be the dominant system. Northward it occupied the West Indies and the Caribbean side of Central America, but did not enter the Southeast United States at all and Mexico in a minor way for a short distance and with only a few plants. Of these the sweet potato found largest acceptance in Mexico, but was nowhere a major crop and, by its Nahua name [*camote*], appears to have been of late introduction. It seems not to have reached the northern limit of high native culture even in the tropical lowlands, excellently suited to it. Spaniards took it into northern Mexico and Englishmen into the United States; the Indians knew it not. This failure to spread in North America is in striking contrast to its dissemination across the Pacific into high latitudes. The peanut had a similar story; the form in native Mexican cultivation resembles the type in Peruvian archaeology. The pineapple was cultivated somewhat in the hot country as far north as the

Tepic area [Mexico]. The slight attention given to yuca is somewhat surprising, considering how well it grows and yields; it too got a descriptive name, huaucamote, "the woody plant with edible roots." None of these plants approach any climatic or other physical limit; they petered out in non-receptive cultures.

This assertion that northward (the Mesoamerica of anthropology) the cultures were non-receptive to the system of agriculture by vegetative reproduction is based on two considerations. The first is the mentioned fact of the fading out of such cultigens northward for no physical reason. The second is that no native plants, useful to man and suited to such reproduction, were developed in the north. I should hesitate to say that such were less available or that they remained unappreciated. An edible [wild] tuberous *Manihot* (*M. carthaginesis?*) grows on the west coast into southern Arizona and is collected. *Camotes del cerro,* one of which is a Euphorb (*Dalembertia*), are common and appreciated boiled tubers, sold in markets and on streets of interior Mexico. The Mexican highlands have a number of species of wild potatoes of some current interest to potato breeders. These grow as volunteers in the *milpas,* especially on the slopes of the Mexican volcanoes. They are dug and used, even sold to some extent in markets. They are small, but of good taste, and quite prolific. Some remain in the ground and restock the *milpa* for the next year. The cultivators, Indians who retain much of their culture, appreciate them and give them some protection, but they do not plant them and of course practice no selection. The orientation northward is that of seed planting and away from root or stem cuttings. To the north we enter a differently mined world in agricultural procedures.

TOBACCO AND ALCOHOL

The plants taken farthest north from South America were three seed-produced, nonedible cultigens, a cotton, a tobacco, and the *Lagenaria* gourd. The bottle gourd is common in archaeologic sites well up into the United States. It may perhaps have been introduced northward about as far as it would grow. Cotton of the western American lineage ([*Gossypium*] *hirsutum* complex) became the textile fiber of the Mexican lowlands, cotton cloth (*mantas*) figuring as a chief tribute item out of the hot country to the Aztec state. It was also established at intermediate elevations in the highlands of Mexico and seems to have been taken to the American Southwest in early Christian times (Hohokam). It never reached the Mississippi Valley or Southeast United States.

Of the two cultivated tobaccos *Nicotiana tabacum* remained closely associated with the vegetative planters; if it got into Mexico before the Spaniards, it did so only slightly, in the lowlands. The other, *N. rustica*, became second only to maize in its agricultural dispersal. That it was the more widely accepted may be attributed in part to its lesser need of attention but also to its much higher nicotine content. Its original home is placed in the highland border of Peru and Ecuador. It was the great tobacco of Mexico, picietl, and is known to us as Aztec tobacco. It was generally grown by the Indians of our eastern woodlands as far as the lower St. Lawrence Valley. Man's interest centered in its alkaloid, nicotine, of all narcostimulants the one of widest appeal for ritual use. In its original home *N. rustica* was taken by drink; it was thence spread into areas where narcotics were used by chewing and as snuff, and still farther, as in North America, to be smoked in pipes. The physiologic effect having been recognized, the plant was adopted into different, probably preexistent ceremonial practices, passing from culture to culture.

Tropical American planters concerned themselves, as did men in few other parts of the world, with plant poisons, with piscicides and poisoned projectiles, with stimulants and narcotics, with medication by effective drugs as well as by magic. Valid observational taxonomy and experimental biochemistry were joined in appropriating plant resources in rather sophisticated ways. The tropical growers of food were also skilled manipulators of potent drugs and poisons and some of these are cultigens. Thus the Solanaceae (Nightshade Family) were pretty fully exploited in domestication; potatoes for food and dye, pepino, naranjilla (*Solanum quitoense*), *Cyphomandra*, *Physalis* as fruits and vegetables, *Capsicum* as seasoning and medicine, and *Nicotiana* and *Datura* as drugs.

Alcoholic beverages were common, in so far as I know, to all the vegetative planting folk. Drinking was ceremonially restricted, especially to feasts, and varied as to moderation or excess in different parts. Spanish chroniclers thought some of the tribes of the Cauca basin of western Colombia [were] given to extreme drunkenness. Sweet fruit and palm sap were fermented but the main employment was of starch food plants, the fermentation aided by mastication in South America. Only farmers brewed. Northward, alcoholic beverages were common to all the farming peoples of Mexico but were not used by the Pueblo people of the Southwest nor by the farming tribes of the interior and eastern United States. Roughly the boundary between Mexico and the United States was the northern limit of alcoholic beverages. The area in which alcoholic drinking was established is one and continuous. To the north it included only the Mesoamerican culture, into which it is concluded that it was introduced from the south, but beyond which the farming tribes, though growing the same complex of crops, did not adopt the practice.

THE SEED FARMERS

The northern hearth of plant domestication, where the process was done by seeds and therefore by sexual selection, lies in southern Mexico and northern Central America. The wild relatives of the cultivated plants grow here; the cultivated forms are here in greatest diversity. Consensus favors this area and I know no reason to disagree. The basic complex is simple, maize, beans, and squash, with grain amaranth as a relic still widely occurring in small cultivation. For the hearth area, chile (*Capsicum*), chia (*Salvia hispanica*), and tomate (*Physalis ixocarpa*...) are included. The classical form of *milpa* planting is maize, beans, squash set into the same mound or hole, the beans climbing up the corn stalks, the squash spreading over the ground, the three together forming effective utilization of sunlight and rain and giving protection against rain wash. The *milpa* system resembles that of the *conuco* in these important particulars: (1) The seeds are set by hand into the exact spot where they are to grow; they are not sown broadcast or in rows. (2) A number of different plants are grown together in the same ground. (3) Where soil permits, the planting is in mounded earth, "hilling" in the terms of the American farmer. (4) Selection is practiced by saving seed from desirable individual plants, as by ear of corn, vine or pods of beans, or a particular squash that is set aside to be used for seed. It is still a planting system.

The food orientation of this system is however very different from that of the vegetative planters. The object here is dietary balance, a proper and adequate supply of protein, fat, as well as carbohydrates. This adequacy was achieved here by plant foods as well and as economically as anywhere in the world. The need of animal food is small, the implication that, when agriculture began here it began on the basis, not of one kind of supplementary food as was the case in the south, but of a complete diet, quite [adequate] in protein content. This interest in protein was marked in the origins of domestication. The seeds of squashes are still as important a food as their flesh. Tzilacayote (*Cucurbita ficifolia*), an ancient domesticate, is of little account for its flesh except in making fermented drink (or now as stock feed), but its large, numerous, and tasty seeds are appreciated. There still are races of *C. moschata* that are grown only for their oily, nutty seeds. The development of squashes for starchy and sweet flesh came later, as did that of the large mealy beans and floury maize. The dietary bias in domestication was initially toward protein and fat, with later breeding to increase carbohydrate yield.

It may be more than a coincidence that the domestic bird of this agriculture is the turkey, inhabitant of temperate oak and pine woodlands, whereas a tropical river duck, the muscovy, was taken into the households of the root planters. There is an air of temperate Mexican and Guatemalan uplands about these cultivated seed plants, as though they had moved from open woodland into clearings, which is exactly what may well have happened. The cultivated beans are all climbers, the Indian runner ([*Phaseolus*] *multiflorus*) being a vine of astonishing growth. Of the *Cucurbita* a single plant of chayote (*Sechium edule*) will provide a bower (*ramada*) for the open air living of a household. Is this plant, useful alike for its fruits and starchy roots (as is the runner bean) a link between vegetative and seed planting ways? Wild cucurbits are common in open spaces such as old fields, as are amaranths where pigs and cattle do not destroy them. The wild kin of the plants in question are vigorous colonizers of clearings and their margins. The surmise may be kept in mind that the first seed cultivators were inland dwellers in temperate lands, of limited resource for fishing and waterside hunting. As they turned to planting, their attention shifted from roots to the seeds of *Cucurbita*, beans, and amaranths that came up as volunteers in clearings. I am not competent to comment on the origin of maize, but both *Tripsacum* and *Euchlaena* [wild relatives] are weeds of fields.

THE SPREAD OF MESOAMERICAN SEED PLANTS

The variety of plants grown became reduced north of the line of volcanoes that stretches across Mexico. Only a few kinds reached [the] eastern United States, and they of course came from Mexico. An attractive task for archaeology is the exploration of rock shelters for the cultivated plant remains left by early occupants, such as those already recovered in Tamaulipas, New Mexico, and the Ozarks, in the present state of our knowledge the most ancient on record.

Southward maize, beans, and squashes filtered throughout the agricultural parts of South America, as associates of root crops. Maize rarely became the staple crop in South America. It was planted, especially across northern South America as tropical flint varieties, for the making of beer, and was boiled or roasted as immature ears (*choclo* of the Andes). Beans and squashes mostly were minor additions to *conuco* planting. However, on the arid west coast and beyond into the higher latitudes of Chile, they became dominant crops; here the lima bean (pallar) acquires its greatest importance. Junius Bird's discovery of the preceramic agriculture of the arid west coast brought as one of its greatest surprises the knowledge that the little esteemed tzilacayote (*C. ficifolia*) was grown there before maize or true beans were introduced. Also, at this early time *Cucurbita moschata* was beginning to be planted in coastal Peru. Farther south *C. moschata* later apparently gave rise to the only South American squash, *C. maxima,* by crossing with some wild *Cucurbita.* (The knowledge of *Cucurbita* domesticates is due mainly to the work of Dr. Thos. Whitaker of the U.S. Dept. of Agriculture at La Jolla, California.)

New World agriculture, it is here proposed, had its beginnings in tropical lowlands by vegetative reproduction and selection of starch food plants. It made the ascent to high altitudes in the Andes by applying the same art and attention to other tuberous plants. At the north in Central America, it changed to seed growing that satisfied nearly the whole range of dietary needs. These seed plants in turn found entry into South American root planting, probably as population increased and animal sources of [protein] food became inadequate or depleted. Aboriginal agriculture occupied one continuous area in the New World, beyond which lie lands equally attractive as to climate, fertility, and plants, such as western Cuba, parts of the western United States such as California, and the central Argentine. Multiple independent invention of agriculture is therefore less acceptable than the spread of an art and its artifacts from a common hearth in low latitudes with derivative changes, substitutions, and exchanges over a long time and large distances.

NOTE

Reprinted with permission from *Agricultural Origins and Dispersals,* by Carl O. Sauer, 113–134. Copyright © 1969, The MIT Press. Originally published in *Actas del XXXIII Congreso Internacional de Americanistas,* San José, Costa Rica, 1958, 1:215–229. Lehmann, San José, 1959.

20

Maize into Europe (1962)

Carl O. Sauer

TURKISH CORN

A new natural science, uncommitted as to doctrine, followed quickly upon the Reformation and turned from classical authority to direct observation. In particular a plant science took form that collected, described, and classified. The new interest was in the identity of plants and where they grew in contrast to the elder attention to their virtues as simples [medicinals] according to Greek medical doctrine. Botany was freed from traditional pharmacognosy. The break came first in southern German lands that were strongly involved in the religious dissent. [Otto] Brunfels' *Herbarum Vivac Icones*, published in 1530 at Strasbourg, was the first of the new plant descriptions that came to be known as the Great Herbals. In these the recognition of all kinds of plants largely took the place of medicinal use and so previously unrecorded plants unknown to Greek prescriptions were given attention. Thus we find the first systematic accounts of maize, not as of the earliest moment of its appearance in Europe, but as of an emergent taxonomy.

A competent study of the growing knowledge of maize by the herbalists of the sixteenth and seventeenth centuries has been made by John Finan in his *Maize and the Great Herbals* (1950). He has the interesting finding that not until 1570 was an American origin and introduction by way of Spain attributed to it in the herbals (by the Italian Matthioli, who had read Spanish chroniclers of the New World).

Hieronymus Bock, collecting in various parts of the Upper Rhine Valley, was [the] first of the herbalists to describe maize. The work was first published in 1539, but was under way in 1531. He called it "the great *Welschkorn*, without doubt first brought to us by merchants from warm lands of fat soils . . . seed of three or four colors, some red, some brown, some yellow, and some pure white . . . in form like wild hazelnuts . . . yields a beautifully white meal." Perhaps, he thought, it should be called *frumentum asiaticum* from what he inferred to be its source. The designation "*Welsch*" suggested to southern Germans an immediate Italian source, *asiaticum* Asia Minor as farther derivation. This vagueness as to its introduction in the south of Germany also indicates that this took place prior to his own time.

The first use of the name *frumentum turcicum* [for corn] (1536) was ascribed by De Candolle [1959:386] to Jean Ruel(lius) of Paris, author of an old style translation of the *materia medica* of Dioscorides, with additional notes supplied by Ruel. Since he described a plant in part reminiscent of buckwheat, we may conclude only that he had heard of but had not seen Turkish corn, and hence that it was probably unknown in northern France.

The most famous of the herbals, that of Leonhart Fuchs, was first published in 1542, profusely and superbly illustrated by full folio page woodcuts that show in close detail the entire plants including roots. The costliness of the work delayed its publication, the formidable task of preparing it requiring from 1532 to 1538.[1] Fuchs was born in Bavaria and became professor (and repeatedly rector) at the University of Tübingen (1534 to his death). His observations appear to have been gathered mostly from parts of Suabia, then in active trade with Venice. The fine woodcut of maize bore the names *Turcicum Frumentum* and *Türckisch Korn*, he said, because it was first brought from Greece and Asia (both then under Turkish rule), adding that it was found *passim* in all gardens. Like Bock he named red, purple, yellow, and white seeded races and said it gave an excellent white meal.

Two illustrations earlier than the one by Fuchs are known. The first is in an Italian translation of Oviedo's first book, published in Venice in 1534 as *Sumario de la naturale historia* [1959] and probably [is] drawn from maize growing in Venezia. The second may be a reduced copy of the former and appeared in the Seville edition of Oviedo in 1535 [*Historia natural y general de las Indias*]. (I am indebted for this information to Francisco Aguilera of the Hispanic Foundation, Library of Congress.) At mid-century Ramusio [1967–1970] in the first volume of his *Voyages*, also published at Venice, gave an exact reproduction of an ear of maize in its husk, showing plump, large, smooth kernals.

Finan has pointed out that Cordius (1561) was [the] first to distinguish a kind [of corn] with [a] prop of brace roots, and that l'Obel (1581) first illustrated this sort and distinguished it from a sort lacking prop roots, limiting the term Turkish corn to the older type.

The name Turkish corn has remained in use to the present, in Italy, for instance, as *grano turco, sorgo turco,* and *sorturco.* To dismiss such "Turkish corn" names as ignorant inventions seems prejudicial. There is no justification for saying that the new grain was attributed to Turkey because no one knew whence it came and Turkey was a casually convenient name for an alien and unknown source. A common argument used in support of such explanation is the English name for the Meleagrid fowl, which, however, in reality is not an accidental term. Nor is the inferred or explicit sequence sound that the New World was discovered by Spain and Portugal, that there had been no prior contact between the hemispheres, that maize could have been disseminated only by way of the Iberian Peninsula, and that this happened only in the decades following the discovery by Columbus.

We may therefore consider the alternative that the German herbalists were competent describers and did know what they were talking about. The south German towns were intermediaries in the great trade from North Italian ports, especially Venice, across the Alps to the Rhine and its tributaries; knowledge as well as goods flowed mainly out of the south into High Germany. Venice in particular was built on the Levantine trade, through which it rose to greatest wealth and power in the fifteenth century, at which time there were Venetian factories and colonies in number extending from the Adriatic to the easternmost Mediterranean. Venice had the closest contacts with the Ottoman Empire both before and after the fall of Byzantium. Nowhere in Europe were things "Turkish" so well known as in Venice. Nowhere north of the Alps was Venetian knowledge as well disseminated as in South Germany. The Turkish ascription of maize suggests further an introduction in the fifteenth or late fourteenth century, prior to which time another Levantine attribution would have been more likely.

The Po Valley long has been known for its large cultivation of maize and for a diet strongly based on dishes prepared from maize. Nowhere in Europe, except to the east of Italy, is maize of comparable importance. Maize growing and eating are similarly characteristic of the Balkans and Hungary, long and early under Turkish domination. (A study of the origin and meaning of the name *kukuruz* for maize, widely used in Eastern Europe, might be revealing.) The early establishment of maize as a human staple in Venice, Lombardy, and Emilia, where it is often simply known as "corn" (*formento*) is not to be attributed solely to the advantages of soil and climate, as appears to be true also for the Balkans and Hungary. An early culture historical element also is involved. Edgar Anderson in an article he called "Anatolian Mystery" (*Landscape*, Spring 1958) has outlined the genetic problem for plants of American origin found in the Near East.

SPAIN AS INTERMEDIARY

There is strangely little evidence in support of the theory that maize, pumpkins, paprika, even tobacco, in fact almost [all] New World plants, were disseminated eastward through the Old World from Spain. Cultivated seed plants originating in the New World are more significant in the eastern end of the Mediterranean and in Italy than they are in Spain, and seem to have been so as far back as there is knowledge of them. Nor are the contacts between Spain and the German lands known to have been productive in plant introductions. Southern areas of German speech and political control came into close relation with Spain through the marriage of Juana and Philip. These contacts were strongest during the long reign of their son Charles V, German emperor and Spanish king. During his reign Spanish soldiers and officials came to Germany in numbers. Persons, posts, and knowledge were in exchange between Spain, Germany, and parts of Italy as at no other time. Witness the role of German printing presses in Spain, as well as in Germany, in communicating the news of the New World, the employment of German clerks, and factors, [such] as by the Welsers and Fuggers, both in Spain and its colonies overseas. Yet in that first half century I know of scarcely any mention of New World plants brought to Central Europe by way of Spain (*tagetes* [marigold] an exception?).

Spanish colonists in the New World did not take readily to native foods if they could provide themselves with the familiar Spanish food items. A familiar illustration is the effort expended to grow wheat in suitable and unsuitable places and the official care to get for every administrative unit a record of all plants "de Castilla" (the so-called *Relaciones geográficas* from 1579 on). Doctor Francisco Hernández [1945] sent in 1570 by Philip II to make a botano-medical study of the plants of New Spain, wrote concerning *tlaolli* (maize): "I do not understand how the Spaniards, most diligent imitators of what is foreign and who also know so well how to make use of alien inventions, have not as yet adapted to their uses nor have taken to their own country and cultivated this kind of grain," the many admirable qualities of which he then proceeded to set forth. "This aliment," he continued, with reference to New Spain, "is beginning to be liked by Spaniards, but chiefly by those born of Spanish and Indian parents, or of Indian and negro, or of negro and Spanish origins." These comments were made late in the sixteenth century and still hold in good part for Spanish America. Maize is for the poor, the Indians, and the mixed breeds, wheat for the better classes in the majority of Hispanic areas. The food of the natives did not find favor with their masters; foodstuffs still mark social status. Columbus brought maize to Spain, but only here and there, as in remote Galicia adjoining Portugal, did maize become a common food. (The historical geography of maize growing in Spain, for food and for feed remains to be studied.)

PANIZO, COLUMBUS, AND PETER MARTYR

Possibly the oldest name for maize employed in Spain is *panizo*. Neither the name maize, nor any other native name of the New World was popularly adapted early in Spain, Portugal, elsewhere in Europe, or for the most part anywhere in the Old World (except in the Philippines), though the name mais, taken from the Island Arawak, came into early use all over Spanish America.

The *Journal* of the first voyage of Columbus (known through Las Casas' *Historia de las Indias*) under date of Nov. 6, 1492 from the north coast of Cuba, made bare mention of *panizo* as cultivated by the natives. On his return to Spain the following spring he brought samples of the grain, which as we shall see, brings Peter Martyr into our theme.

Panizo (Latin *panicum*) is still a provincial name for maize in parts of Spain. According to Corominas[2] "at the discovery of America *panizo*, [the] name of an ancient European grass, became widely applied to maize in many parts of Spain, both of Castilian and Catalan speech." Panicum (common millet, Hirse in German) is about as unlike maize as a grass may be; how the name was transferred from a thin-stalked, small-seeded millet to the robust Indian corn remains unexplained. Nor do I have an idea why Columbus at first called maize *panizo*. By the time of the first edition of the *Dictionary of the Real Academia Española* (1734–37), *panizo* had become just another name for maize.

Columbus gave his account of the discovery overseas to the Court at Barcelona. This took place in May of 1493, Peter Martyr d'Anghiera being in attendance. The latter, Italian cleric and tutor to the royal princes, had been writing letters on contemporary events to high Roman clergy and Italian nobles. Letter 130 of his *Opus Epistolarum*, written in May 1493, took up the arrival of Columbus at Court. Columbus and his companions, the view of the strange goods and trophies and the appearances of an unknown breed of men of wholly alien culture impressed him strongly and gave new direction thereafter to his life. He was henceforth the great reporter on the newfound world, and seemingly was never taken in by the notions of Columbus that he had found the Farthest East that Marco Polo had visited. Though Peter Martyr never saw the New World he was perhaps the first to realize that such it was, and he asked the searching questions of the returned voyagers, of high and low degree, to set down the answers in his letters, some of which later were assembled into his *Decades* [1970]. We should know a lot less of the newly discovered lands and their life but for the prompt and acute interviews he had with captains, soldiers, clerks, any who came back from overseas. First historian of the New World, he recorded by personal interviews and he was a most able examiner.

In a letter of September 1493 (no. 133) he noted the dependence of the island natives on root crops as their staples and how they prepared *cazabe* bread; there was still no mention of any grain. Of similar date and content was the next letter (no. 134) to his old friend and patron at Rome, Cardinal Sforza, a fellow Milanese. This was followed by a mid-November letter, also to the Cardinal, in which he wrote:

"*Panem et ex frumento quodam panico, cuius est apud Insubres et Granatenses hispanos maxima copia, non magno discrimine conficiunt. Est huius panicula longior spithama, in acutum tendens, lacerte fere crassitudine. Grana miro ordine a natura confixa, forma et corpore pisum legumen aemulantur: albent acerba: ubi maturuerunt, nigerrima efficiuntur: fracta candore nivem exuperant: maizium, id frumenti genus appellant*" (*Decade I*, Book I).

It is apparent that Peter Martyr had observed well the maize plants as grown to maturity from seeds brought by Columbus and had recognized the plant. The following spring he wrote once again to Cardinal Sforza (*Decade I*, Book II) that "the carrier will give you in my name certain white and black seeds of the *panicum* from which they (the island people) make bread."

Since Peter Martyr wrote in Latin the *panizo* of Columbus became *panicum*, but also he set down for the first time the name maiz(ium). The observation is competent and constitutes the earliest description of the plant known; it notes the *panicula*, not the panicle of modern botany, but the ear, in modern Spanish in places still called *panoja*. According to Coromínas *panoja* "lives on in many parts of Spain, in place of *mazorca*, more favored in common speech." Also, Coromínas shows its derivation from the classical Latin *panicula*. After noting that the ear tapered to a point, was longer than the span of one's hand, and almost as thick as the human arm, Peter Martyr described the grains as "affixed by nature in a wondrous manner and in form and size like garden peas, white when young." In both letters he referred to black seeds, in the later one he adds a white seeded kind. That the color was in the external aleurone layer is indicated by the remark that broken, across the interior of the seed was whiter than snow.

The plant described differs markedly from the tropical, mostly yellow flint corns of the Caribbean of today. It may be noted that there are agreements in the Peter Martyr, Bock, and Fuchs characterizations of maize.

That bread was made from maize in the West Indies is not mentioned by later writers. Oviedo in his [*Sumario de la*] *Natural hystoria* of 1526 [1959: 13–15] said that the islanders used maize only toasted (*tostado*, popped??) or as roasting ears in the milk stage. When Oviedo knew the natives they were already well advanced in cultural collapse. Peter Martyr on the other hand got his account from the first European contact.

Of highest significance is the identification by Peter Martyr of the grain brought back by Columbus with one which he already knew from two areas of the Mediterranean. Of all grains maize is least likely to be mistaken for something else, unique as it is in its ears, seed, and tassel. The person [Martyr] who wrote the first clear description of maize may not be charged with superficial knowledge. He wrote that this was the grain found in greatest amount among the Insubres and the Spanish of Granada. As one Milanese speaking to another, his old friend the Cardinal in Rome, brother of the Duke of Milan, he named their common countrymen as Insubres. The Insubres had been a tribe of Gauls who once had occupied the area about Milan, and the name was a classicism for the people of and about Milan.

Peter Martyr left Milan in 1478 to live in Rome under Sforza patronage. Thence he was called to Spain in 1487. What he wrote the Cardinal in 1493 was in effect that this was the same grain with which both of them had been familiar years earlier when they were still living in Milan. He repeated this identification twice in later years. In Book Two of his *Eighth Decade* he wrote of the maize on the American mainland as like the *panicum* of Lombardy. In Book Two of his *Seventh Decade*, having interviewed Ayllón and his Indian from the land of Chicora (the Carolinas of Southeast U.S.), he recorded that the bread of Chicora was made of maize as among the islanders and that they lacked *cazabe* bread: This "maize grain is precisely like *persimile,* our Insubrian *panicum,* but is of the size of garden peas."

The other area which Peter Martyr named as having had this grain in cultivation at the return of Columbus was the lately conquered Moorish kingdom of Granada. Again he spoke from personal knowledge for he had been a participant observer throughout that campaign which was carried to its conclusion in 1492.

The testimony of Peter Martyr is entered as that of a key witness, competent and trustworthy, to the effect that maize was cultivated in two parts of the Mediterranean well before the discovery of Columbus.[3]

THE QUESTION OF SORGHUM AND MILIUM

As there has been a transfer of name from the small panic grass to maize, so there has been from sorghum to maize. (Currently the American farmer is calling a grain sorghum "maize," shortened from "milo maize," itself a term of confusion.) Name without some mention of a diagnostic quality may be misleading, as in the current terms "Guinea corn" and "great millet."

Of the three grains that resemble each other in their tall, stout growth and large strap-like leaves, pearl millet (*Pennisetum glaucum*) probably can be disregarded. Mainly of tropical African and Indian cultivation, there is no evidence that it was present in Europe. A field of young maize may look much like one of sorghum, but the sorghums bear their seed in a terminal inflorescence, panicle, or "brush" that stands conspicuously above the rest of the plant, whereas in maize the seed is enclosed in ears that are set at nodes well below the upper stalk. The ears of maize are wholly distinctive (cf. Peter Martyr's *grana miro ordine confixa*), and the seeds are many times the size of those of sorghums or other millets. Ear and seed readily distinguish maize and sorghum.

Sorghum is much earlier in Europe than maize. The elder Pliny said that within ten years of his writing a *"milium"* had been brought to Italy from India. His description has been accepted by J. D. Snowden, monographer of the *Cultivated Races of Sorghum* [1936], as sufficient to establish the plant as a sorghum. *Milium* previously had meant *Setaria* or *Panicum,* or both together as millets. Later Roman sorghum appears to have made slow headway through the Mediterranean, as grain for poultry and as stock feed.

I have gotten very little out of the late medieval writers on agriculture, chiefest among whom was Pietro de Crescenzi, Italian of the thirteenth century, who was reprinted well into the sixteenth century and was largely copied in Spanish by Gabriel Alonzo de Herrera of Salamanca. Their main concern was with good farming practices. What little they had to say of particular plants, in particular of forage and feed plants, which included the category of "millets," was rarely sufficient to tell which plant was meant. Botanically they were still inclined to repeat the old Greeks and Romans. Classical plant names were applied to right or wrong plants, original observations only casually and scantily introduced. Names derived from milium appear to have referred at least in part to sorghums (such as *miglio* and *melica*). *Saggina* or *zahina* appear to denote simply fattening feeds.

Sorgo according to Corominas [1954] was "already documented in the Latin form *suricum* in documents of the North of Italy in the thirteenth century" probably meaning "coming from Syria." It is a fair guess therefore that in the later Middle Ages a post-Roman form of sorghum (which has greatly differing varieties) was introduced from the Levant and acquired some popularity as feed in northern Italy, Venetian trade again serving as intermediary. In Friuli maize is called *sorgo turco* or *sorturco*, the later grain having added the locative word "Turkish" to the older grain that was named from Syria. The earliest use of the name sorgo in the literature of discovery in so far as I know is by Pigafetta during the voyage of Magellan. He identified sorgo as one of the grains in the Philippines. That, as native of Vicenza in the Po Valley, he knew where of he spoke, is further substantiated by his recording as its native name "*batat*," which is Malay for sorghum (cf. Malay names in [I. H.] Burkill's *A Dictionary of the Economic Products of the Malay Peninsula* [1935]).

In 1542, Fuchs described and figured a reddish seeded form of sorghum with very long, lax panicles as *sorgi*, or *Welscher Hirsa*, saying that it was an alien grain brought to Germany from Italy, was grown in many German gardens, but was of difficult cultivation (climate). Thus he introduced plant and name to the literature of botany. The varieties with dense and stiff panicles, the grain sorghums, such as the durras of African origin and related forms cultivated in the Orient, seem not to have been known in Europe until considerably later times; it is these that are mainly used as human food in dry parts of Africa and Asia.

The origin of the name sorghum thus may be credited to Italy. The name spread north into German lands, but not west into Iberia. Where present in the latter parts, we should expect some variant of *milium*. A curious note from Peter Martyr is in order here. Puzzling about a grain other than maize reported by his informants from the Carolinas, he says they think it may be "*milium*," but he was uncertain because very few Castilians know what *milium* is, "since it is never grown in Castile."

PLURAL INTRODUCTIONS OF MAIZE INTO THE IBERIAN PENINSULA

Peter Martyr gave positive testimony of maize in Granada prior to Columbus. This may be supported by a Catalan name *blat de moro*, Moorish wheat. Columbus brought maize from the West Indies, and may have given rise to the local name of *panizo*. Finally throughout Portugal (*milho*) and adjacent Spain (*mijo*), especially in Galicia (*millo*) and on into Gascony (*milhoc*) (Corominas) a name derived from milium was given to maize. This is also true of the Canaries and the Portuguese islands off the African coast. This distribution of *milium* names, down the Atlantic coast requires our attention next.

First, however, a gloss on another name, *borona*, of similar but lesser geographic distribution. Corominas [1945] has documented *borona* in use as far back as 1220 and holds it to be a word of ancient origin, applied to some cereal (a millet?) and the griddlecakes made therefrom to be passed on later to maize "bread" and maize. The accounts of the Spanish occupation of the Philippines in and after 1566 refer occasionally to *borona*, perhaps to a millet like the one known in Spain. Distribution and meaning of this name need study. In Portuguese, both at home and overseas, *milho*, with or without a qualifying word, is the standard or only word for maize and was so from early days.

MILHO AND *ZABURRO*

Pigafetta [1969], Italian chronicler of the voyage of Magellan, noted about Rio de Janeiro that the Indians were growing *miglio* (Italian spelling, same pronunciation as the Portuguese *milho*). The crew was partly Spanish, and from some who had been in the Caribbean he learned that there it was called "*mais.*" As *miglio* the grain was familiar; the new thing was that it had an Indian name.

In the Vizayan part of the Philippine Islands Pigafetta noted *miglio/millio* repeatedly. This may be entered as the first records of maize in the Philippines, antedating by a half century the introduction from New Spain. Pigafetta may be accepted as knowing what he saw. He was not only generally a good observer, but (1) he had been correct in Brazil in identifying *miglio* with maize; (2) he recognized sorghum when he saw it in the Philippines, as was cited above; (3) when forced tribute was levied on the Vizayan natives these were required to bring equal quantities of rice and *miglio*, rice and maize being most desirable and familiar as ship stores to those who had sailed African shores.

These early *milho*/maize terms from Brazil and the Philippines lead us to the Portuguese on the coasts of Africa and their early knowledge of maize. This subject has long been the special concern of Professor M. D. W. Jeffreys of the University of Witwatersrand.[4] I thought to have a look on my own through the early Portuguese writings on overseas and found the Jeffreys theses confirmed that maize in Africa was pre-Columbian and that the Portuguese took it to Portugal from Africa, not from America.

Soares de Souza wrote a detailed and competent description of the natural history and geography of northeastern Brazil in the second half of the sixteenth century, *Grandeza de Bahia de Todos os Santos*. He came to Brazil a generation after its first colonization, but had long and intimate knowledge of the colony and a special interest in the cultivated plants. He described the maize plant and its uses, noting that there were white, ochreous, black, and red seeded races and also a soft-seeded (flour) form, adding "the Indians [Tupi] call *ubatim* [*abatí*] what is the *milho de Guiné*, which in Portugal they call *zaburro*." Maize in his time was the staple food of the negro slaves on the plantations. Guinea corn and *zaburro* he recognized as synonyms for *milho* (maize).

João de Barros in his *Historia* of the mid-sixteenth century described the Jalofa (Wolof) negroes living between the Senegal and Gambia rivers, as practicing a peculiar manner of sowing "*millios de maçaroca a que chamamos zaburro*" (the cob-bearing *milho* which we call *zaburro*). This "was the common sustenance of these peoples." Ramusio [1967–1970] added to his Italian version the marginal comment "the maize of western Indies, on which half the world is nourished and which the Portuguese call *miglio zaburro.*"

About the year 1530 an anonymous Portuguese pilot wrote about navigation from Lisbon to the Island of São Thomé in the Gulf of Guinea, telling of "the grain which is called *miglio zaburro* and in the western Indies is called maize, of the size of chick peas and common to all the [Cape Verde] islands and all the coast of Africa and upon which the inhabitants sustain themselves." It was made into bread "which then was sold throughout the coast of Africa, or land of the negroes, and was traded for black slaves." It was also grown on the lately colonized island of São Thomé for and by negro slaves who had been brought from the near mainland.[5]

The Portuguese crown sent a mission to seek out Prester John in 1515; it traveled through Ethiopia from 1520 to 1526 and returned home in 1527.[6] In this account I found reference at seven localities to *milho* or *milho zaburro*, such as meeting cattle people, very black, naked, and claiming to be Christians, who were guarding their fields sown to *milho zaburro* and had come from afar to sow it on very high and steep mountainsides (ch. 8), of bread made from a mixture of *milho zaburro*, barley and a small black seed called tafa (*Eragrostis teff*) (ch. 13), of traveling through canes of *milho* as thick as those used in staking grapes (ch. 33), of passing through *milharadas* as tall as sugar cane (ch. 49).

A German in Portuguese services, known as Valentim Fernandes, has left an early and informative account of African coasts as known between 1505 and 1508.[7] About Quyloa (Kilwa in southern Tanganyika) there was much *milho* like that of Guinea, all the gardens surrounded with wooden stakes and canes of *milho* like *canaveaes* (sugar cane fields), the stalks as tall as a man (pp. 14–16). The Gyloffa (Wolof of Senegal), as Barros [no reference given] described them later, had much *milho zaburro*, and as their chief food *cuscus* made of *milho zaburro*, their manner of grinding and baking same being described. The Mandingo, southern neighbors of the Wolof, consumed much rice and *milho zaburro*. Of São Thomé he wrote that *milho zaburro* began to be planted there in 1502, about ten years after its colonization, this grain having previously been brought in by ship from mainland Guinea. An important sentence explains the nature of this *milho zaburro*: "E nace propio como ho daca se non q nace grande e o milho en hua maça e non espalhado como o nosso" (p. 128): "It grows like ours, except that it grows large with the seeds in one mass and not spread apart like ours." This is a good comparison of maize and sorghum, the massive ear as against the lax panicle of the sorghum then known in Europe (cf. the plate of *sorgi* in Fuchs [Stuler, 1928]). Also this is the nearest approximation we have to a time when maize was not yet established in Portugal, for the *milho* of Portugal to which he compares the *milho zaburro* was certainly not maize.

Milho has long since become the common name for maize in Portuguese, but *zaburro* also continues to be its name, or the name of a variety.[8] *Milho zaburro* seems to have been the earliest Portuguese name for maize and the name came out of the Guinea Coast. It was in use years before the Portuguese colonization of Brazil began. *Milho de Guiné* became a Portuguese name for maize in Brazil and elsewhere, to be applied in error in later years, especially in the English "Guinea corn," to sorghum. . . . Jeffreys [1957] makes a telling point, saying that in twenty years of residence on the Guinea coast he never saw sorghum grown there, nor is it at all climatically suited to the tropical rain forest; the sorghums belong to the dry African margins.

The origin of the name *zaburro* and of its appearance on the Guinea Coast I must leave to Jeffreys and others who know African languages and early history. [J. M.] Dalziel in his *Flora of [West] Tropical Africa* supports Jeffreys as to *zaburro*/maize names for tribes of the Gold Coast and Dahomey. (In reading Dalziel I wondered whether there might be a connection with the *Digitaria iburua* cultivated as a grain, as by Hausa tribes in northern Nigeria.) Jeffreys is of the opinion that maize was communicated by Arab contacts and pressures out of the north and northeast in the late Middle Ages. He may well be right and if so there may be a common source for Guinea and Turkish corn.

The identification of *milho* as maize, which the Portuguese first learned to know on the Guinea Coast, throws some light on the African slave trade. Maize "bread" from the coast was used for trading in slaves (food shortage in the interior?). Maize was shipped to the African islands to feed the slaves until the islands grew their own supply. Maize was the staple food of the African slaves in Brazil. The inference is that maize was taken with the negro slaves wherever the Portuguese went, to Brazil and to Portugal. It was available on the coast of Guinea, the leading source of slaves; it was [the] accustomed food of the coastal negroes; it was more readily transported than yams and other roots; it was a main food crop on the sugar plantations. It may be therefore that some of the older maize varieties grown in Brazil were introduced from Africa.

NOTES FROM LEO AFRICANUS

Leo Africanus, the Moorish house slave of the Medici Pope Leo X, may have something to add for the Sudan. Born in Granada he was taken to Morocco when the Kingdom of Granada fell to the Spanish arms in 1492. In early manhood he traveled widely about North Africa, crossing the Sahara into the Sudan. These southern travels are dated around 1514; his capture, sale, and coming into the hands of the Pope [were] between 1518 and 1520. As the letters of Peter Martyr informed the Pope about the New World, so the Moorish captive was prevailed upon to write or indite a geographic handbook of Mohammedan Africa.

The Seventh Part of Leo's *Description of Africa* [1550] dealt, somewhat briefly, with the Land of the Negroes, by which he meant peoples more or less converted to Islam and living about the Middle Niger, both to the west and east of Timbuktu. His knowledge barely extended to forest negroes farther south, and not at all to the Guinea Coast. For the "kingdom" of Gaulata (Oualata, in the edge of the desert to the west of Timbuktu) he said there was "little grain and this is miglio (here probably *durra* Sorghum) and another sort of grain, round and white like *cece* (*Cicer*, chickpeas) which is not seen in Europe." Far to the southeast of Timbuktu, in Guber (Gober, a district in the new Republic of Niger) Leo observed "a great quantity of *miglio* and rice and another grain which I have not seen in Italy, but believe that such is to be found in Spain."

In Sudanese lands, where *durra* has long been a staple as well as rice where the local situation was favorable, he noted at two widely distant localities "another grain," which may well have been maize. The seeds, and he only noted grain, were round, white and large like chick peas. Only maize would fit these characteristics; comparison to chick peas has been noted above in Portuguese accounts. He had not seen such a grain in Italy, but he did not know Lombardy or Venice and he had limited occasion to travel in Italy. He thought the "other grain" of Guber was to be found in Spain; Peter Martyr had said that maize was grown in Granada, the birthplace of Leo, at the time of the conquest.

CONCLUSION

This study has been based in the main on printed documents of the time of the Great Discoveries. None of them supports the notion that Columbus was first to bring maize to Europe, nor that it came to be disseminated to other parts of the Old World by way of Spain and Portugal. Nor do I know evidence for such dramatically rapid spread of this grain as Laufer [1938] inferred in his view of its carriage by land in a few decades from Spain east to China. A new food plant finds its way more gradually into an economy. The grower must learn how to plant, cultivate, harvest, and store it, how to fit it into his cropping practices, and how to meet its requirements of weather and soil; the housewife needs to learn how to prepare it in dishes that gain acceptance. Such learning and change of habit take time. At what times and by what routes maize was carried into the Old World still remains to be determined in large part. Its entry into Europe is indicated as pre-Columbian (Balkans,

Italy, and Granada), Columbian, and post-Columbian (*milho* into Portugal out of Guinea). Further light may be expected from documents (Turkish tax lists, Arab travels, accounts of Venetian and Portuguese trade), vernacular names for maize, and perhaps most of all from the phylogenetic study of old local races of maize that survive in Europe, Africa, and Asia.

NOTES (REVISED BY EDITORS)

Reprinted with permission from *Agricultural Origins and Dispersals,* by Carl O. Sauer, 147–167. Copyright © 1969, The MIT Press. Originally published in *Akten des 34. Internationalen Amerikanisten-Kongresses, Vienna, 1960* (1962), 777–788.

1. Stuler (1928:231).
2. Corominas (1954).
3. Edgar Anderson has called my attention to the *Notebooks of Leonardo da Vinci* (1939 [ca.1490–1495]) for two references to maize. Discussing the liver, Leonardo makes a comparison with "maize or Indian millet, when their grains have been separated" (p. 116). In the *Miscellany* [?] (p. 1184) there is mention of "beans, white maize, red maize, panic-grass, millet, kidney beans, broad beans, peas." [These] jottings are thought to have begun in 1508. Leonardo died in 1519.
4. A late exposition is by Jeffreys [1957].
5. Ramusio (1967–1970:1:125–128).
6. Alvares (1889).
7. Fernandes (1941), esp. 67ff.
8. The custodian of our quarters at the University of California, Antonio Trindade, native of Madeira [Island], when asked what *zaburro* was, answered that it is corn of large purplish black seeds, which are white inside.

REFERENCES (INCOMPLETE, MOSTLY ADDED BY THE EDITORS)

Alvares, Francisco. 1889 [1540]. *Verdadeira informaçiõ das terras do Preste Joâo das Indias.* Lisbon.

Corominas, J. 1954. *Diccionario crítico etimológico de la lengua castellana,* 4 vols. Madrid.

da Vinci, Leonardo. 1939. *Notebooks of Leonardo da Vinci,* ed. and trans. Edward MacCurdy. Reynal and Hitchcock, New York.

De Candolle, Alphonse. 1959 [1886]. *Origin of Cultivated Plants.* Hafner, New York.

Fernandes, Valentim. 1941. *O Manuscrito "Valentim Fernandes."* Academia Portugués História, Lisbon.

Hernández, Francisco. 1945. *Antigüedades de la Nueva España.* Trans. Joaquín García Pimental. México.

Jeffreys, M. D. W. 1957. "The Origin of the Portuguese Word *Zaburro* as their Name for Maize." *Bulletin de l'Institute Français d'Afrique Noire, Series B. Sciences Humaines* 19 (1–2): 111–136.

Laufer, B. 1938. *American Plant Migration.* Field Museum of Natural History, Anthropology Series, vol. 28. Chicago.

Leo Africanus. 1550. *Descrittione dell' Africa.* Rome.

Martyr d'Anghiera, Pietro. 1970 [1493–1525]. *De Orbe Novo: The Eight Decades of Peter Martyr d'Anghiera.* Trans. Farncis A. MacNutt. Burt Franklin, New York.

Oviedo, Gonzalo Fernández de. 1959 [1526]. *Natural History of the West Indies.* Trans. and ed. Sterling A. Stoudemire. Univ. of North Carolina Press, Chapel Hill.

Pigafetta, Antonio. 1969. *The Voyage of Magellan: The Journal of Pigafetta.* Trans. Paula S. Paige. William L. Clements Library, Ann Arbor, Mich.

Ramusio, Giovanni Battista. 1967–1970 [1563–1606]. *Navigazioni e viaggi.* 3 vols. Theatrum Orbis, Amsterdam.

Stuler, E. 1928. *Leonhart Fuchs.* Munich.

VI

MAN IN NATURE

21

Introduction

William W. Speth

> The biologist uses the term symbiosis to describe the balanced or harmonious living of a group within its habitat. We are much concerned with this idea in geography.
>
> Carl O. Sauer and John B. Leighly, "The Field of Geography," 1927

1

Carl Sauer had acquired the essentials of a disciplinary world view that sustained his efforts at Berkeley years before he moved there in 1923. His broad and diverse learning at Warrenton, Chicago, and Ann Arbor is an indispensable prologue to an understanding of his later achievements. The first selection among the four included in this section appeared in 1936, thirteen years after he arrived in California. The last of the four was published in 1950, twenty-seven years after he reached the West Coast. These articles on "Man in Nature" are rooted, variously, in the cumulate knowledge that he gained in the Middle West. Ideas and interests that he appropriated before 1923 extend manifestly into the Berkeley phase of his work (Speth, 1987).

Early in his formal training, Sauer began to doubt the received environmentalist definition of human geography. His growing familiarity with the literature of European geographers, which began at Chicago and intensified at Ann Arbor, enabled him subsequently to invert the Chicago doctrine. As to the "materials" of geography, he argued for the visible signs of human occupation of area (chorology) and against the rationalistic linkage of physical cause–human consequence. Late in life, in the last seminar he gave at Berkeley in 1964, he indicated the essence of his geography: what is the "ecological relationship of people to the land" they live in, and is this relationship "harmonious" (1987a:156)? Mechanism (force) was, in time, rejected and organicism (thought) embraced. No mere creature of tropisms, man had become an integral, yet threatening, part of the biosphere. He left Ann Arbor recognizing humankind as a major agent of terrestrial change, yet he apparently did not learn of George Perkins Marsh's study, *Man and Nature* (2003), first published in 1864, until the 1930s.

2

The Warrenton period (1889–1908), including three years of schooling in Calw, Germany, and two baccalaureate degrees (classics, sciences) from Central Wesleyan College (now defunct), was basic to the formation of his disciplinary world view (Kenzer, 1985:265–266; Kenzer, 1987). At this time, he assimilated a Romanticist sensibility that was thereafter reinforced through wide reading, persistent reflection, and personal contacts with sympathetic colleagues, within and beyond the borders of geography. Some of the easily recognized features of Romanticism are lodged in his critical commentary and evolving statements about the nature of geography: Sauer preferred the organic metaphor over the mechanical; the concrete or particular over the abstract; nature over culture, convention, and artifice; historical induction over rationalistic apriorism; *Gemeinschaft* over *Gesellschaft*. The list of attributes is suggestive only and goes on (Quinton, 1995:778; Nemoianu, 1993:1092).

As the Romantic movement took form between 1790 and 1840, it enlivened most spheres of European intellectual life, in particular that of the Germans. Donald Worster writes that

> a new generation sought to redefine nature and man's place in the scheme of things. . . . Romanticism found expression in certain common themes, and one of the most recurrent was a fascination with biology and the study of the organic world. Romantics found [in] this field of science a modern approach to the old pagan intuition that all nature is alive and pulsing with energy or spirit. . . . And at the very core of this Romantic view of nature was what later generations would come to call an ecological perspective: that is, a search for . . . integrated perception, an emphasis on interdependence and relatedness in nature, and an intense desire to restore man to a place of intimate intercourse with the vast organism that constitutes the earth. (Worster, 1994:81–82)

To the idea that all earth features are interconnected one must add that nature (biology) and man became "thoroughly historicized" in the nineteenth century. The result was a totalizing view of the planet and man throughout the entire period of hominid evolution (Worster, 1994:421; Schnädelbach, 1984:43–47). This *unifying* impulse was extended by early twentieth-century German geographers and their heirs, including Carl Sauer.

After moving to Berkeley, Sauer immediately recreated the organismic metaphysic and method as his "Morphology." He asserted that "the organic analogy" is a fruitful "working device" in social inquiry—areal phenomena "are not simply assorted but are associated, or interdependent"—and the element of time, largely nonrecurrent, is "admittedly present in the association of geographic facts" (1925:22, 30). This is the heart of the philosophy enlisted by German Romantics against the mechanists on literary as well as scientific fronts. It is the *Weltanschauung* furthered by Ernst Haeckel, who coined the terms "ecology" and "chorology" over fifty years earlier, who included anthropology as a part of zoology, and who insisted on the unity of nature (Uschmann, 1972:8). It is, in a word, Romantic historicism (Rossi, 2001:6757–6762).

3

Sauer entered the University of Chicago in 1910 as a young naturalist, perhaps already infected by the spirit of the conservation movement. The rise of conservation-mindedness, in the years from 1860 to 1915, served to check an era of greed, devastating exploitation, and waste on the frontier (Worster, 1994:261). "Traced back to its ideological roots," Stephen Fox suggests, "conservation amounted to a religious protest against modernity" (Fox, 1981:359). The leading ideal applied to the new national circumstances was that the sciences must be the basis of natural resources planning, including areas of natural beauty. Maturing from eighteenth-century organismic ecology, the "new" field of ecology was one of these sciences (Worster, 1973:2–3). From 1901 to 1909, Theodore Roosevelt was president of the United States and a powerful advocate of conservation ideology. While he was in office, the movement "swelled into a national panic" only to lose momentum and all but vanish by World War I (Worster, 1973:7).

The crescendo of this national protest echoed widely, and Sauer was at its source in his "Chicago days." Reflecting on the time, he noted, "We were . . . beginning to see the ecologic unbalance that came through industrialization, and listened to early conservationists, especially to Van Hise" (1952b:103). Sauer related to a correspondent the nature of his indebtedness to Charles R. Van Hise: He gave "me and many others . . . an entirely new insight into modern history by pointing out to us that much of what we call progress has been dissipation of what we call capital goods and natural resources. Here was the natural scientist speaking with the wisdom of long range perspective, questioning the optimistic attitude of social science, over confident of the triumph of mind over the physical world" (Sauer to M. M. Vance, n.d. [1948?]).

To another correspondent in the same year, he restated this idea as "the Icarus Complex," noting that civilized man has "thrived briefly on destruction and multiplication" and wondering whether "with only this in his record . . . he has any real reason to think that his ill-gotten supremacy indicates his ability to take over the role of recreating the world" (Sauer to J. M. Russell, October 20, 1948). This is the mature Sauer as moralist, confidently judging the growing ecologic disturbance caused by Western civilization in its newly settled colonial lands.

The philosophy and practice of conservation was at its zenith in 1910 when Sauer began graduate study at the University of Chicago. His first regional monograph from the Chicago period reveals early concern with mankind as careless modifier of the soil and plant covers. The *Upper Illinois Valley* (1916a) launched his lifework as a regional geographer with a strong rural bent. Largely an American product, the work reflects the tendency of the time "to regard geography as the expression of physical geography in human activities" (1967a:70). Davisian geological and Victorian ethnological expressions abound. History (settlement and development) and human geography were "really quite different subjects" (1941b:4). Three-fourths of the content is devoted to geology and physiography, the strong suits of Sauer's mentor, Rollin D Salisbury. Sauer ended the chapter on "present active physiographic processes" with a four-page discussion of "man as a factor in erosion" (1916a:140–143). The same year Sauer published a one-page report entitled "Man's Influence Upon the Earth" (1916b, ch. 8 herein). The decision to construe human action as a physiographic process marks the beginning of his perception of man as "a geomorphologic agent," who could degrade or upbuild (1941b:18).

The idea of man as an agent of physical geography would seem to lie in Sauer's (1967:69) familiarity with the three-volume compendium titled *Geology*, written by Thomas C. Chamberlin and Salisbury and published in 1904–1906. Despite backing evidence amassed by Charles Lyell himself in the 1830s, Lyell did not acknowledge "a revolutionary character to human agency" in altering the natural world (Glacken, 1970:177). Less than three-quarters of a century later, Chamberlin and Salisbury questioned Lyell's restraint in weighing the role of human groups as agents of surficial change. They chose the opposite view and cited the degradation of parts of the Mediterranean basin and the contemporary acceleration of land abuse in the United States, including soil erosion. Hoping that human intelligence would secure the future, they wrote that "man may well be regarded not only as a potent geological agent but as dangerously so to himself" (Chamberlin and Salisbury, 1904–1906:1:620).

What Sauer saw and recorded as a fledgling geographer was only a facet—induced soil erosion—of a gathering civilizational process still incompletely known to him. Adopting the geologists' sense of time, he linked the human animal to the same erosional effects produced by wind, water, and ice. The human impact on the earth is geologically recent, yet human activities have influenced erosional processes in three ways: they stimulate or increase soil erosion by deforestation, over-grazing, and cultivation of slopes (1916a:140). Sauer reasoned:

> Savage man did little to destroy the soil cover, or to invite the gullying of the surface in other ways. In his time, the upland was covered by the thick sod which prairie grass formed, and on the slopes of the valleys grew trees and brush which protected the soil from wash; man was an unimportant factor in erosion. Early travelers have left accounts of this country before the coming of the white man, of thick prairie grass which stretched, an unbroken sea of waving blades, over the upland, interrupted here and there by tongues of woodland along the streams, and of streams that flowed clear and pure. Today the scene is much altered. The grassy prairies have been converted into tilled fields, and the soil is bared to the action of wind and water. Much of the timber has been removed from the valleys, gullies are cutting back into the prairies in many places, and the streams run murky with their load of sediment washed in from plowed fields and denuded slopes. (1916a:140; see also 83, 123, and 124)

This passage foreshadows later ideas in Sauer's intellectual growth: the thesis of his school book, with its unitive title *Man in Nature* (1939), is effectively stated; the heedless white man is censured and the (communitarian) savage is held blameless; Romantic-aesthetic categories and diction are used, as in reference to the billowing prairie grasses and to the historic "scene"; and the eye alone (Goethe) sufficed to apprehend the symptoms of ecological disruption. From this early, dispirited moment at Chicago, Sauer nurtured the theme of soil wastage at Ann Arbor and Berkeley. In the middle years of the great dust storms, he submitted a short article, "Soil Conservation," to the departmental newsletter, the *Geographical Error* (1936b, ch. 22 herein). In this compact piece, he advised: "If the major end of geography is to find the realization or failure of symbiosis of man and nature, the conservative use of natural resources is at the heart of all human geography and the use of soil is one of the major topics with which we must deal" (1936b:3, ch. 22 herein).

Soil wastage was not the only subject of lasting concern that Sauer considered in the *Upper Illinois Valley* monograph. The vast upland prairie surface of northern Illinois—treeless, nearly level, and undifferentiated—elicited comment as a condition of pioneer life. The "problem of the prairie" was discussed, not as possible artifact of aboriginal burning, but as a barrier to pioneer homesteading. Fire was an impediment because "the danger from fires was great to the first prairie homesteads. . . . In numerous instances, houses and crops were destroyed by such fires." On this matter alone, timbered zones were preferred to the prairie (1916a:155).

From the prairies as a problem for settlers, Sauer came to think of them as topographic features created and sustained by humans. In his *Starved Rock* study, he held that the Indian "did little or nothing to alter the natural conditions of the surface" (1918c:44). But in his published dissertation on the Ozark Highlands, he concluded: "Of the various influences that caused prairies on the uplands, man was chief. Indians and other hunters were wont to set fire to the grass in fall or spring in order to improve the grazing for the buffalo, elk, and other big game. Fires were also set to drive game toward the hunters. Through this practice, sprouts and tree seedlings were killed, and thus the grasslands were extended at the expense of the forests" (1920b:53).

Finally, the preservation aspect of the conservation movement drew Sauer's attention. The Missouri Ozarks was attractive recreation ground, already known for its hunting, fishing, and resorts. The state was poised to initiate a park program, and he urged early action to ensure that "the many idyllic spots in which the region abounds may be preserved forever" (1920b:237; cf. 1918c:81–83). Sauer also encouraged the creation of wildlife sanctuaries. Combined with rural economic reform, the procurement of recreation sites would advance "the progress of the state" and thereby better contribute to the welfare of our national life (1920b:237). Optimism and hope suffuse these passages, more perhaps than they ever did again in the course of his intellectual development.

4

Sauer accepted appointment at the University of Michigan in 1915 and remained there for seven years. At this time, his activities were concentrated on land use survey, classification, and policy, the goal of which was to decide the best use of the land. He was honored by the American Geographical Society in 1935 for his "Morphology," for his "contributions to the study of land utilization," for his organizing of the Land Economic Survey of Michigan, and for his report (1934g) on "Land Resource and Land Use" (Anon., 1935:487). Despite the accolade, he believed the Ann Arbor experience belonged to the "errant years," when he and other geographers had moved away from their heritage (1941b:3). At Berkeley, however, Sauer's practice of geography became driven by a search for new knowledge: it was a pure geography that gave "free play to curiosity, restricted only by competence" (Sauer to R. Hartshorne, June 22, 1946). Late in the Ann Arbor phase, Sauer faced a methodological crisis. The "Survey Method" article (1924a), his last at Michigan, reflects this and is philosophically riven and discordant. He faced two contending muses, one pragmatic (non-genetic chorography) but attractive for its ideal of the conservative use of natural resources, the other scholarly (genetic chorology) and increasingly meaningful for the intellectual freedom and scope of learning that it promised.

This reversal of values is clarified by his remarks at the 1944 Huntington Library Conference on "The Relation of Man to Nature in the Southwest" (1945, ch. 24 herein). When asked how we should deal with the depletion of natural resources in the Southwest, Sauer replied, "I am not a land planner or an expert on government." He asserted, "I am more interested in speaking for the documentation of a process that is going on than in undertaking to prescribe the cure." He added, "I do not object if the investigator can also prescribe, but he is not obliged to" (1945:140–141, ch. 24 herein).

During the Ann Arbor years, Sauer had "begun to read seriously what German, French, and English geographers were learning about the world as long and increasingly modified by man's activities." His "Morphology of Landscape" was "an early attempt to say what the common enterprise was in the European tradition" (1974:191). The studies of Norbert Krebs, Joseph Partsch, Siegfried Passarge, Albrecht Penck, Gustav Braun, Jean Bruhnes, and (through the latter) Ernst Friedrich were his models for geographic inquiry (1917b:84; 1920b:52–53; 1924a:22–24). Before he moved to Berkeley, self-instruction in the geographical tradition of Europe formed much of the basis of his rejection of contemporary environmentalism. It served as well to acquaint him with Jean Bruhnes's *Human Geography*, which introduces the reader to Friedrich's concept of *Raubwirtschaft*, or destructive exploitation (Bruhnes, 1920:330–350).

Sauer used this resounding term for the first time, borrowed (evidently) from Bruhnes, in an article on soils geography. Its use is integral to a plea to incorporate "soil geography" into the larger subject. Soil geography would "inquire into the whole matter of the economy of the soil region ... [and] register the experiences of past and present use as to destructive exploitation or sustained yields of more intensive production" (1922:189, ch. 9 herein). Before 1922, he wrote of "exploitation," but the latter lacks the pointed meaning of Friedrich's concept (1918b:83; 1921a:4). However, Sauer came to prefer the designation *Raubbau*, used earlier by Friedrich Ratzel. "This was," Sauer wrote, "a generation before Friedrich and Bruhnes directed the attention of geographers to destructive exploitation, both having been strongly influenced by Ratzel" (1971a:251).

In "Destructive Exploitation in Modern Colonial Expansion" (1938a, ch. 23 herein), Sauer applied this concept to the process of European expansion overseas. Colonial development, largely responsible for the wealth of the modern world, occurred "by the impoverishment of the lands [and peoples] colonized" (494). He continued: "This phenomenon of the deliberate commercial exploitation of the land with deliberate disregard for the permanence of the communities is properly to be called *Raubbau*. It is unfortunate that the term has been applied to two entirely unrelated economies, one commercial, one primitive. The latter use is improper" (496). Again, Sauer defended primitive economies—his "savage man"—as "permanent" or avoiding "destruction of the productivity of the land" (496). He believed that European colonial culture was a colossal example of "bad stewardship of the land." He did not hesitate to judge the pathogenic "realities" produced around the world by modern civilization (494).

After providing the historical context for the phenomenon of exhaustive exploitation, Sauer devoted the remainder of this paper to wastage of soil consequent on commercial plunder. The theme of soil erosion—nurtured now for twenty-two years since its inception at Chicago—should occupy first place of importance among all geographers and should aid in the realization of a comparative understanding of soil wastage (497).

Sauer's concern at Chicago with the problem of prairie origins dovetailed at Berkeley with interest in climatic limits of vegetation, including the Clementsian climatic climax (Worster, 1994:240). In his article "Grassland Climax, Fire, and Man" (1950c, ch. 25 herein), he challenged the argument for a climatic grassland climax, preferring to explain this plant formation in terms of, perhaps, hundreds of thousands of years of deformation "from fires, chiefly a cultural phenomenon." These ecologic assemblages, neither climax nor in equilibrium, are due, he argued, to recurrent burning, a process that has lasted long enough to allow for speciation of new grasses and to affect the characteristics of prairie soils (20). In contrast to this bold conjecture is Sauer's dawning doubt, in the late 1920s, that the American Indian may not have been "the negligible factor in the landscape commonly supposed," that by burning they may have induced the "native" grass cover encountered by early settlers (1927a:202 n. 69). To these earliest livelihood activities, he applied the name "fire economies" (1950c:19, ch. 25 herein). Fire is one of two great geographic forces (the other is climatic change) at work in the late geologic past. In addition to climate, Sauer reasoned: "The second great agent of disturbance has been man [as wielder of fire], an aggressive animal of perilous social habits, insufficiently appreciated as an ecologic force and as modifier of the course of evolution. Man has been in existence throughout the Pleistocene, ranged very widely very early, and during it became the dominant animal over many climates and, it seems, all continents. The earliest human records we have show the familiar use of fire, and they range from England to South Africa" (18).

Thus, an insight buried in Sauer's first monograph (1916a) grew into a piercing aphorism about the planet's "dominant animal"—"modifier of the course of evolution" and threat to symbiosis.

5

The impression that forms in this sketch of when and how the ecological perspective appeared in Carl Sauer's thinking prior to Berkeley is one of hard-won, incremental learning about humanity's growing physical power in nature. His awareness deepened, persisted, and profoundly disturbed him; it could be argued that its expression reached a crescendo in the Dust Bowl years. His Romantic sensibility precluded mere recording of *Raubbau*; he was moved to morally judge the loss of biological patterns and traditional ways. Concurrently, he had to forget the creed of rationalistic environmentalism taught at Chicago. He began formulating a fresh meaning for geography, based on the European tradition that he found congenial. Building on his Warrenton (Goethean) *Naturphilosophie* by reading in the German literature, Sauer aligned himself with the *Landschaftskunde* school and its focus on areal morphology. Also, Sauer leaned on Bruhnes's epitomizing study of human geography. The French geographer was increasingly appreciated at Ann Arbor and drew praise from Sauer throughout his life for classifying cultural forms, including destructive economies. By first reading Bruhnes, Sauer became a lineal descendant of George Perkins Marsh (Whitaker, 1940:157–161). "There is everywhere," wrote Bruhnes in 1910, "evidence of man" (Bruhnes, 1920:32). Human geography was being born, and interest was centered on human *activity*, its nature and extent. Sauer recognized the fire-dependent primate as the last morphological factor to be added to the earth organism. "Man is the latest agent in the fashioning of the landscape." The chorological view of humanity, "areally significant by [its] presence and works," was a "unitary and attainable objective" (1927a:186). Before leaving Ann Arbor, Sauer revealed interest in the human modification of area and, more specifically, in "the quality of [man's] stewardship over his natural inheritances" (1924a:25). At Berkeley, he elaborated upon the idea of "man as an agent of physical geography" as the leading concern of geographic research and tended to give disproportionate attention to it in his writings and seminars (1941b:18–20; 1987a [1964]:156–160). Well into the Berkeley period, Sauer wrote: "I can define human geography as the natural history of mankind, as an historical ecology centering about the skill or lack of foresight with which he has made use of the materials at hand. This is a field that properly belongs to the geographer" (Sauer to L. S. Wilson, April 6, 1948).

REFERENCES

Anonymous. 1935. "Honorary Corresponding Members." *Geographical Review* 25: 486–487.

Bruhnes, Jean. 1920 [1910]. *Human Geography: An Attempt at a Positive Classification, Principles, and Examples.* Trans. I. C. LeCompte. Rand McNally, Chicago.

Chamberlin, Thomas C., and Rollin D Salisbury. 1904–1906. *Geology.* 3 vols. Henry Holt, New York.

Fox, Stephen R. 1981. *The American Conservation Movement.* Univ. of Wisconsin Press, Madison.

Glacken, Clarence J. 1970. "'Man's Place in Nature in Recent Western Thought.'" In *This Little Planet*, ed. Michael Hamilton, 163–201. Scribner's, New York.

Kenzer, Martin S. 1985. "Milieu and the 'Intellectual Landscape': Carl O. Sauer's Undergraduate Heritage." *Annals of the Association of American Geographers* 75: 258–270.

———. 1987. "Like Father, Like Son: William Albert and Carl Ortwin Sauer." In *Carl O. Sauer: A Tribute*, ed. M. S. Kenzer, 40–65. Oregon State Univ. Press, Corvallis.

Marsh, George Perkins. 2003 [1864]. *Man and Nature*, ed. David Lowenthal. Univ. of Washington Press, Seattle.

Nemoianu, Virgil P. 1993. "Romanticism." In *The New Princeton Encyclopedia of Poetry and Poetics*, ed. Alex Preminger and T. V. F. Brogan, 1092–1097. Princeton Univ. Press, Princeton.

Quinton, A. 1995. "Romanticism, Philosophical." In *The Oxford Companion to Philosophy*, ed. Ted Honderich, 778. Oxford Univ. Press, Oxford.

Rossi, P. 2001. "Historicism." In *International Encyclopedia of the Social and Behavioral Sciences*, ed. Neil J. Smelser and Paul B. Balter, 5:6757–6762. Elsevier, Amsterdam.

Sauer, Carl O. Correspondence (cited in text). Carl O. Sauer Papers, Bancroft Library, University of California, Berkeley.

———. For references, see the Sauer bibliography at the end of this volume.

Schnädelbach, Herbert. 1984. *Philosophy in Germany, 1831–1933*. Cambridge Univ. Press, Cambridge.

Speth, William W. 1987. "Historicism: The Disciplinary World View of Carl O. Sauer." In *Carl O. Sauer: A Tribute*, ed. Martin S. Kenzer, 11–39. Oregon State Univ. Press, Corvallis.

Uschmann, Georg. 1972. "Haeckel, Ernst Heinrich Phillipp August." In *Dictionary of Scientific Biography*, ed. Charles L. Gillispie, 6:6–11. Scribner's, New York.

Whitaker, J. Russell. 1940. "World View of Destruction and Conservation of Natural Resources." *Annals of the Association of American Geographers* 30:143–162.

Worster, Donald. 1973. "Introduction." In *American Environmentalism: The Formative Period, 1860–1915*, ed. D. Worster, 1–10. John Wiley, New York.

———. 1994 [1977]. *Nature's Economy: A History of Ecological Ideas*, 2nd ed. Cambridge Univ. Press, Cambridge, U.K.

22

Soil Conservation (1936)
Carl O. Sauer

The wastage of soil by man has won attention tardily though it is perhaps to date the most destructive form of exploitation of which man is guilty. European geographers and other social scientists have given little heed to this occurrence, and the German term *Raubbau* is not necessarily concerned with the destruction of the soil profile at all. The science of soil erosion is principally American; and though it is getting public attention only lately from us, attentive observers have been concerned about it for a long time. [Warren] Thornthwaite's division in the Soil Conservation Service is expecting soon to publish a study that will utilize largely the observations of Thomas Jefferson and his contemporaries. [Eugene] Hilgard and [W. J.] McGee were concerned with this problem. Hugh Bennett in a modest way carried on field and experimental erosion studies for many years before the present national administration launched the Soil Erosion Service, now recognized and enlarged as the Soil Conservation Service. Bennett chiefly has made the American people "erosion conscious" and has developed a national bureau which is attacking the problem on all fronts.

It may not be amiss to point out the connection between soil erosion and commercial exploitation of the land. Soil erosion is far less a problem of the old areas of the world than of the new ones. It is an abnormal feature of man's expansion over the earth, based primarily on specialized economies directed toward maximum cash returns. Where men have settled with a primary regard for permanent living there has rarely been such trouble. Neither agriculture nor grazing *per se* leads to soil erosion. It is principally where men have considered land in terms of quick and large monetary returns that this ill has arisen. Land as a speculation and soil erosion are closely related. Hence cash crop systems and year to year tenant contracts play an important role in the incidence of soil erosion. All of this means that soil erosion cannot be studied alone as a physical condition—though the physical conditions thereof must be studied—but that it must be regarded as an economic maladjustment. Good farming does not lead to soil erosion, bad farming does. Perhaps the anthropologists would speak of a pathologic acculturation.

Hence, any scientific study of the processes and expressions of soil erosion must regard both the physical and cultural factors and forms. The conditioning may lie in nature, the cause is man operating unnaturally, and in the long run unsocially and uneconomically.

This summer [1936] I had occasion to participate in the work of the Soil Conservation Service in the Piedmont, where the pathogenic role of man is most marked. This is one of the great endemic hearths of soil erosion, the disease dating in the north well back into colonial times. The summer was spent as a field seminar in which various members of the Division of Climatic and Physiographic Research participated. The principal base was South Carolina, but the party ranged from Virginia to Alabama. Naturally the work fell into the two parts, erosion morphology (physical) and erosion history (cultural). The individual students are specialized, but their observations and field association is interlocking. The erosion historian must know his way about in soil profiles and erosion forms, and the erosion morphologist must be aware of the history of misuse that has loosed the agencies of destruction which he is studying. The observations and recommendations for study made this summer and during my two previous periods as consultant in the field are being issued in mimeographed form by Thornthwaite's office and indicate somewhat the manner in which the problem of soil erosion may be broken into manageable themes by a geographic approach. Forms, rate, causes, and effects of soil destruction constitute a melancholy major theme in the United States for the geographer who is concerned with seeing "man's relation to the earth." I know of no theme that better illustrates the unity of physical and cultural geography, the necessity of competence and curiosity in both, and the impossibility of doing economic geography without historical geography. If the major end of geography is to find the realization or failure of symbiosis of man and nature, the conservative use of natural resources is at the heart of all human geography and the use of the soil is one of the major topics with which we must deal.

NOTE

Reprinted from *The Geographical Error*, September 11, 1936, 2–3. Department of Geography, University of California, Berkeley.

23

Destructive Exploitation in Modern Colonial Expansion (1938)

Carl O. Sauer

If we take as [the] basic theme of anthropogeography the dynamic relation between culture and habitat, we may organize colonial geography precisely in these terms. The habitat must be appraised in terms of the colonial culture that intrudes itself and of course must be reappraised with every important change of the structure and function of that culture. But also, the colonial culture must be evaluated by the use it makes of the land which it occupies. We may not shirk the question: has the process of colonization been beneficent or malignant in terms of a long-range view of culture history? More specifically, does the manner of colonial development represent good or bad stewardship of the land?

Only in part has the development of modern civilization been based upon more intensive use and more sustained yield of natural resources. We have accustomed ourselves to think of ever expanding productive capacity, of ever fresh spaces of the world to be filled with people, of ever new discoveries of kinds and sources of raw materials, of continuous technical progress operating indefinitely to solve problems of supply. We have lived so long in what we have regarded as an expanding world, that we reject in our contemporary theories of economics and of population the realities which contradict such views.

Yet our modern expansion has been effected in large measure at the cost of an actual and permanent impoverishment of the world. The development of our present civilization has depended in considerable part on the actual consumption of its capital, the natural resources of the world. Economics unfortunately has become restricted increasingly to money economics, instead of embracing the study of *Wirtschaften*, and largely has missed this ominous fact. Destructive exploitation has contributed so largely to the growth of "wealth" [in] the modern world, that it is accepted commonly as a normal process, excused and even approved as a "stage" in economic "development," which is supposed to give way in due time to balanced use and a permanently higher level of production.

In so many cases, however, has the process of European expansion been achieved by impoverishment of the lands colonized that we must consider such as the rule rather than the exception. The thesis may well be set up for trial that the Industrial Revolution itself and the tremendous growth in population and wealth of [the] eighteenth and nineteenth centuries are based on a gutting of colonial lands. For centuries colonial wealth has poured into the entrepreneurial lands about the North Atlantic. This wealth has been only in part normal yield or interest and in part loot or capital dissipation in the larger sense. There have been few to ask how this wealth was got. Spain has been charged often enough with having plundered its possessions in the New World. Yet after the first anarchic years the colonial policy of Spain was far more concerned with the conservation of its possessions than is true of most colonial lands.

De facto the New World is that part of the world which has been recast in its economy to serve as the supply area for the industrial lands of the North Atlantic. In this sense it includes not only the Western Hemisphere but also Australia and New Zealand and at least the south of Africa. The process of Europeanization of these lands has meant generally: (1) The extinction, hybridization, or subordination of native stocks and cultures. In spite of an occasional excess of anger, [Georg] Friederici's indignant volumes on this subject are a much needed corrective to our romantic self-approval as to the process of European colonization. (2) Normally, the permanent productive capacity of the land has been diminished. It is a sardonic quality of modern economic geography that the most severe and widespread destruction of resources is associated with these new countries, not with the Old World.

Hence the more lately settled and often more thinly peopled lands characteristically have today limited or no opportunity for population increase, if living standards are to rise or even be maintained. In the United States we are familiar with areas which after one or two generations of occupance have become areas of emigration and of failing prosperity. One of the worthiest attempts of the present federal administration of the United States is the resettlement of people from areas of growing distress into regions of opportunity. There is no difficulty in locating the distressed regions, but there is grave difficulty in finding even small bodies of land suitable for colonization with modest expectations of comfortable living. In this country distressed areas are as characteristic of the youngest states as they are of the older ones, or even more so. California, Oregon, and Washington are receiving a large and continuing drift of homeless folk from farther east, many of whom come from areas of failing resources. An internal migration has been under way for years in the eastern states, by which workers in the northern industrial centers have been recruited from the rural South. This whole movement is strongly conditioned by declining productivity of the areas in which the emigration originates. The phenomenon of internal migration which has reached such large proportions in the United States is becoming apparent in other lands of white colonization of similar or lesser age. One by one these "new" countries have proceeded to raise bars against immigration from overseas. Is not this change in attitude toward immigration itself a realization that there have developed within these young countries population pressures, though they may be lands of low population density? Informed Americans, and Australians, are well aware that their days of colonization are past and that the time of population pressure has begun.

The situation is not one simply of a momentary dilemma, when production has not yet made the shift from extensive to intensive economies. The Old World holds no proper parallels to the assault made on the basic natural resources of the new lands. It has been said repeatedly and with justice that the exploitation of the new countries has been characterized by cheap land and dear labor. Land has been a cheap commodity, from which the maximum return has been secured with the lowest input of labor. It has been most profitable to use up the land and reinvest the profits in new land, there to repeat the process of exhaustive exploitation. The history of cotton growing and of lumbering are alike in this respect. Too largely men have not settled on land expecting to build their permanent homes there and to have their children's children enjoy the acres which they brought under cultivation. The first waves moved on to newer fields. The residuary legatees increasingly have had left to them the problem of living in economically devastated areas.

This phenomenon of the deliberate commercial exploitation of the land with deliberate disregard for the permanence of the communities is properly to be called *Raubbau*. It is unfortunate that the term has been applied to two entirely unrelated economies, one commercial, one primitive. The latter use is improper. In lands of hoe cultivation there exists a system of land cultivation which involves the deadening of forest growth for the planting of crops. Such a field is maintained only for a short number of years. When sprouts and seedling trees make it difficult to continue the growing of food crops, the field is allowed to revert to forest growth, and is replaced by another clearing. This is the Aztec *coamil* or *milpa*. The author has seen this system in many places where it has been practiced from time immemorial, but has never seen it result in destruction of the productivity of the land. It is simply a permanent primitive economy, in which the per capita area of land needed for a community is rather large and in which the forest growth is to be considered as a part of a long term rotation of fields. Mostly it is found in hilly lands and actually it is an excellent device for maintaining permanent productivity. Such lands are adequately protected by being in cultivated crops only ten to twenty percent of the time and by being allowed to revert to wild growth during the remainder.

The remainder of the discussion will be restricted to one form of destructive exploitation by commercial exploitation, namely the wastage of the soil by improvident use. All remarks pertain to lands of recent settlement. A distinction needs to made between soil exhaustion and soil loss. By the former is meant the removal of plant food in form of harvest, whether this be crops, wool, meat, or wood. Actually, such calculations have little importance except for humid lands of negligible relief. In the smooth parts of our Corn Belt for instance the abstraction of phosphorus and potassium by long continued cropping may become important. Generally however the rate at which soil is washed away is so much greater than the rate at which it is abstracted by crops that the latter is a very minor economic factor. In restricted areas the deterioration of soil fertility may also result from the development of adverse physicochemical conditions in the soil, as by the accumulation of alkali or the formation of an undesirable structure of the soil. But predominantly the problem of soil wastage is one of actual removal of soil, for which the term soil erosion has gained currency. Soil erosion is of course determined by the amount of truncation of the original soil profile.

We may well consider whether the theme of soil erosion should not be moved up to the first category of problems before the geographers of the world. It is very important for the future of mankind. It has critical significance for certain chapters of historical geography. The physical processes involved are poorly observed and generalized, and their study will undoubtedly shake somewhat the rather lethargic present position of geomorphology. Either the geomophologist or the anthropogeographer may apply his special discipline to the study of soil erosion, but, best of all, the subject is suited to a "hologeographic" approach in which the development of surface conditions of specific localization is examined as to interaction of identified physical and economic (i.e., *Wirtschaft*) processes. Out of such localized studies there should emerge shortly a comparative knowledge or general theory of soil erosion.[1]

The physical study of soil erosion has three main descriptive tasks: (1) The determination of the full original soil profile. It is a great pity that geomorphology has been so much concerned with surface form and so little with the soil on which the surface is developed. Given the parent material and also the climate, exposure, and vegetation cover as constant, it follows that there must be determinable a definite relationship between a "denudational" slope and its "residual" soil. In other words it should be possible to construct combined slope-soil profiles which will show for any point on a representative slope how much and what kind of soil and sub-soil should be found there. Walther Penck has directed the attention of geomorphology to slope studies. It is to be desired greatly that the science be enriched by a large series of observations on slope-soil relations, always with the proviso that the basic profiles must be secured from undisturbed slopes. Hence the search for type profiles with the native vegetation intact. Where that is impossible, the slopes may be substituted whose history indicates that a vegetative cover has existed throughout which is approximately equivalent to the original cover in its ability to cover the slope and to absorb precipitation. For any soil erosion studies we must acquire these datum profiles of the intact slopes.

(2) Then we are ready to determine the amount of truncation of [the] soil column that has taken place by human exploitation. Soil erosion has often escaped notice unless it has taken the form of gullying. Yet the major damage may be accomplished by film or sheet removal. Even in our soil surveys in years past many tracts were mapped as clay phases of certain soil types, when actually these lands were the exposed subsoils of sheet-stripped fields. Specifically we want to know how far down into the A or B horizon the surface has been lowered.

(3) We need to know the general drainage pattern of linear erosion in-

duced by man, and the specific form, in cross and long profiles, of the rills and gullies as compared to the natural drainage forms appropriate to such surfaces.

These descriptive materials provide the basis for analysis of physical process. Wind and water segregate out in terms of greater climatic limitation of the former and of its prevalent association with low slopes or even flat surfaces. We do not yet have enough observational data to say where wind will and will not cause erosion. We know very little of the extremes of weather that are most significant in resulting in disastrous wind or water erosion. We can only surmise at present that arid and semi-arid lands are especially vulnerable and that mesothermal climates with seasonal rain are more hazardous than the microthermal lands. We know little of the relative toughness of climatically determined soils. Only fragmentary data are at hand by which to judge the erosivity of soils in terms of their colloidal qualities. We know in general that limestone lands tend to be extra hazardous, but I know of no published study on soil erosion in limestone lands that has comparative value. Morphographic and morphologic literature on soil erosion alike are virtually nonexistent.

Approximately nothing is known of the time required for the development of A horizons. Here, it is possible that observations on dated archaeologic dwelling sites may give bits of evidence as to the span of regeneration for soils.

The anthropogeographic approach to soil erosion again is descriptive and analytic. First of all we need studies in the history of erosion. When did soil wastage become noticeable in a given location? How rapid was its development? What stage of destruction was reached? Has the process been terminated? How do yields of crops and changes in population reflect the waning fortunes of the land? These basic observations are not simply a matter of library and archive, but depend on field studies as well. The remembered experiences of the people on the land are to be collected. The land may itself give datable testimony, as in the case of our forested "old fields," in which the tree rings tell an important story of dates of abandonment of cultivation.

In particular we need records of the form of abuse, of the vicious cycles in which destructive economies get caught and of the attempts of the population to extricate itself from the situation it has brought on. Fundamentally the economy is always to blame. Erosive soils or extreme weather conditions may condition the rate and severity of the damage, but at bottom the agent of erosion is man acting shortsightedly or recklessly. The greatest damages have resulted from monoculture. The crops originating in hoe culture economies, and benign as there produced, may become veritable destroyers when shifted to row plow culture under monocultural practices. Tobacco, cotton, or corn have thus caused the ruin of millions of acres of American farm land and menace a great deal larger upland area in this country. More lately the damage, however, by extensive small grain growing and grazing in the drier parts of the country is mounting at [a] similar rate.

Cultural cause, physical process, rate, and stage of soil destruction are the elements of this geographic pathology which needs to be developed. Then only can we speak sensibly about population and productive possibilities, for then we shall be at grips with the whole grave problem of land rehabilitation which is far more than sound engineering or agronomy.

NOTES

Reprinted with permission from *Comptes Rendus du Congrès International de Géographie, Amsterdam, 1938*, Vol. 2, Sect. 3c, pp. 494–499. E. J. Brill, Leiden. Copyright 1938, International Geographical Union.

1. Although the thesis stated above is concerned with the lands of modern commercial colonization, a series of supplementary theses can readily be set up for examination in the Old World. They would include the following:

[a.] That the sedentary populations have generally maintained highly conservative economies with regard to their land (question of a man in symbiosis).

[b.] That an increase in population is possible by which, without the incentive of export, and under conditions of excessive scarcity of land as opposed to the New World situation of excessive abundance of land, man may engage in progressive land destruction (question of China).

[c.] That aggressive nomadic populations may become very destructive of the land. The vegetated character of New World deserts, in contrast to those of the Old World, and occurrence in the Old World of sand and rock surfaces in broad areas whose climate is not extremely dry, support the theory that the steppe and desert lands of the Old World may have suffered ancient and long continued exploitation at the hands of herdsmen to such an extent that they have become morphologically differentiated from the similar dry lands elsewhere in the world which did not have pastoral nomadism.

[d.] That the Mediterranean lands in the later days of Rome suffered a colonial exploitation similar to that which has happened overseas in the immediate past. An alternative thesis is that the deterioration of the Mediterranean lands came rather as the Roman state began to break up.

24

The Relation of Man to Nature in the Southwest (1945)

Carl O. Sauer

Dr. Robert A. Millikan (Chairman of the Board of Trustees of the Huntington Library): ... The main subject to be presented this morning is an illustration of the fact that we cannot have humanistic studies without mixing in the sciences. In fact our civilization differs from the ancient civilizations primarily in the development of the sciences, so that it is not too much to say that the key to the understanding of modern history is found in the sciences. These were well-nigh non-existent in ancient Greece and Rome.

This morning we have with us Professor Carl Sauer, who has been since 1923 at the University of California at Berkeley as Professor of Geography and who, like myself, had the privilege of working at the University of Chicago with both [Thomas] Chamberlin and [Harlan] Barrows. I now take pleasure in presenting Dr. Sauer as the main speaker.

Dr. Millikan, and members of this conference: Somewhat to my dismay I found that Dr. [Robert] Cleland had put down in a very conspicuous manner that I was to give an address. I shall simply attempt to tell you what is close to my heart and mind, and try to feel my way into the intersects of all of us in a field of study in which we are trying to do something.

The social sciences in particular have been very much agitated in recent years by regionalism and the desire to do regional jobs. Howard Odum and his associates in the South have started this fire and kept it going. We have seen in the programs the Army has imposed on us an attempt to make some meaning out of regions. All this raises a problem of the meaning of "region" to which there is no one answer. I speak as a geographer. So many geographers think they can study a "region" and tell you what it is about. At once we run into the difficulty that there are many categories of regions. We cannot cut the knot by saying "physical regions"; regions considered in terms of climate are different from regions of vegetation. We do not know the common denominator of the thing we call a physical region. A region based on a biotic association—e.g., man—may be quite a different sort of thing. Can you speak of a culture area, or is that a fiction? Much time has been lost in boundary discussion as to types of region and as to where one region ends and another begins. The only general answer is that the region is defined by the thing that is "enregioned," and, on the whole, the people who are most interested in regions are interested in biotic associations, whether it be the study of plants, or lower animals, or man.

We will assume that we here are interested in the area which used to be said to lie "beyond the Pecos"—looking at the West from the East. I shall spend no time in characterizing that area. I shall ask a few questions as to what conditions are present there that represent fields of inquiry with some coherence.

I think by and large, biotic regions have a rough way of being climatic regions. Climate ties more things together and sets up more contrasts with other parts of the world than anything else. A very simple expression can be used to define this Southwest area: all life forms in this part of the world survive and develop because of the fact that they find effective means of economizing in water—the "limited good," whether for the plant world or advanced society, throughout the Southwest. Water is the first, the general physical link.

The second point is that the distribution of water is extremely localized. And again I do not care whether we talk of a family of plants or a civilized society. It is the oasis versus the waste which is the significant antithesis throughout the country between the Pecos and the Pacific. That means we have an extremely powerful operating isolating factor in the presence or abundance of water as against its absence.

I am going to speak of a science perhaps not represented here this morning, the science of genetics. Genetics has done some work which I suspect is most significant for the social scientists and humanists to take on in the future. I am not referring now to the work that is most commonly associated with the findings of genetics but to population genetics. The figure outstanding in this field is Sewall Wright of Chicago. He has been thinking of evolution in terms that begin with the "lower" biota and have been carrying this tentatively into social evolution. A participant and associate is [Theodosius] Dobzhansky, who worked here at the California Institute of Technology for a while. His volume on *Genetics and the Origin of Species* is a "must" for any person concerned with hierarchies of organisms and possibly of society. The importance of isolation in evolution was turned up by Darwin; evolution gets frozen with excessive isolation. The remote oceanic islands are the places where quick evolution takes place, then plasticity goes out of the picture and they remain as they were. Evolution also gets stopped in the large populations or areas that are continually freely-communicating. That has been shown in the *Drosophlia* studies. After a time, with extreme isolation, there is no deviation. Too much isolation stops change, too free communication does also. Take the bisons. They came to the United States early in the Ice Age. Shortly, there were many races or species of them, each perhaps dominant in a particular section. Something happened to wipe them out, until finally only one kind survived. This surviving kind increased in numbers to perhaps fifty millions, clustered into huge herds. The probability of change was pretty well lost in these great herds; the aberrant form was eliminated, or would have been suppressed if it sprang up.

The Southwest is interesting in terms of having available the balanced mechanisms for evolution. It contains "islands," in considerable measure, but not completely isolated islands. This balance, which seems to be enormously important, is of groups that are small, mostly self-contained, but with the means of intercommunication. Such a group gets genes, or ideas, from the outside. It can get these things in from time to time, but the group is not subject to a continual, unlimited penetration of things which it can neither absorb nor get rid of, nor turn into something new. The spots of dominant isolation and partial communication are, I guess, the significant and ever-recurring loci of history. There are certain distinct loci in the world that seem to come into operation over and over again. The eastern end of the Mediterranean is famous for the same sort of thing. When you consider our basic elements of civilization, many of them trace back to that part of the world, and it may be the outstanding illustration of the sort of thing I am trying to present.

I now go to the next conclusion, where I may be in difficulty. If you are going to study man, you should study man in terms of his entire period of existence in the area in which you are interested. My guess is that you can begin the study of man in the Southwest twenty-five thousand years ago or whenever he first came; or you can begin fifty years ago, when American civilization began rolling.

We have a curious tendency to focus on contemporaneity. I have heard it said in all seriousness that the social scientist does not need to concern himself with things that began more than fifty years ago or are likely to happen more than fifty years in the future. There is an enormous arrogance as to our ability to know things in terms of contemporaneity. I think it is as ridiculous for a person to be a student of man in terms of what he may experience in his own lifetime, and not beyond, as it would be for a student of botany or zoology to talk about the contemporaneity of his science.

Not many things can be studied by us experimentally. I believe we have been going off in the wrong direction by assuming that the social sciences are quasi-experimental, that there are sequences that recur under given conditions. There are partial sequences, but I have always found it necessary to think of man as being in the field of natural history. There is an impoverishment of any study of man that says, "Beyond this small bit of time are other conditions and other people—we are not interested."

If this Southwest area is, as I have suggested, one of the great loci of history, it would be very interesting to know what recurrent things have happened to people in this part of the world, perhaps a suggestion of things that may be going to happen to us. This is purely an emotional reaction, but I cannot see where we have any right to say that we of today are important as those of the past were not important.

I am pleading for the natural-history view in terms of the Southwest. Do not throw archaeologists out of your door. To most people around the world the Southwest means its archaeological picture. The Southwest as a culture area means little to scholars elsewhere except in terms of archaeology. Southwestern archaeology is an important chapter of human experience, involving problems of use of environment which are still with us.

Is it of interest that possibly the oldest Americans known were inhabitants of the Southwest? There is no spot in America where antiquity is known to go back as far as in a certain valley in Arizona at Sulphur Springs, where the first Cochise culture turned up. Here we are dealing possibly with the oldest traces of human culture in the New World. I think it is barely possible that there may be some survivors of such primordial, primitive cultures that have lasted into modern times, say perhaps the Yuki of northwest California.

The ancient road by which man moved through the New World—regarding the New World as a peninsula hitched on to the northeast end of Asia and ending in Tierra del Fuego, through which group after group moved—turned southwestward from Colorado through the Southwest. When the earliest groups came, climates had not yet settled down to their modern character. Before the Southwest became a desert, but when the desert lay farther south, the road of human passage was a narrow strip, a funnel through which all men and their arts had to pass to get down to Central and South America. The route probably also had a branch west across what are now desert lands to California on the Pacific Coast. I should not exclude curiosity about these primordial folk, or any of their successors, who tried their skills upon this area or learned them here. I should like to see the study of civilization in the Southwest begin with its inhabitants at the dawn of human learning.

Or one may begin about the time of the birth of Christ, when agricultural economy began to appear in the Southwest. Some very significant things carry through from this age. At this time we get very major introductions of culture out of the south. Before, the stream of man and his skills had been drifting from north to south. Now a movement gets way from the south into the Southwest, which effects very many changes there, and in part does not get beyond the Southwest. Work has been done by Edgar Anderson, George Carter, and others on maize, showing a series of introductions out of the south into this area from South America, Guatemala, and Mexico, with alterations to quite different kinds of maize that happened in the Southwest. And presumably the spread of maize went from the Rio Grande to the east, where it becomes the modern field corn. The ancestral route of corn can now be traced with a good deal of assurance through the Southwest back into southern Mexico. So also with beans, squash, and cotton, though routes, times and perhaps agents are different. With regard to cotton, the Indians of the Southwest and north Mexico made an annual plant out of a perennial, woody bush. This is an achievement of no mean importance, to alter a shrub of the tropics to the form which gave us Hopi cotton. Many other items were similarly introduced and altered.

Another thing that I think is significant is the mystery of irrigation in the lowlands of Arizona. Is it not strange that we have showing up here, rather quickly and on a very large scale, an irrigation engineering that is competent? One must go south to Colombia and Peru to find anything like such formal irrigation, for there was none between. How did it come about that, at the geographic extremes of high culture in the New World, without intervening links, we have this formal skill expressing itself—so far as the record goes—not by halting beginnings but by confident laying out and digging of ditches, even of reservoirs, and that the ancient conquerors carried this to the point of having hundreds of thousands of acres under irrigation in the American Southwest? I think we should not be skeptical about letting the Indian into the history of culture of the Southwest.

The Indian and the Spanish story is one, curiously, in the negligible significance of mining in the local economy, in sharp contrast to conditions farther south. We have this antithesis of the Jesuit in the west and Franciscan in the east, both operating with problems of economy of water but operating in different directions. In spite of good research on the missions, there is still much to be done on them as effective centers of civilization. In the Franciscan areas maize continued to be the basic item of crop production. As in Franciscan Mexico, these areas emphasized sheep; as in Mexico they introduced the ass as the principal beast of burden. The Jesuits formed a different organization. Through the west, through Sonora, into Arizona, one could draw a map with considerable precision showing the limits of the wheat tortilla as against the maize tortilla, and the wheat tortillas would show the limits of the Jesuits' occupation. They did interesting things in wheat growing as an irrigated crop. They went in for fruit growing, they emphasized cattle rather than sheep, and they used horses rather than the mule. This antithesis runs beyond the period when the missions no longer represent the organization of man in relation to his natural opportunities and limitations, into the recolonization when the secular government takes over.

What kind of people formed the New Mexicans and the mestizos of Arizona? This has a great deal of importance. The Chimayo community, with its extraordinary weaving skill, or any other New Mexico or Arizona non-Pueblo communities—where did they come from? We observe new factors, such as the use of the chili in cooking, and a new form of irrigation brought into the Southwest—because the Pueblo people had only the crudest form of irrigation and the great irrigation works of ancient southern Arizona were not maintained by the historic Pima.

Look at the American period, that incredible time that begins less than a hundred years ago. We have here for the first time in the enormous span of human history the introduction of a culture to which the water economy becomes secondary. Mineral exploitation now comes dominantly into the picture. This introduces another general theme. Until this time man had lived in the Southwest (whatever time he had been there) in balance with nature, in what the biologists call a symbiotic relationship. Our modern civilization has been built on the destruction of symbiosis, and I want to question, with due apologies to the chemists and physicists, whether it can continue indefinitely. The magnificent "development" that has given us the modern Southwest of course got rid of most of the gold, pretty well got rid of the silver, is scraping the bottom of the bin on copper, has extremely limited expectation of life as to the last of the minerals, petroleum; and from Silver City to Los Angeles the entire picture of the wealth and volume that we have accumulated is unthinkable except in terms of this harvest of all geologic time in two ordinary lifetimes.

And, going further, we have introduced for the first time significantly the destruction of surface by overgrazing. If you will turn to the accounts of Father Escalante and the Pacific Railroad Surveys and to photographs of the surveys of [John Wesley] Powell, you will see how different uplands and valleys were before the American period. The San Simeon Wash and the San Pedro are now gulches cut into the floors of flat-bottomed valleys. They got started in the 1890s [1880s] when grazing was heavy and a series of dry years occurred. The entire drainage quality of the Gila River has changed since [William] Emory's trip down that river. The Navajo learned the game of raising too many sheep from the white man.

You are familiar with the results of overwatering and its alkali problems. We have scientists who can tell you the decade when lands in the Southwest will go out of use because they have passed the danger point of alkaline accumulation. The use of the water in the upper part of the Rio Grande reduced the amount of the discharge below, but set up silting, overpumping, and other disturbances which raised hob with communities in New Mexico. In California the Santa Clara Valley has at last undertaken a conservancy program after a melancholy history of pumping out its underground water.

At the University of California the geographers have the basement—all my life I have lived in basements—and above us are two floors filled with entomologists. They are having more trouble all the time; the farmers are also calling for more fertilizer experts all the time. Monoculture (which nature abhors), needing more and more artificial measures to keep retribution from overtaking it, is the story of California and Arizona agriculture. Our monoculture systems are making more and more problems. Fertilizer demands are increasing all over California and yields are still tending to go down in most parts of California.

Another thing we have done is overcapitalize and overassess lands in terms of commercial products. I am told that dreadful things are happening to Spanish communities in New Mexico. Benefits of reclamation works are assessed against them and these communities are being caught by what is pressing on the California farmers—more and more intensive commercial production, calling for greater financing and operating skills. The New Mexicans are perfectly good farmers in terms of their own corn and chili patches, but they do not know how to work this accelerating squirrel cage of more intensive production to meet mounting charges against the land. These are illustrations of violations of symbiosis. Yet, on the other hand, the Mormons, coming from New York and New England, by some strange wisdom founded their communities in an almost perfect symbiotic relationship.

This is the manner in which I look at the Southwest scene. The point I hope I have made is that a student of regional culture should say *nihil humanum mihi alienum* and that we shall not throw out as alien anything that will throw light on man's fortunes in the Southwest. The archaeologists are pretty nice people, anxious to learn and anxious to associate. They are not simply technicians of limited skill. Their desire is to see relations between culture and area, and they have a lot to offer.

On the side of social dynamics, you are dealing with ecologic factors which operate in terms of natural selection—not in the old Lamarckian sense, but by making certain things possible in a given situation and denying the possibility of life to them in another. Most of all I think the Southwest is interesting as a workshop for considering what happens in this isolating and recombining mechanism of population, not in the physical sense but in the cultural sense.

I think something that may be vital to progress is nonconformity. The institution cast to type is dead. The melting pot that has worked can be thrown away. Only where you have the permission for something or somebody divergent to arise and to survive do you have the chance for something else than a hundred and thirty million people who are under increasing pressure to hear the same thing, see the same thing, think the same thing. I do not think the great mass, freely intercommunicating in every intellectual and social process, has sufficient possibilities of cultural evolution in it. If biologic evolution teaches us anything it is that change takes place by escape from the pressure of the great, dominant population, by the forming of partly detached colonies that permit variants to survive and establish themselves. The Southwest has been and is today characteristically an island cluster, and [an] area of escape. I think perhaps everything from the artist colony at Taos to the many strange religious sects here in Southern California represent perfectly valid attempts on the part of the groups of men not to become cast into one indefinitely repeating pattern.

NOTE

From a conference held at the Huntington Library, San Marino, Calif., August 22, 1944. Reprinted with permission from *The Huntington Library Quarterly* 8:115–125. Copyright 1945, The Huntington Library, San Marino. Sauer's comments are followed by "Discussion," pp. 125–151, an interesting interaction between Sauer and an interdisciplinary group of social scientists and historians.

25

Grassland Climax, Fire, and Man (1950)

Carl O. Sauer

Ecology has instructed us that plant societies may strike such happy balance with their environment and between their members as to form a stable, indefinitely reproducing order, called a climax vegetation. The concept may have merit at an introductory level of study, when it is good to make the simplest designs of species association. Such simplification of vegetation type and geographic area may be misleading, however, like other regional classifications in which habitat and habit are unified. It becomes seriously misleading if a static view of [the] relation of organisms to environment is substituted for attention to continuing, developing changes in environment, in competition and accommodation between organisms, and in continuity and pace of organic evolution. Both in organic evolution and earth history change is pretty much continuous, and its tempo, direction, and divergences are the real object of inquiry.

As a part-time student of climate I have long been interested in climatic limits of vegetation and have become increasingly doubtful of broad climatic-ecologic generalization. In particular, I have misgivings about equating climate and climax vegetation, leaving aside for the moment the question of the reality of the climax. In the first place, so-called climatic regions admittedly are only an introductory device for inspection of moisture and temperature distributions. We make maps on which graduations and irregularities in the weather across the Earth are greatly conventionalized by showing them as a series of areas, delimited by convenient numerical values secured by instrument recordings. Little has been done with an areal climatology that is based on the dynamics of air masses.

The usual classifications of climates derive largely from the old premise that climate and vegetation are pretty much coincident. Actually, therefore, there has been a lot of fitting of isotherms and isohyets to generalized maps of dominant vegetation. The fit is empirical and may be coincidental. Where meteorologic data is wanting, vegetation lines often have been used to delimit climates. Little progress is made in this circular operation of identifying climate by vegetation, and vegetation by the most closely agreeing set of lines that may be drawn for temperature or precipitation. For the most part, critical climatic limits remain unknown as to physiologically limiting weather values for individual plant species. Nor have we much knowledge of the significance of frequency of extreme weather conditions in plant reproduction or survival. It need hardly surprise, therefore, that the cruder the climatic and plant data the more satisfactory the coincidence of the two. One begins, for instance, with the premise that grasslands have a climatic origin, sorts them into tropical and extratropical, semi-arid and humid, and high altitude grasslands, and selects the most closely agreeing climatologic values as their limits.

Also, this sort of manipulation disregards whatever lies behind the contemporary scene of weather or plants. A present plant complex is very much more than a mere reflection of short-term weather, or of plant physiology related to weather alone. Any assertion that under a given climate there will form a stable self-perpetuating plant complex is likely to perpetuate an assumption that arose before there was modern biology and earth science. Plant associations are contemporary expressions of historical events and processes, involving changes in environment and biota over a large span of geologic time. A real science of plant ecology must rest not only on physiology and genetics, but on historical plant and physical geography.

I should like, therefore, to dwell briefly on the last million years, one of the biologically most critical periods of Earth history. We should also regard the possibility that present time still falls within the overall term "Ice Age." It is by now fairly sure that there were within [the] Pleistocene four successive great glaciations, separated by intervals of deglaciation, perhaps longer than the times of glacial extension. Roughly twenty million square miles of land surface were buried beneath ice masses during each glacial stage, and mostly were uncovered during interglacial times. Sea levels fell as ice sheets grew, and rose as they were melted away; continental connections between Siberia and Alaska were made and broken repeatedly.

Far reaching, major climatic changes initiated each glaciation and deglaciation and involved middle and, to some extent, low altitudes, far removed from any ice mass. The Pleistocene was a succession of wide and deep withdrawals and readvances of plants and animals, involving tens of millions of square miles of land, areas far greater than those covered by ice. During ice advances pluvial climates spread across large parts of the deserts and semi-arid lands of the world; during ice retreat, we think that arid and semi-arid conditions resembled the present.

Mountain making, especially about the rim of the Pacific Ocean, was in major swing in the early Pleistocene, diminishing toward the present.

The over-all picture is one of maximal environmental disturbance, with a succession of climatic swings in opposite directions by which perhaps only the central tropics remained unaffected. This rarely equaled time of great and rapidly changing tensions for the major part of the organic world inevitably brought unusual opportunities for natural selection of variant organisms.

An important faunal change needs to be noted. Whenever there were land connections with Asia, dispersals of land animals took place between the Old World and New. Apparently mainly during the second (Kansan) glaciation the New World was colonized by major groups of mammals. At such times there trooped into the New World our great herbivores, the bison, elk, deer, sheep, and elephants, trailed by bears and wolves. The great herd animals became very numerous during the later Ice Age and can be credited with important ecologic shifts in vegetation, as by disseminating seeds along trails and about their bedding grounds, and by exerting selective pressure favoring plants tolerant of browsing, grazing, and tramping. Our later continental history witnessed a stocking of the land with herd animals, perhaps equivalent to present livestock populations.

The latest glaciation (Wisconsin) is thought to have ended about twenty-five thousand years ago, having lasted perhaps three times that span. During this time there were mountain glaciers as far south as the San Francisco Mountains of Arizona and pluvial conditions blanketed the now arid and semi-arid parts of the West and Southwest, with an extension into central Mexico. The dry lakes of our West and their conspicuous marginal terraces were then freshwater bodies, about which lived herbivorous animals requiring better than desert browse. These lakes in the main were not caused by glacial meltwaters from high mountains, but represent the result of local precipitation. At this glacial time it would appear that aridity has been reduced to minor, detached, low-lying rain shadow basins. The modern general pattern of atmospheric circulation, and hence of modern climates, probably was established with the beginning of the last, great ice retreat. The change from pluvial to arid conditions may have well been rapid, and also the territorial expansion of the organisms capable of living in lands of drought. Shifts in comparative advantage for different species affected most of the United States, least of all the Southeast. Mesophytic types took over most of the deglaciated area, drought tolerant forms much of the West. In each case the race has been to the swift; plants having rapid means of dispersal and pioneering vigor took advantage of the frontier rush.

We are not sure that we are at present in a period of full climatic equilibrium. The general ice recession appears to have stopped about seven thousand years ago. From the states bordering on Canada we are getting, mainly from pollen studies, indication of a sort of climatic "optimum" shortly thereafter. At present, there appears to be in progress a further recession of glacial ice, recorded not only from northern and mountain glaciers, but by a slow, general rise of sea level taking place along the entire Atlantic Coast. At least we have some reason for being cautious about assuming static contemporary climates.

Great changes in climate have dominated the physical world for at least the last million years. It has been a long time for succeeding crises and opportunities for that part of the organic world situated outside of the central tropical areas. It has been, and probably continues to be, a time of marked advantage for mobile and labile forms, for those that are tolerant of environmental change, and those that can change. Evolutionary plasticity pays off at such times. Certainly it has been a time of opportunity for many grasses and herbs as against many trees and shrubs. Fast and heavy seeders, vigorous vegetative reproducers have been favored and thus annuals and plurannuals have had added colonizing advantage. Changing environments have favored survival of mutants. Enforced and invited migrations have brought together previously disjunct forms and multiplied opportunities for successful hybridization. In most respects the balance has been tipped in favor of the lowly and short-lived. The climatic history of the late geologic past supports strongly the view that [the] Pleistocene has been a major time in evolution and dispersal of many grassy and herbaceous forms.

The second great agent of disturbance has been man, an aggressive animal of perilous social habits, insufficiently appreciated as an ecologic force and as modifier of the course of evolution. Man had been in existence throughout the Pleistocene, ranged very widely very early, and during it became the dominant animal over many climates and, it seems, all continents. The earliest human records we have show the familiar use of fire, and they range from England to South Africa. In fact, the most ancient human sites have usually been discovered by finding hearths, broken or crazed stones, baked earth, or accumulation of charcoal. Possessing fire, man was free to move into cold climates, to keep the other predators at a distance from his camp, and to experiment with foods unpalatable in the raw. Fire and smoke became labor saving devices for overpowering, trapping, and driving game, from small rodents to great cattle and elephants. Fire aided the collecting of fallen fruits, nuts, and acorns, and much later was the principal way of preparing land for planting. The earlier human economies collectively may be called fire economies. Often, of course, fire escaped from control and roamed unrestrained until stopped by barrier or rain. The fire-setting activities of man perforce brought about deep and lasting modifications in what we call "natural vegetation," a term that may conceal long and steady pressure by human action on plant assemblages.

In the New World, human agency has been discounted because of dogmatic views that man entered America very late. A controversy has been going on for 50 years as to the age of man in the New World. Even the partisans of the late coming of man have been forced to admit an age of at least 15,000 years. It is my considered opinion, however, that the available evidence and the overall probability favor presence of man in the New World for the last third of glacial time. Work is proceeding in so many places and with new criteria so that it should not be long before we have unequivocal evidence instead of emotionally biased guesses. We may have had man in the New World for two or three hundred thousand years with plenty of time for increase in his numbers to the limits of the resources he was able to use. Whenever and wherever man came to live the prospects of a static ecology soon were reduced. With steadily continuing and increasing disturbances of plant associations by him, areas of plant tensions, already existing by reason of climatic shifts, were accentuated and enlarged.

Grasses have fared especially well in the late geologic past and have expanded as dominants in many parts of the world and in many climates. The explanation does not lie as a rule in the unsuitability of such tracts for woody growth. The normal history of vegetation is an accommodation into the plant society of increasing diversity, of forming and filling of ecologic niches by more and more diverse forms. In grasslands on the contrary we have a simplification of plant structure and scale toward minimizing differences in habit and size, simple and ephemeral stems, shallow and fibrous roots. This is a most curious plant sociology, from which the philogenetically most varied woody plants are mainly or even wholly excluded.

Where shallow and fibrous rooting is an advantage, floristic and morphologic impoverishment is understandable, as by adverse soil aeration and temperatures in marshes and moors (though even here woody perennials may be common though dwarfed). Another case is where a tight subsoil at inconsiderable depth stops root penetration. The clay pans of mature soils on flattish surfaces give numerous illustrations of edaphic grasslands. Long continued weathering on old terraces and other old surfaces that have developed senile soil profiles are notoriously disadvantageous for woody growth, especially in pedalfer soils which appear to have less penetrable basal pans than pedocals. These and other special cases of surface and soil formation do account for a number of edaphic types of grasslands, but are quite incompetent to explain the great savannas, steppes, and prairies of the world, commonly above the average in fertility and in invitation to plant diversity.

For these latter, the explanation has commonly been sought in climate, especially in ecology, which from its beginnings has sought climatic explanations for the presence of grasslands and thus arrived at the climatic climax principle. The more we learn of climatic data the less success is there in identifying climate with grassland. There are grasslands with as little as ten inches of rain a year and with as much as a hundred; with long dry seasons [and] with short dry seasons; [and] with high and low temperature ranges. In this country they occurred from the drier parts of the Great Plains to the markedly mesophytic Pennyroyal of Kentucky and [the] Black Belt of Alabama. Every climate that has been recognized in which there are grasslands has elsewhere dominance of forests, woodlands or brush, under the same weather conditions.

Grasslands are found chiefly (a) where there are dry seasons or occasional short periods of dry weather during which the ground cover dries out, and (b) where the land surface is smooth to rolling. In other words, grasslands are found in plains, subject to periods of dry weather. They may also extend into broken terrain adjacent to such plains. Their occurrence all around the world points to the one known factor that operates effectively across such surfaces—fire. Recurrent fires, sweeping across surfaces of low relief, are competent to suppress woody vegetation. Suppression of fire results in gradual recolonization by woody species in every grassland known to me. I know of no basis for a climatic grassland climax, but only of a fire grass "climax" for soils permitting deep rooting. For millenia, and tens of them, fires, for the most part set by man, have deformed the vegetation over the large plains of the world and their hill margins. The time of human disturbance is probably long enough, even in the New World, to allow also for evolution of new grassland species and for the development of soils of characteristic profile and structure that are known as "prairie soils," a secondary product rather than [the] cause of grassed surfaces.

A word of reserve may be introduced concerning the assumption that grassland soils are more fertile than forest soils. Grassland soils may be fertile because they lie on plains, are composed of favorable minerals, and are favored by surface as to organic accumulation. They do seem to develop physical-chemical qualities congenial to cultivated grasses, or cereals. By their history prior to cultivation, they may be a self-contained revolving fund of fertility if they are still sufficiently youthful. But, short and fibrous roots in the longer run are at a disadvantage with deeply penetrating roots in carrying up mineral salts from depth. We may over-rate the durability of grassland fertility in our short-term civilized view of time and neglect the importance of deeply feeding trees and shrubs in maintaining a productive Earth.

A drift of vegetation toward more grassland, including probably increased variation in grasses and herbs, has been going on for many thousands of years. This deformation has probably derived its driving force from fires, chiefly a cultural phenomenon. Ecologic assemblages, though not climax nor in equilibrium, have thus been established and maintained. The question now is whether civilized man can or should undertake to maintain grass cover and "forb-herb" ratios about as he found them; in other words, whether the desirable usufruct of such lands will justify or require his use of fire, as it did for his aboriginal predecessors.

In a world increasingly pressed by human needs the symbiosis of stock, grass, browse, and multistory plant cover is an increasingly important problem. Largely the world has lost the better plains for grazing and its shrunken range lands are in mountain, hill, and foothill areas, critical for storing and spreading water for lowland populations. These grazing lands were, I think, primevally strongly modified by fire. In that condition they served well to ease the runoff gently onto the valley lands. What place will be left in them for palatable grasses and herbs under full fire protection? What happens to runoff if grasses yield to shrubs? What is the change in fire hazard as full protection is attempted? What are the risks and returns under controlled burning? Fire, or its elimination, is still a main problem in the applied total ecology by which western foresters and graziers must attempt to work out a durable modus vivendi for man in a non-static environment. The layman can only hope for them that they may resolve this difficult and delicate situation by being permitted to take the longest possible view of beneficial activity by man.

NOTE

Reprinted with permission from the *Journal of Range Management* 3:16–21. Copyright 1950, Society for Range Management.

VII

HISTORICAL GEOGRAPHY

26

Introduction

W. George Lovell

> Columbus cannot be discharged of guilt in perverting an existing code of human rights. Whatever others might have done, his was the authority in these first critical years, his the example, and his therefore the major responsibility. He left an evil legacy that was to vex Spanish colonies for many years.
>
> Carl O. Sauer, "*Terra Firma: Orbis Novus*," 1962

The works featured here under the rubric of historical geography span the second half of Sauer's long and fruitful life. All seven selections reflect him at the height of his abilities as, in turn, (1) a trenchant analyst of population trends and prospects; (2) the innovative leader of a team of educators charged with preparing an elementary school textbook that was light years ahead of its time in terms of pedagogic content, ideological bent, and thematic thrust; (3) a geographer able to discern how a distinctive world region, Middle America, came to be; (4) a dogged researcher prepared to grapple with tricky sources in order to illuminate the circumstances under which the name "America" first appeared in print and thereafter anchored itself in the cartographic imagination; and (5) a scholar whose intellectual curiosity and desire to share knowledge saw him maintain, in late retirement, a momentum that resulted in the posthumous appearance of an ambitious synthesis.

Of the seven selections, "The Prospect for Redistribution of Population" (1937d, ch. 27 herein) is perhaps the most intriguing, for while it deals with established and enduring interests of Sauer's, the piece resonates with a palpable political edge, a trait not commonly associated with Sauer. In this case, however, the matter is of considerable portent, for Sauer is writing two years before the outbreak of World War II for the Council of Foreign Relations. He was commissioned to do so, especially pertaining to Europe, by fellow geographer Isaiah Bowman—Canadian-born but by then well-entrenched in

U.S. public life. Sauer does not mention Adolph Hitler by name but alludes to Nazi notions of *Lebensraum* when he observes: "Political concern about population outlets appears from time to time, but perhaps has never been so urgent as at present." Who can fail to conjure up the specter of an invaded Poland and an occupied France in what Sauer says next. "Whenever there is a population problem," he notes with characteristic economy of expression, "there is insufficient room for an expanding body, which seeks release in directions of low resistance and sufficient attraction" (7).

Despite his evident preoccupation with contemporary events in Europe, Sauer looks back and reckons, "We may very well have come to the end of what we have been pleased to call modern history, the expansion of western peoples and civilization over the thinly peopled or weakly peopled spaces of the earth" (8). For Sauer, the period between 1492 and 1918 "saw the greatest migrations of man since Neolithic time and was marked by his most rapid increase" (8). Population movement is linked inexorably to resource exploitation, from silver in Mexico and Peru in the sixteenth century to the oil strikes of the early twentieth century in other parts of the world. Exploitation of resources, however, carries with it enormous destructive costs, be it the "decimation of native populations . . . not only of the New World proper but of Australia and of Australasia" (18) or, closer to home, in Dust Bowl America, where "as yet uncalculated areas of land that have been under use at most for a few generations have been gutted or damaged in such a fashion that they can no longer maintain their economy or take care of their present population" (16–17). It all comes down, in Sauer's eyes, to "a process of skimming the cream from the earth" (16).

He ends this prescient rumination on the past, composed in free-flowing prose without recourse to one single footnote or reference, by returning to the troubled condition of Europe, for which he foresees a future as a peaceful and united entity half a century before it actually transpired:

> In a reasonable world, is not the average European likely to benefit most by elaboration of his skill in his own land? The inhabitant of Europe has at his immediate disposal not only the apparatus and stimulation of an advanced civilization, but he enjoys a unique advantage over the rest of the world. It is that he lives at the center of the world; he has the shortest distance to the ends of the earth and their way stations. This advantage is as permanent as the design of the continents and oceans. If the European can deal freely and amicably with the rest of the world, he is in the best position to do so in his own land.

If he sees the advantages to himself of the greatest possible natural relaxation of barriers, Europe is perhaps of all parts of the world best suited to support a larger population. (23–24)

From demography and geopolitics we move, in a short piece entitled "About Nature and Indians" (1939:8–9, ch. 28 herein) to a glimpse of Sauer very few know about. In the prime of his forties, over a six-year period when professional demands were heavy upon him from every constituency of academic life, Sauer responded to another overture from Isaiah Bowman. Writing to him in early 1934 as director of the American Geographical Society, Bowman approached Sauer to solicit his leadership in coordinating the production of "A First Book in Geography," which would be geared to the instruction of grade-four school children. The result of the endeavor is *Man in Nature: America before the Days of the White Men* (1939), a forgotten classic that engages its subject matter with pedagogic flair, conceptual clarity, and the organizational skills of a seasoned communicator. The book's focus is hardly what most elementary school teachers (certainly in the 1930s, but most likely even today) would expect an introductory text in geography to constitute. Sauer pulls no punches and refuses to spare his young audience the moral indignation he feels when telling them about the Native American past:

> Our people, in settling America, have . . . changed nature a great deal. We have let . . . soil wash away on hills that we have farmed. In many places we have made rivers muddy that once were clear. We have cut down forests and plowed up grasslands. We have killed off many animals, like the buffalo. We have built towns and roads and have done many things to change the country. Some of these changes are good. Some of them may be bad for us. . . . Before the white men came all the land belonged to the Indians. This book is about Indian days. The Red Man . . . was much more part of nature than we are. By learning how and where the Indians lived, we shall learn what kind of country the white man found. We shall then know better what he has done with it. (8)

Though the collaboration was not without its problems and setbacks (see Lovell, 2003, for fuller discussion), Sauer saw the project through to completion, leading by example. "We have gone on a new trail in this book," he wrote late in 1939 to C. F. Board of Charles Scribner's and Sons, the publishers of *Man in Nature*. "It is a trail that readers of all ages, not just nine-year olds, will find a rewarding one to follow."

The bold evocations of "Middle America as Cultural Historical Location" (1959a, ch. 29 herein) feature Sauer at his synthesizing best, delineating the emergence of Mexico, Central America, and the Island Caribbean as a hearth of human growth and development. It all began, Sauer speculates, in ways that recent radiocarbon dating now confirm, much earlier than we thought. A "first peopling of the New World before the beginning of the last (Fourth or Wisconsin) glaciation is no longer a fantastic notion" (196). In terms of settlement flow and direction, Sauer envisions Mexico as "the cone of the funnel, and Central America as the narrowing tube through which poured all but the later migrants that peopled South America." He favors "Yucatán as point of departure" for initial occupation of the Antilles, arguing that "Western Cuba lies northeast of Yucatán, that is, in a favorable position for men at sea to drift across at an easy angle" (195, 197, 198). Geography shaped the destiny of the region as "corridor and crossroads," even though "the peoples living about the Caribbean were more separated than connected by the Carib Sea" (199, 200). Commonly held "elements of culture" in the Antilles and the Central American littoral "came almost wholly from the northern mainland of South America" (199). Sauer asserts, in a memorable turn of phrase, that the Antilles by the time of European penetration "were quite simply colonial South America" (199). He urges that we view Middle America, indeed the New World as a whole, as a dynamic *Gestalt*:

> The cultural content of an area is an accretion and synthesis by different and non-recurrent historical events and processes of people, skills, and institutions that are changing assemblages in accommodation and interdependence. Few human groups have lived in isolation, excluding persons and ideas from outside; the more they have done so the less they have progressed. Isolation after a while stifles innovation; this is perhaps the major lesson of the history of mankind and also of natural history. An advancing culture accepts new culture elements without being overwhelmed by them; it adapts as it adopts and thus change leads to invention. (200)

We stay in Middle America for the fine-grained discussions of "*Terra Firma: Orbis Novus*" (1962e, ch. 30 herein), a precursor to arguments that Sauer would propound at length four years later in *The Early Spanish Main* (1966a). The imaginary geography of Columbus—"he was strangely untouched by the attitude of inquiry that invigorated his time"—receives much of Sauer's attention, and his critique of the Admiral of the Ocean Sea is categorical. His trust was not in observation but in old authority, Sauer writes, pointing out

Columbus's fixation on the reports of Marco Polo. "His was the medieval mind seeking confirmation in writings of the past" (258). Sauer's conclusion is unequivocal: "Did ever any one discover so much and see so little? The original bold and genial venture became the *idée fixe*, bringing confusion to his mind and suppression for others, ignoring the facts of newly known land and life, and ultimately helping to destroy him. He became unable to observe or to think rationally, to the point of being disordered as to mind, temper, and conduct. After the original voyage of discovery he never again rose to greatness or to face realities and make sensible use of them" (263). The power of conviction proved too much for the discipline of observation.

In sharp contrast, Peter Martyr and Amérigo Vespucci are treated with the respect that Sauer demonstrates they deserve, particularly the latter: "A proper name was needed for the great Southern land as its continental proportions and position became known. Influenced, it is true, by the 'four navigations' ascribed to Vespucci, the authors of the *Cosmographiae Introductio*, composed at St. Dié in 1507, invented the name America. The legendary Atlantic isles of Antilia and Brazil had already been applied to parts of the New World. It could hardly be called after Columbus, who had made every effort to deny its existence. To adapt the assonant name of Amérigo was an innocent and rather appropriate conceit" (270).

We close with three brief excerpts from *Seventeenth Century North America* (1980), the manuscript that Sauer was working on at the time of his death in 1975. More a draft than a refined and fully referenced final product, this book has much to commend it, not least the autobiographical "Chart of My Course," with which the volume begins. Sauer recalls his six-decade trajectory as a student of land and life, from his doctoral dissertation at Chicago under R. D Salisbury in 1915, published as *The Geography of the Ozark Highlands of Missouri* (1920b), to his reconstruction of *Sixteenth Century North America* (1971b) during his final Berkeley years. He reminds us of some of his more controversial claims, which include "Vinland as having been in southern New England, the climate as at present," and that "Irish monks settled on shores of the Gulf of St. Lawrence in the tenth century or earlier" (1980:11, ch. 31 herein). Engagement in his last project, he reveals, "has drawn me back to the experiences of my early years, going back to 1910, when I was beginning to learn geographical observation in the Illinois Valley" (11).

Sauer conceived of *Seventeenth Century North America* "as an introduction to the condition of land, nature, and Indian life as seen and influenced by French and Spanish participants" (1980:12, ch. 31 herein). Its last two chapters distill the outcomes of confrontations between natives and newcomers from New Spain in the south to New France in the north. In "Decline of Indian Population," Sauer reviews Indian numbers and demise in Mesoamerica, the Pueblo towns, the East Coast, and the St. Lawrence River Valley. Everywhere Sauer looks he sees unprecedented disaster in the form of "terrible epidemics," "mortality," "great sickness," and "nations destroyed" (1980:248, ch. 32 herein). The vignettes he leaves us with are of entire regions laid to waste, their inhabitants diminished if not extinct, the achievements of the past no guarantee for survival in the future. In the concluding chapter, "The End of the Century," Sauer's final published words, the situations of European New Spain and New France at around 1700 are briefly reviewed. Native peoples survived, but how they lived was a far cry from Verrazzano's image of a "Golden Age," Lahontan's portrayal of Indians as "exemplars of a good society," and Rousseau's depictions of the "Noble Savage" (1980:253, ch. 33 herein). The geographical consequences of empire, as Sauer's lifework on the theme tragically attests, turned out to be markedly different from what the first Europeans to reach and write about the New World could ever have imagined.

REFERENCES

Lovell, W. George. 2003. "'A First Book in Geography': Carl Sauer and the Creation of *Man in Nature.*" In *Culture, Land, and Legacy: Perspectives on Carl O. Sauer and Berkeley School Geography,* ed. Kent Mathewson and Martin S. Kenzer, 323–338. Geoscience and Man, vol. 37. Geoscience Publications, Department of Geography and Anthropology, Louisiana State University, Baton Rouge.

Sauer, Carl O., to C. F. Board, December 24, 1939. Sauer papers, Bancroft Library, Berkeley.

——. For references, see the Sauer bibliography at the end of this volume.

27

The Prospect for Redistribution of Population (1937)

Carl O. Sauer

POPULATION CHANGE MUST BE CONSIDERED HISTORICALLY

The population problems of the world can be defined only in so far as we have knowledge of growth or decline of human densities as distributed about the world in time. Population change has been very unevenly expressed spatially. There are areas that maintain, for a long time, a terrific rate of increase; there are others that, neglected for a time, have filled with a rush; there are still others that continue empty or nearly so to the present. In the first place, we need far more accurate descriptions of the distribution and "movement" of population over the world from period to period. Much more significant population maps could be made than any that we now have, and it is from them that the dynamics of population growth are to be read. The science of population, which has scarcely been begun, can answer questions regarding potential populations only by the most careful work in historical geography, in the phenomena of human distributions, and areal exploitations. As yet we lack mostly the evidence for projecting population trends areally. The following remarks, therefore, are inferences based on fragmentary knowledge and should be considered simply as a working thesis.

Political concern about population outlets appears from time to time, but perhaps has never been so urgent as at present. Whenever there is a population "problem" there is insufficient room for an expanding body, which seeks release in directions of low resistance and sufficient attraction. These areas of attraction then become the "frontiers" with reference to the expanding group. We who are of middle age remember the covered wagons that trailed westward across the Mississippi Valley. As we watched them go by, there was only slight interest in their specific destination. Westward lay the frontier, and somewhere on that frontier they would find a place. The "movers" themselves often did not know where they would establish their homes. We were looking at a drift that had been going on for generations. There had been new lands for so long that people scarcely thought of a time when there would be no more such land. Yet in the present century the American frontier has about vanished. There was one last great burst into the Canadian Northwest and the Great Plains of the United States and, almost suddenly, the "mover" faded out of picture. In our continent we have finally come to the end of the "new" land. If we have reached this condition elsewhere in the world, or are approaching it, we may very well have come to the end of what we have been pleased to call modern history, the expansion of western peoples and civilization over the thinly or weakly peopled spaces of the earth. For purposes of this discussion the thesis is set up that we have just ended one great period of history which may be dated as extending from 1492 to 1918, the latter date being used because of the war-time expansion of agricultural settlement into certain areas, such as our Great Plains. This period of four centuries saw the greatest migrations of man since Neolithic time and was marked by his most rapid increase. The expansion was due principally to the invasion of lands of lower and weakly resistant cultures by overflow from higher, aggressive cultures. The population growth in this time, however, depended also on migration of culture as well as on migration of people.

In order to regard the question of room for colonization, we may appropriately review the great population shifts of modern history.

Numerically we are most ill-informed until the nineteenth century. Since 1800 we estimate that the population of the world has more than doubled. We know that in the four centuries of modern history an enormously greater growth of population has taken place than in all the thousands of years of prior human existence. We know that approximately half of the growth of the human race has occurred in about one century. The period immediately behind us is therefore a time that is unique in all history.

In terms of production, what has happened is that all of the major physical resources of the world have been brought into use, resource by resource and area by area, at first slowly, then with a rush, until only minor possibilities remain of the discovery of great additional centers or forms of raw materials. It may be shown similarly that, whereas in prior history cultures were in large measure self-contained, in modern history cultural resources have been diffused until it is difficult to find areas that have not received and tried the cultural goods of the other parts of the world. In both respects, that blocking out of the world as to productive capacity, and hence as to population, has been realized as never before in history. To a degree previously unparalleled, population differences reflect the economic potentials of areas. Perhaps we can even say that, by and large, the people of the world today are where they belong under their present standards of living and cultural organizations. At any rate, a marked stabilization of the population of the world has come about, and marked shifts in relative density will become more exceptional and restricted as to areas involved. The physical world has been pretty well explored economically.

THE DECLINE IN MINERAL PROSPECTS

Modern history began with the great mining strikes of the New World. Perhaps this chapter of history terminates with the oil strikes of the early twentieth century. It is not at all likely that there will again be a discovery of minerals in any way comparable to that of the four centuries that lie between.

In little more than a century after their establishment on the mainland of America, the Spaniards had located and partially exploited every major district of precious metals in their territory that is known to us today, with the one exception of California. Latin America, as to gold, silver, lead, and copper mining, still depends mainly on districts found and opened up in colonial time. In particular, the silver production of the world comes principally from ore bodies discovered by Spaniards during the first half century of exploration (except for silver as a by-product). For precious metals, both free-milling and the metallic sulphides with copper or lead in association, the major possibilities for discovery apparently are no longer in the young mountains of the Pacific but on the margins of the ancient shields in Arctic and sub-Arctic latitudes. We do not, however, expect the Canadian Northwest or the [Siberian] Lena Valley to bring about population movements comparable to those of the mines of Potosi, Zacatecas, or California.

On the common metals there is no impending shortage, nor is there any great prospect of discovery of such ore bodies or processes as might revolutionize population distribution. Continued advance in the solution of metallurgic problems is not likely to cause any great relocation of industry or of population. In the case of iron ore, there are certain areas of large reserve that are now not producing, or are scarcely doing so. Their utilization in a world of reasonable freedom of commerce, however, will involve no important change in population. For iron and steel the tendency will continue to be to smelt the major part of the ore at or near the centers of power, though additional centers of hydro-electric power and "lower grade" fuels may emerge. The majority of the metallurgic areas of the world are very well established with regard to the most favorable combination of raw material assembly and distribution of products. It is likely that heavy industry will continue to be centered as at present, with probable additions in Brazil, China, and at ore bodies near the sea, with power reserves, such as the gulfs of St. Lawrence and Biscay.

The coal fields of the world are quite well reconnoitered, and the best ones are not only in use but are likely to be experiencing chronic population ills, as in the Appalachians. One may dream of coal-mining centers in Antarctica and of great chemical industries about the brown coal beds of Montana, but such visions are not likely to be realized for a long time to come. North China, now densely settled, holds the world's best prospects of increase in coal production.

Perhaps there still are deserts or tropical forests where petroleum booms will establish clusters of towns. The life of oil wells, however, is hardly such that they offer population outlets. Moreover, a pessimistic tone dominates the explorations for petroleum. It would require an extraordinary optimist to predict that we shall find new oil fields for the present generation that will make good the exhaustion of the ones now in production. Search for new oil reserves has narrowed to more and more limited districts in increasingly remote and difficult locations. At best a small number of workers will be able to produce such oil for export to refineries.

Though great mineral discoveries began our age of history, we now rely increasingly on technology rather than on discovery to satisfy our changing mineral needs. Hence, the mining boom is pretty well eliminated from major significance in future population shift.

THE SPREAD OF OLD WORLD GARDEN CROPS

From the Orient, orchard and garden crops, dominantly of hand tillage, early penetrated the Mediterranean and thence passed to the New World, there to give rise to a great development of plantations. Of these, sugar cane was earliest and chiefest. It principally was responsible for establishing an African population on the west side of the Atlantic. Sugarcane has continued to be one of the greatest agencies in distributing contract and peon labor through the margins of the tropics. A less equivocal gift from the Old World were the plantains, which have in modern time largely increased the food supply of tropical areas in the New World and in the islands of the Pacific. It is interesting that the banana, though it is not aggressive in reproducing itself, spread more rapidly than did the Europeans through the tropics of the New World and soon came to be considered a native plant.

Within the Orient a great trade has developed in the principal native starch, rice. The younger and less populous Indo-Chinese lands of monsoon rains and of ample flood plains of late years have poured a great stream of rice into the lands where population outstripped the local food supply. This inter-Oriental rice trade itself is the result in considerable measure of the large world demand developed for the more intensive garden cultures of the Orient, principally tea and silk. The ancient spice lands of the East have continued to supply the rest of the world with greater and greater quantities of agricultural luxuries, depending on larger use of skillful labor on small areas. It is here the commodity and not the crop that enjoys an increasing spread. Increase in these goods rests on their steadily increased absorption by the western world, and this has made possible intensification and extension of settlement within the Orient.

A belated migrant from the subtropical Old World is coffee, which has found its home primarily in the New World, where a fortunate combination of climate, terrain, and freedom from disease made possible the large-scale appropriation of new lands at the time when the demand for coffee was increasing most sharply in our urbanized populations.

In general the migration of the crops from the monsoon lands of the Old World is about finished. They are known and grown almost everywhere where they will succeed. They are not thus grown on a large scale except in limited areas of superior advantage. There is available a great deal more subtropical land on which they might be largely produced; but they are garden crops of high yield from small areas, requiring mostly intensive application of skillful labor, and the demand for them is limited. Some of them, like coffee and sugar cane, are now available in surpluses sufficiently large to cause distress.

Additional ricelands, well situated for export, could be used at present by the terrifically growing population of the rice-eating Orient. It is difficult to find such lands, however. The nearer valleys, such as those of Indo-China, are pretty well engaged in such cultivation. Rice does best in flood plains of monsoon climate. Where might one find similar large fertile lowlands with plenty of moisture in early summer and a marked dry season following? A few thousand hectares here and there, especially with the aid of irrigation, can undoubtedly be developed, but the great good ricelands are occupied. The rainy tropics may perhaps be invaded by rice, but it is significant that they have thus far resisted such invasion on any significant scale, though plantings have been made in them for many years.

THE DIFFUSION OF NEW WORLD CROPS

Another great phase of modern expansion of population is associated with the spread of New World crops. The role of the potato in revolutionizing North European agriculture and making possible a large increase of population is too well known to need elaboration. In limited measure the potato may still pioneer expansion of settlement into the colder and less desirable fringes of far northern lands. Any such expansion, however, will be far more important for subsistence farming than for commercial agriculture. If, for instance, an economic basis for the development of Alaska can be worked out, the potato may make possible the feeding of such a population. It is not likely, however, to cause people to settle Alaska.

Manioc has made possible an important expansion of population in Negro Africa. It seems to have revolutionized living in various savanna lands of that continent by introducing a high-yield crop into lands too dry to grow plantains. Farther east as well, manioc has increased population densities in lands on the dry borders of the tropical rain forests. Maize has contributed similarly to population increase in lands of summer moisture in southern Europe and the Orient and now appears to be well established in agricultural economy all over the Old World to its climatic limits.

The most important extension of New World crops has been in the New World itself. Here the settlement of the United States well into the nineteenth century is primarily a story of forest clearing and "hill" planting of native American crops, carried on by whites. In each case the plant was well perfected aboriginally, but the white man added to the technique of production and developed a large commercial outlet. The white man took the ritually used tobacco of the Indian and made it one of the first great crops of overseas commerce. He has kept it in large production [with] the aid of the fertilizer chemist. He discovered that maize was a far better feed grain than he had possessed before, and he built on it a world trade in lard and pork. He took the upland cotton of Mexican origin and by the invention of the gin gave the world its first cheap vegetable fiber. Each of these cultural adaptations resulted in a great forward surge of population. Together they account in the main for the expansion of American settlement to and beyond the western limit of the eastern woodlands of the United States. Because these processes of adaptation were developed in North America, they are still more significant in the distribution of population with us than in the Old World. An intensified livestock farming based on corn could undoubtedly be developed beyond present conditions in the Argentine, Uruguay, Paraguay, and parts of Brazil. Here lies one of the more hopeful outlooks for population growth, however, provided the world needs more pork products. Tobacco and cotton admittedly can be grown advantageously in many parts of the world in so far as physical conditions are concerned.

In the nature of their domestication the American crops demand a good deal of labor and give high per acre returns. They also lend themselves moderately well to mechanical cultivation. They are perhaps the most attractive group of crops as to expansive possibilities and are likely to be significant in any future expansion of the settled area of the world, as well as in the increase of population of already developed areas.

THE SPREAD OF NONTROPICAL OLD WORLD CROPS AND ANIMALS

Until the nineteenth century the transfer of European crops and animals to the New World meant little more than that the colonists maintained through them their material culture on a subsistence basis in their new homes. The expansion of population in the nineteenth century is principally a matter of the commercial occupation of the great grasslands of the world. These had lain largely neglected previously, except for a small trade in hides and tallow. The industrial revolution made their exploitation possible, and they in turn made possible the full realization of the industrial revolution. They gave us our famous frontier of cattle and sheep ranches, followed by the plow that broke the plains. Steamship, canal, and railroad are necessary to move the bulky products of the grassland farms and to carry in machinery and structural materials. Capital, in large amount, was necessary to provide the means of improvement.

The great grasslands of the world from the beginning of time to the nineteenth century had been left pretty well to the nomads and had thus remained the great relatively empty spots of the world in mid-latitudes. It required modern transport, heavy horses, horse husbandry, and the modern plow which could turn the heavy sod effectively, before they were ready for settlement. The great migrations of the last century everywhere have been into the grasslands, first into the humid prairies and park lands, then into the subhumid lands. The story begins first in south Russia. It reaches its greatest sweep in the United States. It is continued in the Argentine, in western Canada, in Australia, and in South Africa. It ends, and ends for all time, in the Canadian Northwest and the plains of Manchuria and Mongolia.

The critical crops are the small grains. First come the ancient field grains of Europe. As the dry margins of the grasslands were penetrated, drought-resistant varieties of these grains were collected from the drier fringes of Old World agriculture and were further improved in the new lands. Finally, the very desert margins of the Old World were combed to yield the least exacting of all grains, sorghums and their kin, for the last push of farming into the steppe. These plant resources have been well canvassed. We now know that, with all we have of dry country crops and all we know of dry farming, expansion of fields has been pushed into the steppes well beyond the limits of reasonable climatic risk. We now have an ebb of population to claim our attention in these dry margins.

EXTENSION BY IRRIGATION AND DRAINAGE ENGINEERING

The last half of the past century saw important extensions of the habitable area by agricultural engineering. Irrigation and drainage projects have been carried out at a rapid rate in the technologically more advanced lands. The cycle of development, usually, has been brief. The cheaply reclaimable lands soon were thus appropriated. Where expensive works are necessary, crops of high return must be produced. These are in general subtropical crops limited to warm lands of long growing season. It is not likely that additional development of such land is economically feasible on any large scale in a country such as the United States. There do remain, in rather low latitudes, various attractive possibilities of land reclamation [by] engineering. Perhaps the most notable such area in the world is Mesopotamia, but there are numerous lesser possibilities in Latin America.

EXTENSION BY LUMBERING

Lumbering as a form of land utilization leading to large and permanent settlement is hardly in the same category with the forms of extension of population that have been mentioned. It has come very late in economic history and still has modest reserves of land available, mostly in tropical forests of high logging costs.

SHRINKING SUBSISTENCE BASES

This sketch of the filling in of the world with new people who bring new activities must at once be revised and reduced. One of the striking features of this greatest population movement of all time is that it sadly overreached itself. In the first place, while men were prospecting land by trying what they could do with it, many of them settled on land that was unfit for the purposes of the settlement. A notorious illustration of this has been the invasion of the semiarid Great Plains of the United States by grain farmers during recent periods of high grain prices.

Again, the rush of people into new lands for their commercial exploitation had brought about an extent and degree of destructive exploitation of the land which we are just beginning to realize and which is extremely difficult to arrest. A considerable part of the growth of the nineteenth century and of our own time has been by a process of skimming the cream from the earth. As yet uncalculated areas of land, that have been under use at most for a few generations, have been gutted or damaged in such a fashion that they can no longer maintain their economy or take care of their present population. And this is an outstanding feature of the new lands of the earth, not of the old inhabited ones. In this sense large parts of North America, Australia, South Africa, and southern South America have become problem areas as to the possibility of maintenance of their population.

Let us regard briefly the shrinking of the subsistence base in the United States, which has given rise to the so-called conservation movement. A serious population problem exists in our southeastern states, which are the area in which destructive exploitation first became apparent. This area is primarily one of the forested uplands in which tobacco and cotton culture developed. The economy resulted in the development of a relatively dense rural population, but also in the process of long continued clean cultivation there resulted a gradual destruction of the topsoil. In very large measure present agriculture is being maintained only by the feeding of the subsoil, which now is at the surface, with chemical fertilizer and by increasingly costly mechanical devices to retard the washing away of the surface. Both crops, and they are the fundamental crops of our southern economy, have shown themselves to be very hazardous in cultivation of slopes, even of the slopes of low degree. Cotton, for instance, found two areas of superior attraction in very gently rolling prairie plains. The older of these was the so-called Black Belt, located largely in Alabama, a region of faint slopes and initially high fertility. Yet here, on slopes of as little as one degree, the fertile soil has been lost and almost the entire area has been forced out of cotton growing into a more extensive pasture and dairy economy. The great, black, waxy prairies of central Texas are now suffering rapid destruction under cotton growing. Prevalently the southern uplands, even the very smoothest, have been damaged, probably irreparably, by their agricultural economy, and the redevelopment of an adequate mode of living for the rural population of the South is one of the gravest problems confronting this country.

Even the Corn Belt, pride of American agriculture, has not remained unscathed. It first became widely apparent in northern Missouri that the prairies of Corn Belt farming were becoming frayed. We now have evidence that a fourth of the land of Iowa has suffered serious soil erosion. The dust storms originating on the western High Plains are conditioned by the plowing up of those plains for small-grain farming. Still farther west, the range lands have not only lost a large part of their stock-carrying capacity, but permanent damage by washing, resulting from overgrazing, is again generally prevalent on slopes. The agricultural surpluses of the United States have been produced, it would seem, by dissipation of the land capital. Soil conservation, resettlement, and reorientation of rural economy are major current problems of the United States.

The debit column of our land exploitation could be continued indefinitely. The states on the Great Lakes realized years ago that in dissipating their timber resources they had reduced a large part of their land to permanent poverty. Farms cannot follow forests in much of the timber country, and the cutting of trees is not simply a matter of waiting for their regrowth. The history of our lumbering again has often involved permanent damage to the land or at least damage that cannot be made good except by an excessively long period of re-established forest growth. These illustrations may suffice to call attention to the really serious theme of destructive exploitation in lately settled lands in various parts of the world, which brings them very early to a condition of overpopulation.

THE DECIMATION OF NATIVE POPULATIONS AS AN ELEMENT IN MODERN SETTLEMENT

The whole oversea expansion of European population has been based on the feeble opposition presented by aboriginal populations. This condition was true not only of the New World proper but of Australia and of Australasia. In highly polemic manner the fading of the native populations was first presented by Bartolomé de Las Casas in his diatribe on *The Destruction of the Indies*. Whereas the missionary bishop placed excessive emphasis on the cruelties of the Spaniards, the central fact that native populations melted away under the impact of Europeans is increasingly substantiated by our growing knowledge of the early records and the archaeology. At the end of the sixteenth century it was a commonplace with virtually all recorders to say that only a small fraction of the original population remained in Latin America.

The main point is this, apparently: The continental lands of the Old World had had contacts between their populations for so long that their contagious diseases were widely endemic through the Old World and that immunity, prophylaxis, or treatment, but especially the first, restricted the severity of the outbreaks. In America and the island masses of the Pacific Ocean, on the other hand, these contagious diseases were lacking. Perhaps we shall never have a good history of contagion; but this much is apparent, the newly found lands were swept, with great mortality, by disease after disease brought from the Old World. Relatively benign Old World diseases became New World plagues. Death by plague out-traveled the white man and emptied the land for him so that he had an abundance of unoccupied land for all his wants. In good measure the occupation of the New World had been a matter of refilling it with population.

No such event is likely ever to happen again. Should the white man, for instance, undertake to settle in number in Africa, he will not find that disease will clear the way for him. Like other manifestations of culture, the diseases of man have attained the major limits of their possible distribution. On the other hand, the entry of a higher, let us say European civilization, now means a general improvement of sanitation and a lowering of mortality. The introduction of white settlement tends to be reflected in major benefits of sanitation applied to the native population. In so far as the population has higher reproductive rates than the white colonists, there is therefore a tendency for its growth to outstrip the growth of white population and to exert pressure against the European colonist increasingly as time goes on.

REMAINING EMPTY LANDS

Where then, may we ask, are the land areas that still are thinly peopled? The answer has become almost distressingly easy. With two principal exceptions, these are areas of unsolved climatic problems. The first and perhaps most notable exception is to be found in South America. In South America, in the aggregate, large areas of land are still held out of full settlement by the existence of what we are in effect *latifundias*. The other principal exception is in Inner Asia, where pastoral peoples that will not practice farming have held lands capable of agricultural settlement. This situation, however, is breaking down very rapidly, not only in Manchuria where it is well destroyed, but by a similar invasion of farmers under way from the Chinese side in Mongolia and from the Russian side in Turkestan and adjacent Siberia.

On a population map of the world the greatest empty spot is the Arctic fringe. Here everywhere the frontier of civilized settlement is advancing very slowly. Shortness of growing season, hazard of summer frost, limited utility of possible crops, the "raw humus" cover of the soil, and the extreme slowness and difficulty with which trees may be removed are the principal difficulties. Even in Sweden and Finland, where the population is immediately at hand to press against the northern forests, rate of advance has been exceedingly slow.

Less thinly peopled than the polar margins are the tropical forests. Those of the New World are far emptier than those of the Old. It may be noted in this connection that our American tropical forests seem to have far fewer people than they did at the first coming of the white man and that perhaps the present inhabitants live at a lower cultural level. This is as yet one of the unexplained problems of the New World, but the recurrence of the condition is so widespread that generalization appears to be sound. The disparity between the population of the American and Old World forested tropics may well be due to the introduction of diseases from the Old World. At any rate, in some manner the establishment of contacts with the Old World destroyed the forest people of American tropical lowlands, and they have remained poorly occupied to the present. It would appear, therefore, that modern hygiene might have the opportunity of repeopling the New World tropical forests. Whether such lands come seriously into question for colonists from lands beyond the tropics is a fiercely debated question. There seems to be more sound than knowledge in the literature on the subject, yet satisfactory physiologic inquiry should be quite feasible.

Perhaps equally serious is the cultural problem of these forests. Their tillage is prevalently by hoe culture, and both in their crops and physical conditions there are serious difficulties in the way of a shift to the tillage methods of European farmers. The fact that these lands have been so resistant to reorganization of their economy, except in a few instances, indicates that they present unsolved problems—problems, however, which it must be admitted have not been sufficiently canvassed.

The thinly peopled dry lands are of speculative interest chiefly in so far as they have possibilities of irrigation. The expectations of dry farming as to conserving water by tillage have had only limited success, and the crops of minimum water requirement have been pretty well brought into use. The savanna lands of excessively brief but rather dependable rains may have possibilities if experiments to be mentioned below, in reducing the period necessary for germination, are successful.

LATENT TECHNOLOGIC POSSIBILITIES

In terms of the present productive skill of man, the world is pretty populated. Density of population tends more and more to approximate known productive capacity. For future relocations there is involved the solution of a series of technological problems. A few of these are:

(1) The finding of new crops offers minor opportunities. Civilized man has shown little success in the domestication of wild plants. Various reasons may be advanced, but the fact is admitted. We do not know whether he will find new plant resources that he can shape to his ends, in order to extend his range.

(2) The world's resources in domesticated crops have certainly been canvassed pretty thoroughly. There is little likelihood that any valuable domesticated species remains to be disseminated. This is somewhat less true of varieties adapted to special and extreme conditions. Even here, however, the returns of agricultural plant exploration for drought, cold, or alkali resistance are diminishing. The plant breeder is still extending somewhat the range of plant conquests in these directions, but the rate of progress is slow and "miracle" plants are likely to be of the order of frequency of miracles.

(3) In agricultural practice there may be sensational possibilities in the seed treatment known as vernalization. It has been discovered that the growing period of many seeds can be shortened by a simple treatment prior to their planting. In simplest terms, these seeds fall into two groups, those that are cool-starting and those that are warm-starting. If it is possible to shorten the growing period of the cool-starting seeds by several weeks or a month, it may be possible largely, for example, to increase the poleward range of the small grains. This in itself, however, will not mean an easy conquest of an additional great belt of territory along the polar fringe of agriculture. There still remains the stubborn fact that the removal of stumps and roots involves labor to such an extent as to maintain these lands economically submarginal. It is easier to dream of large change in connection with the warm starting crops. If here again the growing season may be shortened markedly, a good deal of the drier savanna country can be occupied by farmers. In any case, the experiments with vernalization are among the most interesting things to watch in connection with possible population shifts.

(4) Potentially a large role may be assigned to the extension of public hygiene. The problem of the tropics is largely a problem of health. If low latitudes can be made more habitable in terms of health, they can and will carry larger populations, but perhaps not people of mid-latitude origins.

It would appear, therefore, that our expectations of colonization must turn principally to low-latitude lands, and in far less degree to high-latitude and arid lands. Most expansion will depend on technologic advance, not on old-fashioned pioneering.

Technologic advance, however, may develop more benefits within the areas of advanced technology than in the outlying parts of the world. It is quite likely to do so, in so far as European migration is concerned. For instance, the problem of the tropics, if it is solved, will be solved step by step. In that case, will not an increase in the native stocks or an influx from nearby areas continue to take up whatever slack there is in the population, in so far as most of the increased production of goods in concerned?

In conclusion, it appears to this commentator that the population of the world has become sedentary permanently, that most of its inhabitants are where they belong. Some possibilities of small-scale colonization have been indicated, though these for the most part involve in some manner questions of land reclamation or of hygiene. A migration by low-standard groups could press on a large scale into lands of higher standards, but the latter have become highly sensitive to such pressure and have raised elaborate barriers. In a world that is less given to dislike and distrust of the foreigner than our momentary world, there is of course still a great deal of room for the dispersal of trained men from the old countries into the younger ones. These professional men and artisans, however, will most naturally find their places as individuals in going communities. That the young European doctor or mechanic should again find place overseas is desirable, but he and his offspring will not be colonists. They will pass into the communities that receive them.

In a reasonable world in which goods may flow freely, there may be expected to continue strong contrasts in population density which reflect persistent contrasts in economic advantage. The less strangulation of trade there is by politics, the less likely areas of dense population to be areas of overpopulation. At bottom, areas of great accumulation of population are areas of great and persistent economic advantage and cultural energy. In a reasonable world, is not the average European likely to benefit most by elaboration of his skill in his own land? The inhabitant of Europe has at his immediate disposal not only the apparatus and stimulation of an advanced civilization, but he enjoys a unique advantage over the rest of the world. It is that he lives at the center of the world; he has the shortest distance to the ends of the earth and their way stations. This advantage is as permanent as the design of the continents and oceans. If the European can deal freely and amicably with the rest of the world, he is in the best position to do so in his own land. If he sees the advantages to himself of the greatest possible national relaxation of barriers, Europe is perhaps of all parts of the world the best suited to support a larger population.

NOTE

Reprinted with permission from *Limits of Land Settlement: A Report on Present-Day Possibilities*, ed. Isaiah Bowman, 7–24. Copyright 1937, Council on Foreign Relations, New York.

28

About Nature and Indians (1939)

Carl O. Sauer

Man in Nature is the name of this book. We call it that because first of all we want to learn about nature. Everyone should know the nature of his own country. We shall learn about its mountains and plains, its lakes and rivers, its forests and deserts.

Our people, in settling America, have made many changes. Some things remain the same. The weather is still the same. Parts of the land and the water are much the same as they were.

But in numbers of ways our people have changed nature a great deal. We have let a great deal of soil wash away on hills that we have farmed. In many places we have made rivers muddy that once were clear. We have cut down forests and plowed up grasslands. We have killed off many animals, like the buffalo. We have built towns and roads and have done many things to change the country. Some of these changes are good. Some of them may be bad for us.

Later you will learn about how we have changed our part of the world. But first you should know how it was before our people came here.

INDIAN DAYS

Before the white men came all the land belonged to the Indians. This book is about Indian days. The Red Man lived in the land much as he found it. He was much more part of nature than we are. By learning how and where the Indians lived, we shall learn what kind of country the white man found. We shall then know better what he has done with it.

[Below] is a picture of land before the white men came. [Below] it is a picture of the same land in our day. The hills and the sky have stayed the same. Nearly everything else is changed.

The land when Indians lived on it.
[Ilustration by Antonio Sotomayor]

The same land with white men living on it.
[Illustration by Antonio Sotomayor]

We think it is a good thing to know about Indian days. We could not live like the Indians, even if we wished to do so. We have our own ways of living. But we did not need to cut down so many forests, and we did not need to destroy so much wild game. Often we have made the land poor and ugly. The land was natural and beautiful in Indian days. Perhaps we should make parts of it look once again as it did in Indian days.

Look again at the two pictures.... Do you think that it would be good to have more of our country as it was in Indian days? That is why we now have national parks and state parks and forest reserves. We like to see such places where the natural beauty of the land is left as it was. We still need to set aside more places where wild trees and flowers can grow, and where wild animals can live.

CIVILIZED AND PRIMITIVE PEOPLE

This book also tells about Indian ways of living, as well as how the land looked in the days before white men settled on it.

We know many more things to do than did the Indians. We have machines, factories, books, schools, automobiles, and airplanes. We say that we are civilized.

We say that the Indians were primitive for these reasons: They had few tools. They lived in nature in simple ways. They made use of the things that were near at hand. They made everything they needed. Their needs were few.

THE INDIANS WERE NOT ALL ALIKE

Some Indians were very primitive. They had very simple ways of living. Others had learned more ways of making use of nature. Little by little these found or invented new ways by which they could live better. They remembered these ways and taught them to their children. Some of the Indian people had learned many good and skillful ways of doing things.

We may think of civilization as a ladder up which men have climbed by learning more and more things. We are civilized because we have learned a great many good and useful things. Some of the Indian peoples were near the bottom of the ladder of civilization, but many others had climbed a good part of it.

This book begins with the simplest or most primitive peoples. Later it tells of other Indians who knew more and more ways of doing things. We shall learn, therefore, an important part of the story of how men grew more civilized by learning more skill in the use of nature.

INDIAN GIFTS TO US

Have you ever stopped to think that our own people learned a good many things from the Indians? In this book we shall tell now and then of things the Indians taught the white people. Some of our ways are really Indian ways, as you will see.

NOTE

Reprinted with permission from *Man and Nature: America before the Days of the White Man, A First Book in Geography*, pp. 8–9. Copyright 1975, Turtle Island Foundation, Berkeley. First edition, Scribner's, New York, 1939.

29

Middle America as a Culture Historical Location (1959)

Carl O. Sauer

To place Middle America as a scene of distinctive human occupations and experience we begin with its geographic position and configuration in the overall pattern of the inhabited earth. The significance of location needs to be appraised rather than the limits that may be assigned. I shall be concerned only most casually with delineation of area, saying, for the sake of convenience rather than by conviction, that Middle America is bounded by the Bahamas, the Windward Islands, the Gulf of Darién, and the Isthmus of Tehuantepec. This is intended to be only a rough physical identification; it is sufficient for the immediate geographic focus and it defers argument about culture areas and their boundaries. If I thereby dislocate the expected frame of reference, it is to avoid the horns of a contrived dilemma, i.e., Mesoamerican vs. Andean culture area, on which concern with autonomy of cultures would impale us.

POSITION AND CONFIGURATION OF LAND AND SEA

Physically, our area is the space between the two massive Western continents, including the island garland of the West Indies, the Caribbean Sea, and the mainland of Central America, the latter the only isthmus between the world oceans (except for that of Suez). Thus the uniqueness of its position is both intercontinental and interoceanic. Situated on the northern flank of the Tropics, Middle America has similar climatic regimes, whatever the local altitude, as to length of day, march of temperature, and pattern and season of rainfall. In these latitudes also the probability of major change in climate within human times may be discounted. To one coming from the north its plant world, especially in the temperate highlands, appears quite familiar as far south as Nicaragua and to a lesser degree through the Greater Antilles. South American elements of flora and fauna have penetrated northward in mass and diversity, farthest so through the lowlands. Biotically both mainland and islands have been a zone of meeting and mingling of stocks from north and south and centers of endemism only in minor degree.

CORRIDOR AND CROSSROADS

What then of the physical situation and condition as to human attraction, access, and passage? The mainland *Istmo* is an unobstructed passageway, the only way between north and south available to men until they learned to use boats. Mexico is the cone of the funnel, Central America the narrowing tube through which poured all but the later migrants that peopled South America. Perhaps nowhere in the world has there been as narrow, long, and significant a land passageway.

Radiocarbon dates are now assigning some tens of thousand years to human presence from Texas to California. They are also reducing greatly the time of the last phases of the Ice Age, so that a first peopling of the New World before the beginning of the last (Fourth or Wisconsin) glaciation is no longer a fantastic notion. It should not have taken a great time for men to drift south from Texas and California to and through the inviting Isthmian lands. Through most of human time in the New World people, ideas, and goods have been flowing into and across this mainland bridge. At first, and for quite a time, it was a one-way bridge leading men south.

The corridor of Central America is extraordinarily inviting in the diversity of its terrain, soils, climate, and life. Nature offered no deterrent to man to enter, nor barriers of relief or climate to halt him. Unless and until established peoples blocked the passage of newcomers, such migrants could live and move widely with ease and at their leisure.

The general level of the ocean has risen about 30 meters in the last ten thousand years and probably at least 100 meters in all since the last maximum of glaciation. The coastal swamps, estuaries, and widely flooding river valleys of today in the main are the result of recently risen sea levels. The Gulfs of Panama and Darién and the Atrato Valley [Colombia] indicate such progressive late submergence by eustatic rise of sea level, because of the deglaciation in high latitudes. The lowered sea levels during the migrations of early men resulted in stream erosion and good drainage of lowlands; the later rise of sea level brought flooding by both sea and lowland streams.

The rain forest between Panama and the Atrato was no serious obstacle. It is more formidable now than it was before the coming of the Europeans. When Columbus discovered the Portobello coast he was impressed by the well-populated and cultivated country, like a *huerta pintada*. The Spanish colony of Darién lived chiefly off the numerous, large, and advanced native villages to the west. Balboa's discovery of the Pacific was by an easy and well-supplied march. Three centuries later a surveying party of the United States Navy, of which the youthful [Matthew] F. Maury was a member, tried much the same route and nearly perished in a trackless and foodless wilderness. By then the Indian populations had disappeared; the present rain forest has repossessed lands of former abundant Indian habitation.

For passage by sea, currents and winds must be considered and also the range of visibility of land across water; the voyaging is thought to have been by drift and paddle. Access to the West Indies island bridge was easiest from the southern bridgehead of Trinidad and Tobago. Because of the strong westward set of current the island of Tobago was most eligible as a starting point for the Windward Islands, the high parts of Tobago and Grenada also being within sight of each other. Thence north to Cuba the passage from island to island presented no difficulty, except for Barbados, somewhat distantly apart, rather low, and lying strongly upwind and upcurrent from the islands to the west. Discovery and possession of the Bahamas, also on the windward side of the large islands, was most feasible from Haiti. These low islands could only be sighted from canoes at close range; that they were nearly all occupied bespeaks seafaring competence and venturesomeness. Jamaica was most easily reached from Haiti; at both ends of this sea passage mountains tower to two thousand meters, and brisk and steady winds and currents run straight west. (Note the ease of Columbus' discovery of Jamaica and the difficulties he had in getting back from there to Santo Domingo.) Haiti is thus suggested as the common source of colonization of Jamaica, Cuba, and the Bahamas.

Various passages from the Bahamas to Florida offer no great difficulty, but to come from the northern mainland to the islands against the powerful Gulf Stream was quite another matter. Along the Florida east coast the northward current may exceed a hundred kilometers a day; entry to the Bahamas by canoe from Florida seems quite unlikely, to Cuba from Key West only somewhat less so. The passage to the West Indies from North America, however, seems feasible for primitive man coming by way of Yucatán. Thus, speculations would run as follows: Folk moving down the Mexican Gulf coast had become accustomed to knowing and using tropical plants and animals. In Yucatán the access they had to seals and sea turtles along the shores offered an invitation to prowl the shallow seas. Along the east coast of Yucatán a steady drift sets northward, the great outflow of the Caribbean into the Gulf of Mexico. Western Cuba lies northeast of Yucatán, that is, in a favorable position for men at sea to drift across at an easy angle. Also, once such seafarers got well out into the Yucatán Channel the Cuban highlands would be in sight, an advantage lacking for any passage from Florida.

From the east the two equatorial currents, driven by the trade winds, sweep across the Atlantic to and through the West Indies. The equatorial sea drift, deflected by the shoulder of Brazil, running westerly, and picking up the waters of the Amazon and Orinoco, move at a rate of 50 to 80 kilometers a day as far as Trinidad. West of Trinidad the drift along the north coast of South America continues steadily but with a decreasing rate toward Panama. Within the Caribbean the trade winds become almost easterly, as does the drift; both wind and drift reach the mainland coast of Central America with some vigor. (Note that Columbus, sailing eastward on the north shore of Honduras, named Cape Gracias á Dios in relief at getting out of the windward belt.) We may not exclude from possible attention the narrow passage of the Atlantic and its extraordinarily favoring winds and currents out of Africa.

On the Pacific side a weak sea drift sets from Panama to [the Isthmus of] Tehuantepec during most of the year, with winds prevailing out of the south. Navigation along the shore here also may take account of regular daily alternation of land and sea breezes. Below Nicaragua the highland coast is broken by numerous sheltered bays accentuated by submergence. From the Gulf of Fonseca west-northwestward there are long beaches and attractive lagoons behind great sand bars for coastal voyagers.

EARLY MIGRATIONS

This geographical position and configuration, as I have tried to set them forth briefly, are at the center of the New World stage, whoever the actors and whatever the act and shift of scene. As yet we know almost nothing of the long early times, except the footprints of men that trailed bisons in the fossil volcanic ash of Nicaragua, and some early midden sites that are scarcely reconnoitered. The Isthmus lacks offside pockets of isolation as refuges for what Father [John] Cooper [1942] has called "internal margins." Survivals of primitives who passed through are not to be sought here but out in the back woods and far corners of South America; the ancestors of these early peoples must once have inhabited Central America. Nearly the whole of our knowledge of aboriginal Middle America is of later peoples who were agriculturists, in contrast to Mexico and most South American countries.

The exception is presented by the so-called Ciboney, to which name Cuban scholars properly object, as applied to all or any non-agricultural predecessors of the Arawak. Neither the scant historical knowledge nor the scattered archaeologic remains of the islands warrant the attribution of everything that preceded Arawak colonization to one culture or one people. By what route, what sort of craft, and at what times may they have come? Whenever they came, the sea passages, winds, and currents were about as they are now. Notable ability to survive at and move by sea seems an inescapable premise. I have suggested above the superior eligibility of Yucatán as point of departure. A South American entry has been discounted by the restriction of the known primitive and early sites to the northern, great islands. By whatever bridgehead they came, it is worthy of note that all three mainland approaches have long been occupied by agricultural peoples. It is unlikely that such primitive folk [Ciboney] could have forced their way through wide areas occupied by advanced and numerous peoples. Hence the hypothesis may be put forth that these primitive islanders came in before the higher cultures existed. From this it may follow that the early island colonists were already venturesome seafarers, who stagnated in the isolation of their final homes. At the beginning of European contacts they occupied the western part of Cuba and southwestern part of Haiti, both areas of highest suitability for Indian agriculture, an adverse comment on the case of independent invention and, subsequent to their contact with Arawaks, on the readiness to accept new and advantageous learning.

In earlier times, when migrations of peoples were continuing or recurrent by southward displacement and mingling, the sources of population and ways were out of the north. Language is carried by migrants coming as groups, and the historic languages of the Central American corridor should disclose old linkages to Mexico and the north, vestiges of ancient population drift. This is one of the most attractive problems of comparative linguistics and glottochronology, the clarifying of surmises about old linguistic filiations, such as Swadesh [1952, 1954] is investigating. Of immediate interest is the question of older common roots of the Chibchan group with linguistic stocks to the north; might its member groups in Nicaragua and Honduras be in part not merely a western fringe of Chibchan expansion out of South America, but also rear guard remnants from the north? In the cultural substrata of Central America there may also survive techniques and tools of fishing and hunting brought down from the north. On social organization and ceremonial ways I have no competence to make any suggestion other than to urge that such elements, sufficiently identified, be studied as to their overall distribution without regard to the limits of Middle America.

PROVENIENCE OUT OF THE SOUTH

For the later cultural introductions and modifications, a preponderance of South American influences is found, aside from southward drives of late aggressive folk (Pipil). South American crop plants, mainly vegetatively reproduced, dominate lower Central America, giving way in the north to annual seed plants. A large complex of kinds, techniques, and uses of narcotics, intoxicants, body paints, and unguents centers on South America. Metallurgy, I should think, was derived out of South America. Has it been noted that the prizing of gold had a continuous distribution in the New World, greatest in South America, including all of Middle America, and limited in the north by the Mexican highland high cultures, and that it was not merely related to cultural level or source of metal? What are the distributions and their significances of inclosed towns (by palisades or live hedges), ritual cannibalism, idols of stone or wood, hereditary caciques, and many other traits in and beyond our area?

Let us take one of the available lines of passage of men and ideas, the West Coast, as illustrating questions of distant migration and communication, probably from the south. I should like to begin in the north, with the Hohokam of the Arizona Desert. Their appearance by mass immigration, dated by Gladwin [1957] as of the early eighth century A.D., was without any preliminary local antecedents, and they remained without notable connection with neighboring cultures. They came with an elaborate and distinctive culture, at its best in the earliest period, when they colonized rapidly and fully the Middle Gila Basin. It was a nation of implicit social classes and political organization. No comparable ability and enterprise in irrigation engineering is known to the north of Peru and none at all in neighboring areas, except the historic irrigation of Opata and Pima, almost surely derivative from the vanished Hohokam. The lively animation of free hand pottery designs, copper bells, palettes, mosaic, plaques, and other traits also bespeak a distant origin. Could such migrants have come by land northward without leaving trace of their passage or could they have forced their way through already long and well-occupied lands? Or did they come by sea, continuing until they came, at the end of the sea, to a desert sparsely and primitively peopled, which they were able to transform by their irrigation arts and organization of labor?

Farther down the coast, Sinaloa has its southern (?) mysteries: The seemingly aristocratic tumulus of Guasave excavated by Ekholm [1942] has elaborate grave furnishings (including lacquer on gourd) strangely out of place in the known simple archaeology of northern Sinaloa. In the prehistoric towns of the Culiacán Valley are clusters of great urn burials and on the coast are mounds of slag where sea shells were used in smelting complex sulphide ores difficult to reduce to metal [McLeod, 1945]. The nearest source of such ores is well to the east in the mountains; the metallurgy was of an advanced technology. In the State of Colima vault burials have long been sought and plundered for their anthro- and zoomorphic pottery. These are chambers excavated laterally in rock from vertical shafts, the latter subsequently refilled. They are strongly reminiscent of Colombian *huacas*, as on the Quimbaya ridge crests [West, 1959]. Such are a few of the many items in archaeology and ethnology that need to be properly identified and examined as to the whole range of their occurrence. Is it our own agoraphobia and thalassophobia as students of New World culture that have so greatly restricted knowledge and consideration of far connections of culture, of mobility of peoples, of communications of ideas beyond the sheltering limits of so-called culture areas?

LIMITATION OF THE CONCEPT OF CULTURE AREAS AND STAGES

The peoples living about the Caribbean were more separated than connected by the Carib Sea. What is known of movements by sea is that they were peripheral and occasional. Whatever elements of culture the Antilles and Central America held in common came almost wholly from the northern mainland of South America. The islands were quite simply colonial South American. The islanders preserved with little change the culture they brought with them, living largely in isolation from other cultures. Such at least is the picture as it has come down to us. It might be less simple if we knew more. On the Island of Haiti north of Samaná Bay, the Spaniards found the Ciguayos, warlike and differing in customs and speech from the others. The southwestern Haitian province of Xaragua was more aristocratic than the rest and was said to have had intensive irrigation. The Jamaicans were praised for their agricultural skill; here also, elaborately sculptured stone seats are known in some number and seem to be like those from Colombia and Costa Rica. It may be, therefore, that the island Arawak included a number of distinct colonizations, not all wholly Arawakan.

For Central America a southeast-northwest division between Circum-Caribbean and Mesoamerican cultures is commonly recognized. (I do wish that a term other than "Mesoamerican" might have come into use for the Mexican-Guatemalan complex.) As a synoptic picture at the time of European contact it has value in introductory orientation. It is, however, only a rough delineation of status at one moment of history, synchronic, not diachronic, as our colleagues like to say in the fashion of the day.

Geography, which basically is concerned with position and areal extension, holds suspect all simplified and inclusive areal generalizations. Schemes of climatic, vegetation, and geomorphic regions, and even more so those of natural and cultural regions, may be useful elementary conveniences in helping to see major patterns of differentiation over the earth and thus as approaches to the processes of such differentiation or conjugation. As with landforms and vegetation, the cultural content of an area is an accretion and synthesis by different and non-recurrent historical events and processes of people, skills, and institutions that are changing assemblages in accommodation and interdependence. Few human groups have lived in isolation, excluding persons and ideas from outside; the more they have done so the less have they progressed. Isolation after a while stifles innovation; this is perhaps the major lesson of the history of mankind and also of natural history. An advancing culture accepts new culture elements without being overwhelmed by them; it adapts as it adopts and thus change leads to invention. The history and prehistory of the Old World are read throughout in terms of the communication of people and culture traits, of their blending, and modifications into new forms as they are farther removed in time and place from their origins. Why should the New World be different; why should there be construed here a congeries of autochthonous culture areas, each passing through independent parallel stages of development?

No part of the New World has been less isolated or self-contained than has Central America. I return to the initial theme that it has always been corridor and crossroads. It needs study as such with the most accurate identifications of its elements of culture and of the total range of distribution of each, however far this may extend into other areas and times and thus of whatever may be learned of their appearance and movements in actual time, and not by reference to inferred or imagined stages.

Since I am strongly impressed by the role of the Isthmus as a cultural passageway, I should assign it the role of culture hearth only in one instance—as the place of origin, at its northern borders, of New World seed agriculture. Much inquiry remains to be carried out on this problem and it may never be resolved, although I am hopeful that it will be. I like placing the origin of the maize-beans-squash complex in southern Mexico and Guatemala. I share the view that vegetative reproduction, that is, planting pieces of the desired plant in the ground, is earlier than seed agriculture. I am struck by the fact that seed agriculture all over the New World is still by planting, and not by sowing processes as in the Old World. It seems probable to me that agriculture by vegetative reproduction was brought into Central America with domesticated plants out of tropical South America. (Central America has added nothing new to the list of such domesticates.) As planting was carried northward into areas environmentally different as to the rainy season, highlands, and native flora, attractive herbaceous weeds appeared in the planted clearings which became the basis of the new seed crops.[1] This is my tentative surmise of the basis of "Mesoamerica."

As these questions of the geographic range of traits and assemblages and of the derivative modifications or variations into new forms and complexes are diffusionist in import, they ask for an examination of the evidence that geographic distributions may bear, wherever identifiable traits are known. Culture history, and this means culture dynamics, may not build Chinese walls.

NOTES

Reprinted from *Readings in Cultural Geography*, ed. Philip L. Wagner and Marvin W. Mikesell, 195–201. Copyright 1962, Univ. of Chicago Press, Chicago. Originally published in the *Actas del XXXIII Congreso Internacional de Americanistas*, San José, Costa Rica, 1958, 1:115–122. Lehmann, San José, 1959.

    1. For elaboration of these remarks, see Sauer (1950a, 1952a).

REFERENCES (ADDED BY WAGNER AND MIKESELL; REVISED BY THE EDITORS)

Cooper, John M. 1942. "Areal and Temporal Aspects of Aboriginal South American Culture." *Primitive Man* 15:1–38.

Ekholm, Gordon F. 1942. "Excavations at Guasave, Sinaloa, Mexico." *Anthropological Papers of the American Museum of Natural History* 38:23–139.

Gladwin, Harold. 1957. *History of the Ancient Southwest.* Bond Wheelright, Portland, Maine.

McLeod, B. H. 1945. "Examination of Copper Objects from Culiacán," In *Excavations at Culiacán, Sinoloa,* by Isabel Kelly, Appendix II, pp. 180–186. Ibero-Americana, No. 25. Univ. of California Press, Berkeley.

Swadesh, Morris. 1952. "Lexico-Statistic Dating of Prehistoric Ethnic Contacts." *Proceedings of the American Philosophical Society* 96:452–463.

———, et al. 1954. "Symposium: Time Depths of American Linguistic Groupings." *American Anthropologist* 56:361–377.

West, Robert C. 1959. "Ridge or 'Era' Agriculture in the Colombian Andes." *Actas del XXXIII Congreso Internacional de Americanistas.* San José, Costa Rica, 1958, 1:179–182. Lehmann, San José.

30

Terra Firma: Orbis Novus (1962)

Carl O. Sauer

WEST TO INDIA OR TO A NEW WORLD?

This essay is concerned only incidentally with the precise sea routes of the Atlantic explorers. For the most part these are known about as well as they are likely to be.[1] Rather I shall consider conflict and change of opinions as to where the newly found lands lay and what they were like and thereby the emergent questions of Spanish rights and prospects. The time runs from 1492 to 1507, years of gradually clearing geographical uncertainties, when strong and willful personalities operated overseas with little restraint and often without scruple, while authority in Spain had not yet determined how to administer the unanticipated accessions afar.

The Renaissance that developed in northern Italy turned the minds of men to observing and describing, by whatever symbols, the external world. Florence of the Medicis in particular became the seminal center of intelligence, discourse, and innovation in art and science. North Italian mariners and merchants were pioneers in geographic discovery and tutors of navigation to other nations. After Prince Henry, the Great Age of Discovery owed most to Italians, such as [Paolo] Toscanelli, Peter Martyr d'Anghiera, Cà da Mosto, Columbus, [Amérigo] Vespucci, [and] the Cabots.

When the time was ripe Columbus sailed his three ships across the western ocean to enduring fame. Modern history conventionally and conveniently is dated from this voyage. Agent or symbol, Columbus continues to engage our attention as a determinant of the later course of history. Except for his faith that he could reach the East by sailing west, as derived from or confirmed by Toscanelli, he was strangely untouched by the attitude of inquiry that invigorated his time. His trust was not in observation but in old authority; his was the medieval mind seeking confirmation in writings of the past.

Columbus set forth to find a new way to India. To the Europeans of the time India was about the equivalent of what we call Monsoon Asia, divided into two parts by the Ganges River. The India that lay beyond the Ganges included Cathay (China) and Cipango (Japan), with islands unnumbered sown through the Eastern Sea. Maps existed that gave fair approximations of the position of eastern lands, though placed in latitudes that were generally too low. Columbus was acquainted with the accounts of early travelers to the Far East, especially with that of Marco Polo. The latter, returning from Cathay by sea, touched at south Chinese and Indochinese ports before passing through the strait (Malacca) into the Indian Ocean. Thus he gave the first determination of the southeasternmost extent of the "Indian" mainland, thought by cartographers of the fifteenth century to extend somewhat south of the Equator.

The "Journal of the First Voyage" of Columbus is straight-forward and reasonably clear, the account of land, vegetation, and natives fuller and more objective than for any of his later voyages.[2] His first declared objective was to reach Cipango, underestimated as to latitude (entry of October 13). On October 21 he was ready to leave the Bahamas "for another very great island (Cuba) which I think must be Cipango . . . , in which they say that there are ships (*naos*) and many men skilled at sea. . . . Moreover I have determined to go to the mainland (*tierra firme*) and to the city of Quinsay and to give the letters of your Highnesses to the Great Khan." All this was straight out of Marco Polo, Columbus being unaware that the Mongol Khan had been replaced by the Ming Dynasty almost a hundred and fifty years before. Four subsequent entries in October identified Cuba with Cipango. On October 28, having arrived in Cuba, he understood that the ships of the Great Kahn came there and that the mainland was but ten days distant. On October 30 however he had a report that Cuba was the name of a city, that its king was at war with the Great Khan, and that he, Columbus, was actually on a great mainland that stretched far to the north. On Nov. 1 he had further assurance that he was on the mainland and that Sayton and Quinsay lay ahead, one about a hundred leagues from the other (again Marco Polo?).

In his so-called First Letter, which was composed near the end of the return voyage,[3] Cuba had ceased to be Cipango in his mind. He was still somewhat doubtful as to whether it was an island or part of the mainland. He would have thought it part of continental Cathay except for the assurance of the natives that it was an island, in which case he thought it must be greater than England and Scotland combined (actually half as large). He had, he wrote, followed its northern shores for 107 leagues and beyond his farthest point lay two more provinces, in one of which the people were born with tails! As he proceeded west he continued to find only small settlements without means of communicating with the natives, since these ran away. Hoping still to find cities or towns he kept on for many leagues. Finally seeing that there was no change (*innovación*), that the coast had taken on a northward trend whereas his wish lay in the opposite direction (*mi voluntad era contrario*), and that winter was about to set in, he decided to turn about and head south. Such was his resume of the Cuban part of the voyage of discovery.

To return now to the *Journal* entries. He had been showing samples of cinnamon, pepper, gold, and pearls to the Cuban natives and getting the answer that [a] great store of these was to be found to the south in another large island, to which he crossed early in December. Here, in Haiti, he noted on Dec. 11 that the people knew of Carib raids. He wrote *Caniba* for Carib inferring that they "were nothing else than the people of the Great Khan, who must live near by." On December 24 he wrote that "among other places which they name where gold is found they spoke of Cipango which they call Cibao." On January 6 there was information of another large island called Jamaica, abundant of gold in pieces greater than broad beans (*habas*), whereas in Haiti it was found in the size of grains of wheat. Also he was getting ornaments of base gold which were traded up from the south, as he thought from an island called Goanin.

The three months spent in the islands left Columbus somewhat uncertain as to their geography. At first Cipango was Cuba but later he placed it in Haiti, and Cuba therefore should be part of Cathay. What he saw in Cuba of course failed to fit Cathay. The people were simple woodland cultivators of root crops, living in round thatched houses clustered into small villages, using body paint rather than clothing and wearing a few ornaments of gold. At his farthest west he heard of still more primitive folk, born with tails. That this should have been sheer fancy when he thought himself within ten days of the great cities of Cathay seems less likely than that he had some vague report of the Guanahacabibe who inhabited westernmost Cuba, simple collectors displaced into the far end of the island by the advance of the Arawaks. To blend such an account with a medieval myth of people with tails was easy.

For stopping his westward exploration he offered the poor excuses that the coast was trending somewhat to the north and that winter was approaching. Cuba is without winter cold and the trend of coast was favorable for the direction which he thought Cathay should lie. Gold, pearls, and spices, all thought to be associated with tropical lands, became the object of his quest. He thought he had heard and seen enough to retrace his route until he could turn south for the crossing to Haiti. Thus he came to the most populated of the islands, having also a class society and a material culture advanced over that of Cuba. Here his easily stimulated imagination recognized in the name Cibao the Cipango of Marco Polo. Peter Martyr, back in Spain, questioning the returned seamen, made the correct identification of Cibao as "stony mountains."[4] The mountain streams of Haiti yielded gold in grains and nuggets to native workmen. Here also he [Columbus] became aware of a base gold which he referred to a southern source from an island he called *goanin*. Actually *guanin* was the native name for an alloy of gold, copper, and probably silver, originating mainly in Colombia and traded into the island. Finally he heard of the (non-existent) golden gravels of Jamaica.

Much of the information that Columbus reported was by a free reading of the travels of Macro Polo into an incongruous scene. He picked up a few natives in the Bahamas, those of Cuba ran away; the time available to learn verbal communication was very short. This wishful thinking that was to becloud more and more the discoveries of Columbus [thus] became apparent ... early.

Peter Martyr, most acute and objective of reporters, maintained from the outset a position of marked reserve as to the geography of Columbus and did not accept the latter's Oriental identifications as his own. Four letters of 1493, contained in his *Opus Epistolarum* [1670] have such terms: No. 130 (May) Columbus "*rediit ab antipodibus occiduis*," No. 134 (September) "*perrexit ad Antipodes*," No. 138 (November) "*Colonus ille novi Orbis repertor*." A letter, written in mid-November of 1493 and later included as the first chapter of [Martyr's] *Decades*, said: "*de insulis maris occiduis nuper repertis ... Ad orientem* [from Cuba] *igitur proas vertens Ophiram insulam* [Haiti] *sese reperisse refert: sed cosmographorum tractu diligenter considerato, Antiliae insulae sunt illae et adjacentes aliae.*" Humboldt in his *Examen critique* [1836–1839] referred to this statement of Peter Martyr as the earliest application of the name Antilles to the West Indies. The Antilles of course were legendary islands of the Atlantic Ocean, not of the Far East. It would appear that Peter Martyr from the beginning inferred the discovery of lands remote from the Orient. It may also be noted that the standard term in early Crown documents was *islas y tierra firme del Mar Océano*. (The name Indies became widely popularized in the titles of numerous issues and adaptations of the "First Letter" that were published all over western Europe.)

COLONIZATION BECOMES THE REAL OBJECTIVE OF COLUMBUS, THE IMAGINARY GEOGRAPHY PERSISTING

When Columbus composed his letter for the Sovereigns on shipboard as he was returning to Europe he construed a glowing prospectus for Española, as Haiti henceforth was to be called. Wealth unlimited was to flow into the coffers of the Crown as well as into his own. He began by promising Isabella and Ferdinand as much gold as they might need, all the spices and cotton they would require, mastic and linaloe as desired (in both cases he was mistaken as to the plants), and "slaves as many as they would order to be transported and (he added piously) these shall be taken from idolaters."[5] The people he had in mind were Caribs since he exculpated at that time the natives of Haiti and Cuba of idol worship and had only warm praise for the habits and friendliness of the Arawaks.[6]

The First Voyage (1492–1493) brought a radical change of plans. He would still fit his discoveries into the map of the Far East as then accepted, but it no longer mattered that all that he had found or would find differed greatly from what Marco Polo had seen. There was no further thought of diplomatic missions to Oriental states to open the gates of their trade. He now had a grandiose plan of colonial power that would far exceed the Spanish venture in the Canary Islands. Total possession and exploitation was what he came back to offer to the Sovereigns, with himself as their regent. The confidence of Isabella in him had been justified by the successful issue of the voyage of discovery; now he could assure the impoverished Crown solvency and prosperity through the colonization of Española. And he had least difficulty of all in persuading himself that the prospect he offered would become reality.

The Second Voyage was a full scale colonizing enterprise to take possession of Española. Its armada was not equalled again until 1502 when [Nicolás de] Ovando came to assume full charge. The ships carried around fifteen hundred men and a large stock of animals and plants from Spain and the Canaries. The direct down wind route followed became the standard passage from Spain to the West Indies. A rapid reconnaissance was made of the Leeward Islands, with two brief contacts with Caribs that confirmed the prior reports about them as cannibals and warriors.

Having set the main body of colonists at building a town and reconnoitering inland for gold, Columbus set out in April 1494 with three caravels, first to see what Jamaica was like and then to run out [the] south shore of Cuba, which he followed almost to its end. Having all but proved that Cuba was an island, he halted the exploration and required all hands to attest that Cuba was part of *tierra firme*, being at the beginning and end of India, namely part of the Province of Mango (the South China of Marco Polo). Whoever might later say anything to the contrary was to have his tongue cut off and be fined or lashed. Why did he stop almost at the point of proving that Cuba was an island; why the attest under duress with the grisly threat of mutilation; why did he never again look in this direction, in which he thought lay Cathay? The two years following were given over to the exploitation of Española, chiefly in the search for gold.

Peter Martyr had a letter from Columbus that adds otherwise unrecorded items to this expedition along the southern shores of Cuba. Briefer references to it is in No. 142 of *Opus Epistolarum* [1670] written to Giovanni Borromeo, the larger account in No. 164. Columbus wrote that he had followed the southern shore of Cuba for seventy-six days continuously, well into a southward turn of the coast "*ita ut se proximum aliquando reperiret aequinoctio. . . . Nec existimat se duas integras ad Auream Chersonesum orientalis termini metam horas solares reliquisse.*" Passage was reported of many islands, shoals, and narrows (characteristic of those waters). Stretches of the sea were covered by vast numbers of sea turtles (the first notice of their once great feeding and breeding grounds south of Cuba). To such items Columbus added from his imagination that tracks of great land animals, many streams, some cold, some very hot, and so on. This time, in approximately the same latitude in which he had turned back on the north coast before because he thought himself getting too far north, he thought himself nearing the Equator, calculated his distance from the Golden Chersonese (Malaya), and turned back abruptly.[7]

In his Third Voyage (1498) Columbus went well to the south of the by then familiar ocean crossing, it has been surmised in search of the source of the *guanin*. After discovering the Isle of Trinidad he entered the Gulf of Paria and to the west sighted for the first time the *terra firma* of South America. He held the Peninsula of Paria to be mainland, but for reasons of fantasy, not of scrutiny. Here he came up with idea that the world was pear shaped, with Paria lying somewhere up the stem end and the Terrestrial Paradise at the tip. The broad sheet of fresh water discharging over and through the Gulf of Paria (and which came from the Orinoco River) he ascribed to sweet waters cascading down from the mountains that enclosed Paradise. Ending the report of the Paria discovery, Peter Martyr added soberly *"fabulosa mihi videantur."* On the shores of the Paria Gulf members of the crew did a brisk trade in pearls and *guanin*, abundantly in possession of the natives. From the north shore of Paria a direct course was steered to Española, after sighting and naming the island of Margarita, without landing. Columbus himself seems not to have set foot on shore anywhere on this voyage until he reached Española, the reason given being that he was suffering an eye affliction. The discovery stopped short of the mouths of the Orinoco and the Pearl Coast. This time he was right as to *terra firma* but for imaginary and erroneous reasons.

The preconceived geographic views continued through his Fourth and final voyage (1502), in which he remained unaffected by the discoveries made in the years immediately preceding. My uncertain guess is that the last voyage was undertaken to find another colonial venture which might replace the authority he had lost in Española rather than to discover a passage or strait through the land. Off the coast of Honduras he picked up a seafaring group of Indian traders who carried worked, smelted, and soldered metal objects, together with forges, the tools of gold or silver smiths and their crucibles (*crisoles*). He makes no mention of inquiring whence they came or where they were bound. Instead he kept driving east along the Honduran coast against head winds and storms.

Having passed south along the Mosquito Coast he heard (about Puerto Limon?) of the mines of Ciamba (literally out of Marco Polo, the Champa of Cochinchina) and also of Ciguare, at a distance of nine days' journey inland. This was said to be a marvellous land of infinite gold, its people richly and fully clothed and bedecked with gold and gems, and engaged in trade by means of fairs and markets. "Also they say that the sea surrounds Ciguare and that thence at ten days' journey is the River Ganges." There is enough in the account, especially as to the system of trade and the clothing, to accord with the high Mesoamerican culture that extended south along the Pacific coast and which he could have reached readily by a march of a week or little more. He may even have heard there the first report of the Pacific Ocean ("the sea surrounding") which as well as the culture of its people were surely known to the natives of the Caribbean coast.

Instead of going west by land he took again to the sea and ran out the Panamanian coast to the point of its prior discovery by [Rodrigo de] Bastidas and Juan de la Cosa. He also made a brief incursion into the Veraguan rain forest, where he saw "greater signs of gold in the first two days than he had in Española in four years." After grievous off season misadventures with storms he found himself once more (May of 1503) "in the Province of Mangi which adjoins that of Cathay." After ten years of wide experience in and about the Caribbean he was still unshaken in his belief that he had been all the time on the borders of farther India.[8]

Did ever any one discover so much and see so little? The original bold and genial venture became the idée fixe, bringing confusion to his mind and suppression for others, ignoring the facts of newly known land and life, and ultimately helping to destroy him. He became unable to observe or to think rationally, to the point of being disordered as to mind, temper, and conduct. After the original voyage of discovery he never again rose to greatness or to face realities and make sensible use of them.

Columbus was a courageous and able mariner, but he was not of the stuff of explorers. Mostly he stayed shipside and let others bring him the news of the land, which he then recast to suit his theories. On at least six occasions during the four voyages he might have put his geographic ideas to decisive test and discovered their error; instead he turned away or back each time when he should have carried through. With his convictions why did he not once go on when he came to the critical point? The final discovery he thought he made was of the gold country of Veragua. This he said was one and the same with the mines of the Golden Chersonese. Yet the mines of Veragua have no existence; all he got from this claim was the empty title of Duke of Veragua, still borne by his descendants, [the] ironic end of his illusions.

THE COLONIAL LEGACY OF COLUMBUS

The geographical ideas of Columbus had little to do with his colonial administration. The plan he drew up on the return from the voyage of discovery was put into effect only as to gold and slaves. The cultivation of tropical plantation crops, cutting of dye woods, and procuring of aromatic resins came after he ceased to govern.

It made sense to concentrate Spanish occupation on the island of Haiti in terms of its central position, diversity of resources, numbers and condition of natives. What decided Columbus, however, was that it had gold bearing streams. Nothing of the sort was known from Cuba or Puerto Rico; Jamaica, which he had thought to be rich in gold, dropped out [of] his plans after he reconnoitered it. The mountain core of Haiti, however, consisting of igneous and metamorphic rocks, held gold-bearing quartz veins the waste of which accumulated in its stream beds. These placers were worked to some extent by the natives and their gold beaten into ornaments.

At first Columbus thought to bring in placer miners (*lavadores*) from Almadén,[9] but Alonso de Hojeda, at the time his favorite captain, discovered an easier source of labor by rounding up petty chiefs (*reyezuelos*) and making them set their people to such work. The structure of the island society with *caciques* in some hierarchic order made it possible to press native labor into service by control of their chiefs. From its beginning the Spanish colony forced disruptive and destructive servitude upon the natives. Humboldt cited Columbus as saying: "*El oro es excelentisimo: del oro se hace tesoro y con el quien lo tiene, hace quanto quiere en el mundo, y llega a que echa las ánimas al Paraiso.*"[10] Nothing truer or more cynical could be said; such was the basis and goal of the colony. The allocation of individual chiefs and their people to particular Spaniards, under the term of *repartimiento*, formalized the procedure, but hardly tempered the extreme and often cruel use of power.

The initially friendly attitude of the natives broke down quickly. Reduced to hapless and hopeless subjection, diverted from the care of their plantings, and driven from their homes, they were doomed to rapid decline and extinction. Their condition was sized up objectively by Peter Martyr [1970] in the fourth book of his *First Decade* and later in the angry diatribes of Las Casas [1951].

Columbus himself began the business of shipping slaves to Spain during his second stay in Haiti. In his original proposal of the gainful operation of the colony he had included the taking of slaves, but had the Caribs in mind. This he repeated in his letter of January 30, 1494 in which he asked that "each year enough caravels be sent to bring over livestock and other supplies of food and other items ... to make use of the land ... which things then could be paid for in slaves of these cannibals, people so fierce and fit and well proportioned and of very good understanding, who, having been quitted of their inhumanity, we believe will be better than any other slaves. This (inhumanity) they will lose as they are kept away from their own lands and of these many can be used in the galleys of the vessels it is intended to build here." He drew up this statement carefully within the rules of taking slaves as then accepted. However the Caribs had to be overcome and taken in their islands to the south, which he did not do or even attempt. The inoffensive Arawaks could be loaded at will in occupied Haiti. This he did do, an act of duplicity and villainy.

Prisoners of war, enemies of the Christian faith, and slaves bought in African slave markets were within the code of slavery, but Spain, charged with and acknowledging in good faith its missionary responsibilities, could not in conscience accept such treatment of harmless natives who had given no provocation. Protests by jurists and theologians in Spain put a stop to this kind of slave trade by 1500, in so far as their shipment to Europe was involved. In the islands and on *terra firma*, however, the indiscriminate seizure of slaves, especially for shipment to Española, was not controlled until much later.

Columbus cannot be discharged of guilt in perverting an existing code of human rights. Whatever others might have done, his was the authority in these first critical years, his the example, and his therefore the major responsibility. He left an evil legacy that was to vex the Spanish colonies for many years.

THE *TERRA FIRMA* TO THE SOUTH

The year 1499 brought to a head the troubles of Columbus, both in Española and in Spain. Things were in a bad mess. Peter Martyr summed it up: "*Cum reges tot querelis undique conflictati, et maxime quod ex tanta auri et aliarum rerum amplitudine parum, ob eorum discordias et seditiones afferetur, gubernatorem instituunt novum.*"[11] A new government was needed in Española; it was time to review and reduce the privileges and powers that had been reserved to Columbus. Henceforth others were to be granted license to discover and to traffic overseas. Initially such licensed expeditions were required to keep well away from any lands Columbus had discovered. The proviso however was lightly regarded from the outset and soon was ignored.

The year 1499 began a burst of exploring and trading activity, wholly directed to the southern *terra firma* and mainly engaged in by former pilots, masters, sailors, and soldiers of Columbus. All were interested in the region of Paria that Columbus had coasted the year before. Stories of the abundance of pearls and *guanin* had gotten around. We may also infer the guiding hand of Bishop Fonseca, who, Columbus being disposed of, was the power behind the throne in overseas affairs. Our knowledge of the events is lessened by the fact that Peter Martyr saw only two of the returned expeditions before he left Spain for Italy and Egypt in 1500, not to return until years later.

The first party to sail from Spain (May 1499) was that of Alonso Hojeda, Juan de la Cosa, and Amérigo Vespucci. Hojeda was the picaresque soldier of fortune of the second Columbus expedition, who had the first brush with

Caribs and had begun the business of entrapping Haitian chiefs, to be for a decade a bad actor on the southern mainland. Juan de la Cosa had been with Columbus on the first and second voyages and was pilot for Hojeda. These two began their coastal exploration somewhere on the shores of Guayana, continued through the Gulf of Paria, and followed the mainland shores to Cabo de la Vela; in other words they ran out the entire coast line of modern Venezuela.

Vespucci separated from the others well before the end of the ocean crossing. It is probable that as a foreigner he sailed under the license that was given to Hojeda and then, by previous agreement, struck out on his own. As the other turned west Vespucci sailed to the south and east. His first landfall is thought to have been near the northern extremity of Brazil, whence he followed the coast southeastward against the strong South Equatorial Current and the Southern Trades to about three or four degrees south latitude, entering the estuaries of the Amazon and Pará en route. On the return he followed the route of Hojeda and la Cosa as far as Cabo de la Vela. Whether the three rejoined is uncertain.

The other important expedition of 1499 was that of Vicente Yañez Pinzón, which left Spain in December, took a direct course to the shoulder of Brazil, went south beyond it, and on its return also explored the mouth of the world's greatest river. Within one year five thousand kilometers of the mainland coast of South America had been run out. The new discoveries appeared on the *mappa mundi* of Juan de la Cosa, probably drafted in the latter part of the year 1500. The map is signed as of this date, probably correct though it carries some later additions.

URABÁ AND JUAN DE LA COSA

Juan de la Cosa was the subject of the last monograph written by the distinguished historian Don Antonio Ballesteros [y] Beretta, *La marina cantabra y Juan de la Cosa*, published posthumously in 1954. By patient study and large erudition he reconstructed the leading role of this person in the opening of the Caribbean. Cosa was master of the Santa Maria on the First Voyage of Columbus and chart maker on the Second,[12] and piloted and charted the first expedition of Hojeda.

In October of 1500 he left Spain again, as pilot of the expedition licensed to Rodrigo de Bastidas. The exploratory part of the voyage began at Cabo de la Vela, where the prior voyage had turned north. A close reconnaissance was made of the entire Caribbean coast of modern Colombia, including the Gulf of Darién and ending at the northernmost bend (and narrowest part) of the Isthmus of Panamá (Puerto del Retrete). Another thousand kilometers of mainland coast was thus made known. The discovery of the Magdalena River betokened a large land area to the south, as did the Gulf of Urabá, [the] southern extension of the Gulf of Darién, terminating in a maze of great freshwater distributaries and swamps.

The voyage was carried out at a leisurely pace, the return to Española in the summer however considerably delayed by damage done to the ships by wood-boring mollusks, the teredo known henceforth to the Spanish as *broma*. This was the first such experience recorded for the New World.[13] The Darién and Panamá waters thereafter became notorious for such damage. Columbus in the following year had a similar experience. Both parties made a narrow escape, barely getting their sinking ships to Jamaica.[14]

Bastidas and Juan de la Cosa made the first contact with the source of *guanin*, which had been the object of long search. They found large settlements, including a "great city," in which they traded gold objects and saw others of copper and bronze (*latón*), bringing back several chests filled with gold items.[15] The city can hardly have been located anywhere else than in the Sinú country [Colombia], where a high native culture soon was to be looted at great profit.[16] West from Sinú through the country of Darién they were in contact with skillful metallurgists, traders, and prizers of precious metals, whose work still astounds us as to its craftsmanship and beauty.

Soon after, ships in number began to frequent the southern shore of the Caribbean, from Paria and the adjacent Pearl Coast westward to and beyond the Magdalena River. Friederici is probably right in holding that these trading ventures largely turned to looting and slave hunting; native labor was running low in Española; authority [was] remote and indifferent.

The Crown set up the Casa de Contratación in Seville in 1503, in [the] charge of [Juan Rodríguez de] Fonesca, for the regulation of trade and the collection of overseas revenues. That same year Juan de la Cosa, then in Spain, was named *alguacil mayor* (chief constable) of Urabá to serve under a governor to be selected and who was to reside on the Gulf of Urabá.[17] Thus, ten years before such a colony was actually established, Urabá had been chosen as [the] initial base for mainland settlement, the reason apparent being that it was the gateway to precious metals.

The services of Juan de la Cosa had won the esteem of the Crown, which chose him to lead a third venture to the Gulf of Urabá and other *Islas e Tierra*

Firme del Mar Océano to trade, discover, and have authority to settle. The capitulation was signed in February of 1504. He was forbidden to take slaves except for certain named places on the coast east of Urabá, where the natives had been declared to be cannibals.[18] Such was the newly found solution to the question of taking slaves: cannibals, which meant Caribs, were fair game. The order was given in good faith, but there were no disinterested officials to determine who were cannibals and who were not. If the natives fought back at a raiding party, they ran the chance of being recorded as cannibals. Who were the Caribs and who the Arawak or others on the coasts of Venezuela and Colombia is thus still a matter of some uncertainty.

The expedition left Spain in June, 1504, and worked its way westward from the Pearl Coast, as usual, acquiring pearls and cutting brazilwood. It soon turned to looting and the hunting down of natives, disregarding the prescribed restrictions. The Indian town of Urabá on the northeast side of the gulf was captured, with the intent to settle there. Troubles continued to mount, from Indian reprisals, disease, lack of food, and the riddling of the ships by *broma*. The data have been critically collated by Ballesteros [1954]. The small remnant barely made it back home in 1506.

Thus failed the first attempt to colonize on *terra firma*, to be revived after some years in the same manner and place and for the same reasons: the gold-bearing land, the native metallurgy, and the gulf that reached deeply south into a land of which great tales of riches were told, some of which turned out to be true.

Tierra firme was the Spanish name from the beginning for the southern side of the Caribbean from Panamá and Darién east through Paria and beyond, to include whatever southern lands Spain thought lay on its side of the Line of Demarcation. Other names came to be applied as colonial governments were established, but *tierra firme* continued to be the most inclusive term. The English translated it as the Spanish Main and in the course of their raiding activities transferred the name to the Caribbean Sea, which lacked an early Spanish name.

VESPUCCI AND THE NEW WORLD

The Spanish discoveries of 1499 and 1500 established a continuous coast from Cabo de la Vela southeast to and beyond the shoulder of Brazil. The great rivers, Orinoco, Amazon, and Pará, declared the continental proportions of the land. All of it lay well to the east of the original landfall of Columbus, actually extending for about forty degrees of longitude, though longitude then was guessed at rather than measured. However there was little doubt that the Spanish expeditions of [Vicente Yáñez] Pinzón, [Diego de] Lepe, and Vespucci had penetrated eastward well beyond the Line of Demarcation into the sphere reserved to Portugal. Lepe died shortly after his return, Pinzón appears to have turned to his private affairs for a time, and Vespucci was left to be concerned with the question of the southern land.

Amérigo [Vespucci] wrote his Medici patron from Seville (July 18, 1500) an account of the completed voyage, more academic than adventurous in form. The long mangrove shores, the drift of fresh water out of the great rivers, the strength of the sea current are soberly noted as well as the appearance and habits of the natives. He was especially interested in matters of cosmography and mathematical geography, such as the length of a league and of a degree, of the circumference of the Earth, and how one determined latitude and above all longitude by celestial observations. The businessman amateur, become navigator, was watching the night skies thoughtfully and thus, during an enforced stay of three weeks on shore, he worked out a much improved system of finding longitude by the time of conjunction of the moon with a planet. He still accepted the tenet of Columbus that he was on Far Eastern shores (as did Juan de la Cosa), but by asking himself the right questions as to how distance and direction were found he was well on the way to the proper answer.[19]

Although Amérigo had planned a further voyage of exploration for Spain, this was carried out in 1501 by him for the King of Portugal. Relations at the time were cordial between the two states, neither of which wished to trespass on the rights of the other. Cabral had made discovery at Easter of 1500 of the land of Santa Cruz in Brazil. It was rather urgent to have knowledge as to where the Line of Demarcation intersected the southern land mass. This was a question of longitude, the subject in which Amérigo was especially interested and competent.

Two further letters to his Medici patron constitute all that we know as [an] authentic account by Amérigo [see Vespucio, 1951]. The course of the second voyage was down the east coast of Brazil, by his own account as far as 50° S. far down the Argentine coast. His observations on latitude and longitude in general were quite good. It may be noted however that his accounts of natives apply only to those of the Brazilian coast, and that there is no notice in it of marked change in people and culture to the south. (Having experience that copyists were prone to error in transcribing numerals, I wonder whether Amérigo might have written 30 and not 50 degrees.)

At the outset of the voyage Amérigo still hoped to find the strait at the southeastern extremity of Asia by which Marco Polo had returned from Cathay. The Ptolemaic atlases of the time, with the East Asiatic shores skewed well to the south of their true position, gave him the idea that [the] Cape and a Strait of Catigara lay some eight degrees south of the Equator. The progress of the voyage destroyed this concept. He came back with the assurance of a continent stretching far south of the Equator and in longitude far to the east of Asia. This discovery found its way promptly into maps that began to appear in 1502 (Canerio, Cantino, etc.).

The new continent needed a name but Amérigo had nothing to do with giving it his own, as has been known for a long time. The Medici letters were private; somehow they were circulated after the death of the recipient in 1503. In August 1504 a Vienna printer issued in Latin the *Mundus Novus*, a somewhat free rewrite of the genuine Medici letters.[20] The book with the attractive title had enormous and continuing success and made the name of Amérigo known over western Europe. Harrisse [1866] lists seventeen editions by different publishers, all in Latin or German.

In September of 1504 there was published in Italian a long letter purporting to be from Amérigo to Soderini, then gonfalonier or head of the Florentine republic. This bears a strong odor of fraud and forgery.[21] The so-called Soderini letter first came under scholarly suspicion because of its rude language, out of keeping with the style of the Medici letters and with the speech of a cultured Florentine, which Amérigo was. The letter describes four voyages, the first of which [was] purported to have been in 1497 into the region of the Mexican Gulf and thus to anticipate the Columbus mainland discovery. It is patently a fraud with invented accounts of land and natives. The second and third voyages are garbled versions of the only two voyages made by Vespucci. The fourth appears to have been invented to balance the four voyages of Columbus.

Western Europe was aware that Spain was engaged in discovery, trade, and colonization across the Western Sea but its reading public had almost nothing more available for a decade than numerous versions of the "First Voyage of Columbus." This was hardly due to [a] policy of secrecy on the part of Spain but rather to the condition of the Spanish press. The publication of *Mundus Novus* was well timed to satisfy the hunger for news, in particular for readers in the German Empire.

The experience of Peter Martyr was similar to that of Vespucci. As an Italian in intimate touch with events in Spain, he had developed the habits of writing the news, and commenting thereon, to numerous leading persons of church and state in Italy. Somehow copies came into the hand of a Venetian printer and were published there in modified form, also in 1504.[22] This piracy of private letters, "now printed and dispersed throughout Christendom unawares to me," as Peter Martyr said, led to his decision to issue a selection in full original form as the first volume of his *Decades* (1511). (Peter Martyr remained unpublished in Spanish until 1892!)

By 1503 Vespucci appears to have been back in Spain working as cosmographer. Queen Juana conferred Spanish citizenship on him in 1505, and the same year he and Pinzón were commanded to make plans and arrangements for maritime discovery (Junta de Toro). Nothing came of it because of the internal political situation. In 1507 Ferdinand again picked up the idea of a program of exploration and exploitation and named Vespucci first of the four seasoned pilots summoned to meet early in 1508 as the Junta de Burgos, there to fashion the design that ushered in the second phase of overseas empire. In this connection Amérigo was appointed the first Pilot Major of Spain, which office he discharged to the time of his death.

A proper name was needed for the great Southern land as its continental proportions and position became known. Influenced, it is true, by the "four navigations" ascribed to Vespucci, the authors of the *Cosmographiae Introductio*, composed at St. Dié in 1507, invented the name America.[23] The legendary Atlantic islands of Antilia and Brazil had already been applied to parts of the New World. It could hardly be called after Columbus, who had made every effort to deny its existence. To adapt the assonant name of Amérigo was an innocent and rather appropriate conceit that should not have brought the obloquy which followed him through the years.[24]

NOTES (REVISED BY THE EDITORS)

Reprinted with permission from *Hermann von Wissmann-Festschrift*, ed. Adolf Leidlmair, 258–270. Copyright 1962, Geographisches Institut der Universität Tübingen.

1. The first landfall of Columbus is still under discussion (Link and Link, 1958). Also, for parts of the Brazilian coast, Portuguese and Spanish claims of prior discovery are still being argued.

2. The direct source is Bartolomé de las Casas (1951).

3. For the writings of Columbus and immediately related documents, the source used is Fernández de Navarrette (1825–1837: vols. 1 and 2). [Also see Columbus, 1961, for "Letters."]

4. Note also that *siba* is a general Arawakan word meaning stone (Friederici, 1960).

5. Letter from Columbus to King Ferdinand and Queen Isabella (first voyage), version addressed to Santangel, 1493.

6. Letter from Columbus to King Ferdinand and Queen Isabella (first voyage), version addressed to the Royal Treasurer Raphael Sanchez, March 14, 1493.

7. Letters written by Peter Martyr [1970] during 1494 and 1495, it may be noted, still avoid the term "Indies," cf. No. 140, 142, 144, 154.

8. The data are from the Jamaican letter (Columbus, 1961) of July 7, 1503 (fourth voyage), written while stranded there and delivered much later.

9. Letter from Columbus of January 30, 1494, to Antonio Torres (second voyage).

10. Humboldt (1836–1839:1:110).

11. Martyr (1970, Book 7, *Decade I*).

12. I think that Ballesteros (1954) has disposed of the view that there were two persons who bore the name Juan de la Cosa.

13. Ballesteros (1954:263–265).

14. Ballesteros (1945:2:552–558) shows that Columbus, setting out on his last voyage, met Bastidas returning in Española in the summer of 1502. The authority is mainly Las Casas, who had just come to live in Española, and met both men there. This explains why Columbus, five months later (November), ended his exploration near Nombre de Dios; he had linked up with the farthest point of Bastidas.

15. Ballesteros (1954:266, 214).

16. Gordon (1957).

17. Ballesteros (1954:280).

18. Ballesteros (1954:308–309).

19. The revaluation of Vespucci has engaged the attention of scholars in our century, but also earlier, as well set forth by Frederick Pohl (1944). The most searching critique and revision is that of Alberto Magnaghi (1926). Roberto Levillier (1948) has contributed further study of the maps of the period. I have also used the comments of Don Antonio Ballesteros, both in the Columbus (1945) and Juan de la Cosa (1954) memoirs. Germán Arciniegas's (1955) *Amérigo* is of special interest for the Florentine background.

20. Pohl (1944:148 ff).

21. There is a good account in Pohl (1944:150 ff.).

22. *Libretto de Tutta la Navigatione de Re da Spagna.*

23. As to its authorship, see Laubenberger (1959:163–279).

24. Volume editors' note: Much of this article appears scattered in Sauer's *The Early Spanish Main* (1966a), but not his discussion of Vespucci and the naming of "America."

REFERENCES (ADDED IN PART BY THE EDITORS)

Arciniegas, Germán. 1955. *Amérigo and the New World.* Trans. Harriet de Onis. Octagon, New York.

Ballesteros y Beretta, Antonio. 1945. *Cristóbal Colón y el descubrimiento de América.* Salvat Editores, Barcelona.

———. 1954. *La marina cántabra y Juan de la Cosa.* Santander.

Columbus, Christopher. 1961. *Four Voyages to the New World.* Trans. and ed. R. H. Major. Corinth Books, New York. Original edition, 1847, Hakluyt Society, London.

Friederici, Georg. 1960. *Amerikanistisches Wörterbuch und Hilfswörterbuch für den Amerikanisten*. University of Hamburg, Hamburg.

Gordon, Burton L. 1957. *Human Geography and Ecology in the Sinú Country of Colombia*. Ibero-Americana No. 39, Univ. of California Press, Berkeley.

Harrisse, Henry. 1866. *Bibliotheca americana vetustissima: A Description of Works Relating to America Published between the Years 1492 and 1551*. New York.

Humboldt, Alexander von. 1836–1839. *Examen critique de l'histoire de la géographie du Nouveau Continent*. . . . 5 vols. Gide, Paris.

Las Casas, Bartolomé de. 1951. [1875–1876]. *Historia de las Indias*. 3 vols. Ed. Agustín Millares Carlo. Fondo de Cultura Económica, México.

Laubenberger, Franz. 1959. "Ringmann oder Waldseemüller." *Erdkunde* 13:163–179.

Levillier, Roberto. 1948. *América la bien llamada*. Buenos Aires.

Link, Edwin A., and Marion C. Link. 1958. *A New Theory on Columbus' Voyage through the Bahamas*. Smithsonian Miscellaneous Collections, Vol. 135, No. 4. Washington, D.C.

Magnaghi, Alberto. 1926. *Amérigo Vespucci, studio critico*. Instituto Cristoforo Colombo, Rome.

Martyr d'Anghiera, Pietro. 1670 [1530]. *Opus Epistolarum*. Paris.

———. 1970 [1493–1525]. *De Orbe Novo: The Eight Decades of Peter Martyr d'Anghiera*. Trans. Francis A. MacNutt. Burt Franklin, New York.

Navarrete, Martín Fernández de. 1825–1837. *Colección de los viages y descubrimientos que hicieron por mar los españoles*. . . . 5 vols. Madrid.

Pohl, Frederick. 1944. *Amérigo Vespucci, Pilot Major*. New York.

Vespucio, Américo. 1951. *El nuevo mundo: Cartas relativas a sus viajes y descubrimientos*. Ed. Roberto Levillier. Buenos Aires.

31

Chart of My Course (1980)

Carl O. Sauer

Geography, well established at European universities, came late to academic attention in the United States, geologists mainly fathering its introduction. At the University of Chicago R. D Salisbury, Professor of Geographic Geology, formed a department of Geography, independent but closely associated with Geology. Physiography, which now would be called geomorphology, was Salisbury's special field, a field in which all of his students were instructed in the processes that have shaped the earth's surface. We learned to recognize land forms as to their origin within the larger context of Earth history. Learning about the nature of the terrain led to inquiry into patterns of vegetation and human response, as it then was called. Regional studies were proposed of the natural environment and its utilization by man. To this end Salisbury secured support by the Illinois State Geological Survey for a series of field studies.

After a year of graduate study at the University of Chicago I was sent in 1910 to study the Upper Illinois Valley. When I asked Professor Salisbury about the range of observations required, his reply was that this was left to me to determine and defend. The result, published in *Geography of the Upper Illinois Valley and History of Development* [1916a], was in the main physical geography, with some addition to knowledge of Ice Age land forms. The prairie plains of the area were the start of my interest in the origin of grasslands and led to reading what pioneer histories told of their nature. The record of local history began with the French and Fort St. Louis, built in 1683 on top of Starved Rock. This landmark, newly made a state park, its sandstone cliffs and miniature canyons a scenic attraction, had my detailed notice as to physical origin and historical significance, but I paid little attention then to the Indian history of the valley. As to human geography this first study was an attempt to apply the orientation then prevailing of human adaptation to physical environment, with some early doubts that this direction was adequate or proper.

The Geography of the Ozark Highland of Missouri [1915a (1920b)] was my doctoral thesis, the subject area a compromise cutting off the large natural region at the Missouri-Arkansas border. The component lesser natural regions were outlined by belts of rocks encircling the granite Ozark dome, escarpments of beds resistant to weathering. The limitations of such classification were apparent: not all the soils were derived from the weathering of the underlying rock; vegetation paralleled only in part the stratigraphy; kinds of people and their habitats did not sort out by physical environment. It was important to know the different terrains, but it was apparent that these only helped to understand the different ways of life. The people at the north and east were early German immigrants; at the southwest, a settlement of antislavery New Englanders; in the interior, hill folk from Tennessee and Kentucky; and each of these communities carried on the usages of their own very distinct and different traditions. Cultural geography, it had become evident, was more than "response to natural environment"!

Geography of the Pennyroyal [1927a], the field work done while at the University of Michigan, has as subtitle *A Study of the Influence of Geology and Physiography upon the Industry, Commerce and Life of the People*, a definition chosen by its sponsor, the Kentucky Geological Survey. In the Preface I wrote: "The dominant theme is the expression of the individuality of the region as the site of a particular group of people and of their works," the marks of this tenure being the cultural forms inscribed on the land. One chapter treated the problems of conserving land resources, in particular soil erosion, shown by maps of type localities. Man was recognized as a major agent of physical change.

The move to California in 1923 opened larger horizons of place and time. Field studies began in Lower California and continued in Sonora and Arizona. South of the border the land was little known except to naturalists. Villages had been Jesuit missions, in part still had an Indian population, and grew crops of native origin in the native manner. A centuries-old past still survived only partly modified by European ideas and skills. The northwest of New Spain had been a Jesuit mission province comparable to Paraguay, and of similar age, from the late sixteenth to the Jesuit expulsion in the eighteenth century. In their best tradition Jesuit missionaries wrote large and perceptive accounts of the land and its diverse natives, in fact a series of geographies ranging through two centuries. These were manuals that helped to introduce us to a larger and different perspective of cultural persistence and change, allowing us to review the attitudes of native peoples to land and life.

From arid north Mexico we went on to the tropical lands of the Pacific coast, using Spanish accounts of their condition in the past [1932a, 1948a]. By chance we came upon a forgotten area of high Mesoamerican culture well preceding the Spanish conquest. For almost half a century geographers have continued to go from the University of California into Mexico, Central America, and South America to compare past and present conditions and inquire into the activities of man in whatever ways and times human societies have intervened to alter the physical and biotic environments.

After retiring from academic service I had leisure time enough to write *The Early Spanish Main* [1966a], the land about the Caribbean as the Spanish found it at its discovery by Columbus and what they did to it in the first quarter century. I made use of what I had seen during a field season in the Dominican Republic, in travels about Cuba, Jamaica, Puerto Rico, and the Lesser Antilles, and in visits to the mainland coast. With this background of observation in mind I then studied the early reports as to places, plants, animals, and people, items peripheral to the interest of historians of the Spanish Indies. It was possible to trace the routes taken by the Spanish and to identify what they found, beginning with the Greater Antilles of Arawak population and continuing to the Isthmus of Panama, Chibchan in culture. The records give extensive information on the various economies and societies. Starting with Columbus, prosperous and well balanced native ways of life were broken down in a few years, and a population of millions [was] reduced to the point of extinction, all of this competently documented by Spanish sources.

Northern Mists [1968a] took its title from Fridtjof Nansen's classic study of sea faring into the North Atlantic before the time of Columbus. The global shape of the earth, known from classical times, was applied by the Portuguese in the fifteenth century in seeking a western passage from the Azores to the Orient. There was information from fishing ships in the fifteenth century that crossed the North Atlantic to western lands. The Norse settlement of Greenland, begun at the end of the tenth, lasted into the fifteenth century, the abandonment according to Nansen not due to adverse change of climate. Climatic change was rejected also by Vilhjalmur Stefansson.

Vinland was placed in southern New England by early, well informed students. Later, others located it in northern Newfoundland, inferring either a climate much milder than at present or that *vin* did not signify grapes. Reviewing what the sages said of plants, animals, and people, I found additional evidence in support of Vinland as having been in southern New England, the climate as at present.

Norse sagas told of a Great Ireland beyond Vinland. Irish monks with their households were the first settlers of Iceland, which they left after the coming of Vikings. They did not return to Britain, but the way west was open and inviting. There is unnoticed evidence of their having been in Greenland, whence the route to Newfoundland, later used by the Norse, was available. I have suggested that Irish monks settled on [the] shores of the Gulf of St. Lawrence in the tenth century or earlier.

Sixteenth Century North America [1971b] has the subtitle *The Land and the People as Seen by Europeans.* European explorations made known the Atlantic coast from the Gulf of Mexico to Davis Strait, the interior to Tennessee, Arkansas, and Kansas, and at the west the coast of California. Coronado and De Soto went in search of treasure and mines, Verrazzano, Cartier, Frobisher, and Davis to seek the western passage to the Orient. Spanish, French, and English attempts were made to found colonies in Florida, Canada, and Virginia. These earliest accounts give good descriptions of vegetation and animals, of oak and nut trees differing from those of Europe, of kinds of wild grapes, of herds of strange wild cattle on great grasslands, and of great flocks of water fowl including white cranes that shouted in unison like an army of men. The native peoples of the Eastern Woodlands were good farmers, cultivating a kind of tobacco, sunflowers, and a *Chenopodium* in addition to Indian corn, beans, and squash. Some had orchards of plum trees and vineyards. In some parts the coming of white men was followed by great sickness then death among the native population, an affliction unknown among these cultures on anything like the same scale before. The historical geography of the first century of European presence was assembled, therefore, from observations by participants, and from my own knowledge of terrain and biota.

The view presented here of seventeenth century North America, as it was known to the French and Spanish, has drawn me back to the experiences of my early years, going back to 1910, when I was beginning to learn geographical observation in the Illinois Valley. I have known the Mississippi Valley, the Great Lakes, and the Southwest widely and, I think, well enough to recognize the geography as it was before the changes brought by civilization.

Land and life are depicted by selected excerpts, given in my own translations. The Spanish reports for the most part are terse, of distances, places, native numbers, hardships, and frictions, the bare bones of geography and history. The empire Spain had acquired in the sixteenth century, greater than the world had known, was too much for its declining strength. The Viceroyalty of New Spain, charged with the support of New Mexico and Florida, had more pressing problems than those of these remote and profitless lands.

France began settlement in Acadia and Canada in the first decade of the seventeenth century. The profitable fur trade, supplied by Indian purveyors, gave access to the Great Lakes. In the course of the century habitants settled the lowland St. Lawrence Valley, Jesuit missions served the upper Great Lakes, and the Mississippi Valley was explored to the Gulf of Mexico.

The publication of French observations began with [Samuel de] Champlain's *Des Sauvages* in 1603 and continued through the century in volume after volume, a lively and widely read literature of the time, some of which was translated into other languages. More memoirs and documents were found in the nineteenth century, in particular the series published by the French archivist Pierre Margry [1879–1888]. From time to time historians have continued to add more records of early Canada, Acadia, and Louisiana.

I offer this study, then, as an introduction to the condition of land, nature, and Indian life as seen and influenced by French and Spanish participants.[1]

NOTES

Reprinted with permission from *Seventeenth Century North America*, ed. Bob Callahan and Kenneth Irby, 9–12. Turtle Island Foundation. Copyright 1980, Netzahualcoyotl Historical Society, Berkeley.

1. The work [on this volume] has been eased and made enjoyable by the cordial participation of Adrienne Morgan who has designed the maps, Marijean Eichel who has assisted me in numerous ways without requiring direction, and Margaret Riddall who has done the bibliography and checked the text.

REFERENCES

Margry, Pierre, ed. 1879–1888. *Découvertes et établissements des Français dans l'ouest et dans le sud de l'Amérique Septentrionale (1614–1754).* 6 vols. Paris.

Sauer, Carl O. For references, see the Sauer bibliography at the end of this volume.

32

Decline of Indian Population (1980)
Carl O. Sauer

NEW SPAIN: KEYSTONE OF THE SPANISH EMPIRE

In the course of the sixteenth century Spain gained a greater empire than the world had yet known, its keystone the land that [Hernando] Cortés took and presented as New Spain to the King. This was the land of high native cultures (Mesoamerica), from Pánuco at the north to Honduras at the south: central and southern Mexico and Guatemala. By mid-century the discovery of great silver mines was under way in the interior north. Florida was attached to the Viceroyalty of New Spain in 1557 to guard treasure-laden ships sailing to Spain. Cortés sent the first ships across the Pacific from the Mexican west coast, foreseeing that this would provide the shortest route between Europe and the Far East. In 1564 the Orient trade became regular sailing, to be known by the Manila galleon. By 1590 the Philippine Islands were made a dependency of New Spain. New Spain supported an empire and did so at a heavy cost of native life.

DECLINE OF NATIVE POPULATION IN MESOAMERICA

Old World contagious diseases came with the Spanish as terrible epidemics. Smallpox spread ahead of the Spanish. Tarascans visiting Cortés in 1521 at Mexico [Tenochtitlán], devastated at the time by smallpox, took the infection back to Michoacán, where it killed a multitude, including the Tarascan King and many of his household. Pedro de Alvarado, marching to the conquest of

Guatemala, found it desolated by smallpox. Tropical lowlands of Vera Cruz and of the Pacific Coast were depopulated, malaria being thought largely responsible. Subsequent to the mass mortality of the first years, outbreaks of lethal epidemics took place from time to time throughout the century.

The subjugation of Mesoamerica by force and disease raised the concern of the Council of the Indies, which enacted the New Laws of the Indies (1543–1546) for the protection of the Indians. The abuses were gradually abated; the decline of the native population slowed. The monumental demographic studies of Sherburne Cook and Woodrow Borah [e.g., Borah and Cook, 1963] documented the decimation of native numbers in central and southern Mexico during the century, as to manner, rate, and place.

The Rodríguez-Chamuscado party set out in 1581 on northern exploration, three Franciscan friars and a small guard of soldiers, the friars to look for a new mission field of which they had heard, the soldiers hoping to find mines. The result was the first record of location, number, and size of the Pueblo towns, the *Relación* of Hernán Gallegos [1927].

The *Relación* recorded sixty-one pueblos in the order visited, counting or estimating the houses in each and their size. The first, entered as they went up the Rio Grande, had forty-five houses of two to three stories. As they continued up river the number and height of houses increased, Nuevo Tlaxcala (Taos) being the greatest, with five hundred houses ranging to seven stories. Snow and cold prevented exploration further north, where they heard that there were more pueblos. East of the Rio Grande Valley they visited five pueblos of the Salinas (Estancia Basin), and at the west they got to the five Zuñi pueblos, these totaling 386 houses of two to five stories. The pueblos visited in the Rio Grande Valley added up to about five thousand multi-family houses; those entered farther east and west added another thousand. Gallegos gave an objective account of size of pueblos in terms of houses, without estimate of the number of inhabitants.

Soldiers who had taken part in the expedition were called to testify in Mexico. They agreed that the sixty-one pueblos visited contained more than a hundred thirty thousand souls. In terms of the house numbers of Gallegos this would assign a score or so of inhabitants to a house, a reasonable number for the multi-family, multi-story Pueblo dwellings. The estimate is moderate, I think, for a people of high skills, industry, and thrift, living in ecologic balance and undisturbed by external pressures. Parts of the Pueblo country to the north, east, and west were not included. Reexamination of the Pueblo condition before the Spanish occupation is indicated.

The ruin of the Pueblos began with the first governor. [Juan de] Oñate threw the people of one town out to make it his seat. Going to the Pueblos of the Salinas he punished one for "insolence," destroyed a part, and took away captives. The Pueblo of Ácoma, he informed the Viceroy, was of three thousand *indios*, who killed his lieutenant, and were therefore punished. Ácoma was a large town, the term *indio* perhaps meaning adult males, an exaggeration that underscored the danger he claimed to have overcome. His report to the Viceroy of the first year told of having visited all the provinces, with seventy thousand *indios* living in pueblos, whatever he may have meant by that word. Grave charges against Oñate were made in Mexico, of terror, flight of inhabitants of pueblos, hunger by seizure of their food, decrease in population. Inquiries as to their number gave estimates of fifty thousand or more, one being twenty-two to twenty-four thousand men and women, another of twelve thousand men not counting women and children, a third of fifty to sixty thousand inhabitants of a hundred and thirty pueblos. The continuing abuses led to Oñate's recall and exile in 1608.

Franciscan review of Pueblo provinces began with the Benavides *Memorial* of 1630 [1916], numbers of souls baptized given in round thousands adding up to about fifty thousand and not distinguishing between the living and the dead. The Zuñi and Hopi pueblos were not included. The revised *Memorial* was given to the King in 1634, indicating 34,380 Indians then living. The Prada report in 1638 was of eight thousand households paying tribute to soldiers, a Pueblo population of forty thousand or less after two recent epidemics. The friars' interest in numbers was mainly in baptisms. The pueblos east of the Rio Grande Valley were in marked decline by the 1660s. The Governor monopolized the trade of the Apaches who had taken to rustling livestock from pueblos and ranches, the connection unexplained. The Pueblos of the Salinas had been directed by officials to furnish salt, were taken south to the mines, and therefore became unable to give needed attention to their crops. A great drought began in 1666 and continued for three years, four hundred fifty persons dying of hunger in one Salinas pueblo. There was great pestilence among people and cattle in 1671. The last of the Salinas pueblos was abandoned in 1679. The fixed yearly tribute required having eliminated the practice of storing the surplus of good years against the recurring years of drought, poor years brought starvation.

The complaints of the friars were against officials, against forced labor, injury to pueblo lands by livestock, and above all the payment to soldiers of fixed tribute imposed on pueblos. The abuses were known and deplored, both in Mexico and Spain. New Mexico was a liability with no prospect of mineral wealth and with a declining and impoverished native population. The suggestion was that it should be abandoned. Little was done to ameliorate its condition.

Florida was the other remote and indigent appendage of New Spain. Its natives were Muskhogean tribes, served by Franciscan missions of obscure record. Mexico knew and did little about the Indians of Florida. Gangs were put to work at St. Augustine. There were occasional notices of revolt and of famine. Late in the century the Yamassees, a Creek tribe, got out and settled in Carolina.

INDIAN CONDITION ON THE EAST COAST

Nicolas Denys [1908], describing the declining condition of the Indians, placed the blame on alcohol provided by codfishers. The fishermen camped on land to dry their cod, plied Indians with drink, and got pelts in return. Denys came to Acadia in 1632 and spent most of forty years on its coasts, trading from the Bay of Fundy north to Chaleur Bay.

Father Pierre Biard, Jesuit missionary at Port Royal and Mt. Desert Island from 1611 to 1613, reported that at most nine to ten thousand Indians lived from Newfoundland to Saco (Maine), their numbers greatly reduced since they had intercourse with the French. Biard observed that they had become addicted to drink and were afflicted by dysentery in summer.

Champlain made his surveys of coast and natives from 1603 to 1606. Lescarbot [1907–1914] came to Port Royal in 1605 to remain a year. Neither mentioned disease, drunkenness, or decline; both found the Indians in good circumstances.

Champlain made two voyages along the coast of New England, getting as far as Nantucket Sound. From Saco south the Indians lived in villages, cultivated fields, and were in goodly number. Fifteen years later the Pilgrims found the fields and villages abandoned and attributed the empty land to plague sent by Providence. At the time the bubonic plague carried by rats was ravaging parts of western Europe. Norway rats infested houses and crossed the Atlantic on ships. Lescarbot told how their ship brought rats to Port Royal and that these spread promptly from the French settlement to the Micmac village, a nuisance previously there unknown.

French and English ships occasionally visited the Massachusetts coast. One ship, carrying infected rats, would suffice to introduce the contagion to natives that lived congregated as they did in agricultural villages of southern New England. Ships also carried house flies and other vectors of pathogens, suggested by Biard's reference to summer dysentery.

THE EMPTY VALLEY OF THE ST. LAWRENCE

[Jacques] Cartier [1924] found the St. Lawrence Valley well inhabited by natives living in palisaded villages of long houses and cultivating ample fields. The lower valley was called Canada; at the upper end of the valley was the island that he named Montreal, and the major Indian town of Hochelaga. His description of the Indian society included a vocabulary of their speech that certifies them as Iroquois. Cartier found them prosperous, numerous, and living at peace. Sixty years later Champlain found a greatly different condition of the country, no settled people or cultivated fields, the valley a great woodland with some prairies near Montreal.

Champlain's description of well grown trees of hardwoods and hemlock, and of many great grape vines indicates a vegetation of ecologic balance, the small prairies perhaps being former cultivated fields. Champlain's description suggests also that the change came shortly after the time of Cartier.

Explanations have been offered that the attractive valley, occupied by a people superior in numbers, organization, and skill, was abandoned because of attack by enemies such as Algonquins and Montagnais, lesser hunting tribes. Instead, the introduction of Old World disease was sufficient to cause the disaster. In 1535 Cartier and a company of a hundred ten French went up the St. Lawrence River to Montreal and returned to winter quarters adjacent to the Canada village of Stadacona (near Quebec). The Indians confined to the close quarters of their crowded long houses experienced a great sickness (a new respiratory contagion?), the cause of which neither they nor the French understood. Mass mortality followed Spanish, French, and English in the New World, the more so the more the natives lived congregated in clustered houses of several families. Mexicans, Pueblos, and Iroquois were prone to grave contagion by mode of habitation.

NATIONS DESTROYED

The *tabagie* [festival] in 1603 at Tadoussac celebrated a victory of Montagnais, Algonquins, and Abnakis over Iroquois. Montagnais were hosts; the two allied and kindred nations came from afar for the scalp dance. This was the first notice the French had of Indian warring and alliance.

The French became involved in 1609 when Champlain, ignorant of the cause of the enmity, joined a raiding party of Montagnais, Algonquins, and Hurons going to the land of the Mohawks. This is the first record of alliance of Algonquian tribes and Iroquoian Hurons against [the] Five Nation Iroquois, the Hurons thereafter known as "the good Iroquois."

Champlain in 1610 took part in another savage raid on the Mohawks. He came to Huronia in 1615, was awaited there by a large Huron war party, and went with it across Lake Ontario to attack an Oneida town. The attack was repulsed and the attackers returned home. Champlain took no further part in these raids thereafter.

Missionaries in Huronia recorded recurrent Huron raids across Lake Ontario and the bringing back of Iroquois captives, the men usually tortured to death, the women and children adopted into Huron society. The initiative was Huron; Algonquian bands were minor participants. Occasionally Iroquois made a counterattack. The last Huron invasion of the Five Nations was in 1638.

Informed missionaries estimated the Huron population in the 1620s as thirty to forty thousand. Later they were ravaged by a series of epidemics. In 1634 a sickness, thought to be measles or small pox, broke out among Montagnais at the trading post of Trois Rivières, spread to Huronia, and continued there for two years with high mortality. Another great sickness, probably small pox, came to Huronia in 1639 by way of Algonquins of the Ottawa River, who had brought it from Quebec. By missionary report the Huron nation in 1640 had been reduced to a third of its former numbers. Disease had cost them the initiative in the long and deadly feud.

The Senecas, westernmost and largest of the Five Nations, took the lead in the events that were to change the course of history. In 1648 a Huron party, hunting in the no man's land north of Lake Ontario, was ambushed by a Seneca war party. In 1649 Senecas led the offensive that laid waste [to] Huronia, killed its people, took them captive, or drove the remnant to distant refuges.

Having disposed of the Hurons and the affiliated Tobacco [Putun] Nation, the invaders continued north to devastate the land of the Nipissings, an Algonquian tribe that apparently had no part in the conflict. The Iroquoian Neutral Nation was next to be destroyed, reputedly because it had given refuge to Hurons. The Erie Nation, last of the non-aligned Iroquois, was eliminated in 1658. Eries and Neutrals, occupying lands on opposite sides of Lake Ontario, were the most populous nations of the Northeast, each perhaps greater than the combined Five Nations. Neither had sides in the intertribal and intra-Iroquois feud. The records do not report what became of the survivors.

In the course of ten years the Five Nations eliminated their four kindred nations to the west, leaving an empty land, still thus known at the end of the century on Delisle's [1700–1731] map as *nations détruites*.

Algonquian tribes living to the west between the Great Lakes and Ohio River were next to be dispossessed. Shawnees, Miamis, Potawatomies, Kickapoos, Foxes, and Sacs, drifted west beyond Lake Michigan and the Illinois River to find respite from Iroquois raids. They had not taken part in the feuding and appear not to have suffered great damage. When the French came to know them in their new homes they were numerous, in good condition, and apparently in normal proportion of sex and age.

The French relations of the Mississippi Valley and the western Great Lakes, elaborate and diverse in detail, lack mention of mass sickness and unusual mortality, an exception being [Jean François Buisson de] St. Cosme's finding the Quapaw villages at the mouth of the Arkansas (1698) newly decimated by smallpox and in part abandoned.

NOTE

Reprinted with permission from *Seventeenth Century North America*, ed. Bob Callahan and Kenneth Irby, 245–249. Turtle Island Foundation. Copyright 1980, Netzahualcoyotl Historical Society, Berkeley.

REFERENCES

Benavides, Alonso de. 1916. *The Memorial of Fray Alonso de Benavides, 1630*. Trans. Mrs. E. E. Ayers. Chicago.

Borah, Woodrow W., and Sherburne F. Cook. 1963. *The Aboriginal Population of Central Mexico on the Eve of the Spanish Conquest*. Ibero-Americana 45. Univ. of California Press, Berkeley.

Cartier, Jacques. 1924. *The Voyages of Jacques Cartier*. Ed. and trans. H. P. Bigler. F. A. Acland, Ottawa.

Champlain, Samuel de. For various references used by Sauer, see *Seventeenth Century North America*, 255.

Delisle, Claude, and Guillaume Delisle. ca. 1700–1731. *Atlas de De L'Isle*. Paris.

Denys, Nicolas. 1908. *The Description and Natural History of the Coasts of North America (Acadia)*. Ed. and trans. William F. Ganong. Champlain Society Publications 2. Champlain Society, Toronto.

Gallegos, Hernán. 1927 [1581–1582]. "The Gallegos Relation of the Rodríguez Expedition." Trans. G. Hammond and A. Rey. *New Mexico Historical Review* 2:249–268, 334–362.

Lescarbot, Marc. 1907–1914. *History of New France*. Ed. and trans. W. L. Grant. Champlain Society Publications, Nos. 1, 7, 11. Champlain Society, Toronto.

33

The End of the Century (1980)

Carl O. Sauer

NORTHERN BORDERS OF NEW SPAIN

The thousand miles of California coast were sailed yearly by Manila galleons untroubled by enemy or storm. There was no need of the ports Vizcaino had surveyed. California remained unoccupied and unknown inland.

Santa Fe was retaken in 1693, the last Pueblos brought to submission in 1697. The Hopi pueblos, distant from the rest, were not put under mission management. A number, such as those of the Estancia Basin, were not reoccupied. Colonists were brought from the south to stock lands with cattle and sheep and take over former Pueblo fields and irrigation works. Where Coronado had made his base on the Rio Grande the New Mexican town Bernalillo was settled in 1698 and then farther up river a colony was located at Santa Cruz. Spanish Creoles, Mestizos, and some Mexican Indians thus began to envelop the Pueblo missions.

The Council of the Indies became concerned with Louisiana as a French design against New Spain. The ship sent from Vera Cruz in 1685 to search the north coast of the Gulf of Mexico found no French activity. For reasons undisclosed the selection of Pensacola Bay was deferred to 1693, years after La Salle already had failed. The bay had long been known to [the] Spanish; the professor of mathematics at the University of Mexico was sent to make its survey; Sigüenza y Góngora sounded the channel, found a good site to build, and composed a competent and vivid appraisal of west Florida. Pensacola was built in 1693 on a sheltered harbor, convenient to watch the mouths of the Mississippi and whatever the French might be doing. It was well located with regard to Vera Cruz and Havana. The land inland was judged by its vegetation to be proper for cultivation and livestock, correctly rated by Sigüenza. The new town was the base of Spanish West Florida, St. Augustine to the east continuing to have indifferent attention.

NEW FRANCE

When New France was made a crown colony in 1663 the French population was little more than two thousand, mainly male, impermanent, and engaged in fur trade, civil and military duties, and in the service of the church. The French lived mostly in Quebec, Montreal, and Trois Rivières, with some two hundred about Port Royal.

[Jean-Babtiste] Colbert, as Minister of the Marine in charge of French affairs overseas, held that a viable colony needed to be based on small farmers, *habitants*, who would raise crops, livestock, and many children, thereby making a new France of the Valley of the St. Lawrence. The *habitants* of Canada, as the valley was known, would feed the colony and give defense against the growing numbers of English and Dutch neighbors. Colbert's aim was to populate Canada and keep the French from dispersing into distant parts.

The enduring organization of New France was put into effect in the first decade of the crown colony, the Intendant [Jean] Talon its administrator. At the end of Talon's tenure the French population had increased to seven thousand from Quebec to Montreal, was mainly rural, and provided ample supply of crops and livestock.

In the same decade Jesuit missions were built at the crossroads of the Upper Great Lakes, Indian settlements growing up about them. Talon sent parties to explore and take possession of the extremities of Canada. La Salle went on this exploration, Seneca guides taking him to the falls of the Ohio River. Talon heard also of portages by which to get to Florida and to Mexico. As one of his last acts Talon engaged [Louis] Joliet to start on the discovery of the great river, known as the Mississippi. The hydrography of North America was becoming outlined from Hudson Bay to the Gulf of Mexico.

The simple colonial plan continued to work. At the end of the century there were fifteen thousand French, living mainly along the St. Lawrence from Île d'Orleans to Montreal, the slow doubling of population in thirty years mostly by natural increase. The townspeople were in Quebec, Montreal, and Trois Rivières, the *habitants* in *côtes*, strings of farmsteads along the waterside.

The interior remained in Indian possession. After the death of La Salle, Jesuit missions were added among the Miamis at Chicago and [the] Illinois on Lake Peoria. At the end of the century the Crown ordered *voyageurs* and *coureurs de bois* out of the interior, holding that Indians were capable to take the furs to market, thus reaffirming the policy of restricting French settlement to the east.

The economy of the fur trade was a free partnership of French and Indians. Indians took the pelts and dressed them; theirs were the skills of purveyor and processor, of which there was rarely mention. Indians made, manned, and owned the canoes that carried the furs to market; Ottawas in particular [were] middlemen of trade and transport. Jesuit missions were hosts to French and Indians, maintaining order and sobriety. Indians were not subjected to any of the servitudes imposed by the Spanish. Unlike the English colonies there was no advancing frontier from which Indians were being driven.

The extractive nature of the fur trade brought early depletion of the resource of aquatic mammals. Otter, marten, and other carnivores were soon much reduced in number. The growing demand for beaver was met by finding new beaver dams. By the end of the century the fur business was exploiting the headwaters of the Mississippi and the drainage of the Assiniboine. The extension west was arrested at the prairies. At the north the Hudson Bay Company was moving onto the Laurentian Highland. With declining yields, Montreal and Quebec sought more and more remote sources.

The friendly and equal relations between French and Indian did not have their like in other European colonies. Champlain did blunder into enmity with the Iroquois which continued beyond his lifetime but ended in a peace of consent, not of force. French officials met Indian dignitaries in ceremonies of elaborate ritual, or oratory, dance, feast, and smoking of the calumet [pipe] of peace, and the pledges were kept. Missionary priests ministered to natives without duress or support of soldiers. Traders engaged Indians by mutual agreements. There was peace throughout after the Iroquois were conciliated, a peace of respect and good will.

The Indian way of life was admirable, it was agreed, before it was changed by European influence. Verrazzano [1970] thought these Indians to be living in the Golden Age. Lescarbot [1907–1914] came to Acadia to escape from corrupt civilization. Denys [1908] deplored the loss of morals and customs he witnessed in Acadia during his lifetime. Lahontan [1905] portrayed [the] Iroquois as exemplars of a good society. Rousseau adopted Lahontan, and created the "Noble Savage." At the end of the seventeenth century New France was still Indian country beyond Montreal.

NOTE

Reprinted with permission from *Seventeenth Century North America*, ed. Bob Callahan and Kenneth Irby, 251–253. Turtle Island Foundation. Copyright 1980, Netzahualcoyotl Historical Society, Berkeley.

REFERENCES

Denys, Nicolas. 1908. *The Description and Natural History of the Coasts of North America (Acadia)*. Ed. and trans. William F. Ganong. Champlain Society Publications 2. Champlain Society, Toronto.
Lahontan, Louis Armand de Lom d'Arce. 1905. *New Voyages to North America by the Baron de Lahontan*. Ed. Reuben G. Thwaites. 2 vols. Chicago.
Lescarbot, Marc. 1907–1914. *History of New France*. Ed. and trans. W. L. Grant. Champlain Society Publications, Nos. 1, 7, 11. Champlain Society, Toronto.
Verrazzano, Giovanni. 1970. *The Voyages of Giovanni da Verrazzano, 1524–1528*. Ed. and trans. Lawrence C. Wroth. Yale Univ. Press, New Haven, Conn.

VIII

CARL SAUER ON GEOGRAPHERS AND OTHER SCHOLARS

34

Introduction

Edward T. Price

> Through a wide range of stimulating lectures ... she exerted an influence over virtually the entire living generation of American geographers and recruited to the field a large number of its present representatives.
>
> Carl O. Sauer, "Ellen Churchill Semple," 1934c

Carl Sauer wrote short biographical statements on eight geographers—Karl (Carl) Ritter, Oskar Peschel, Friedrich Ratzel, Ellen Churchill Semple, Ruliff Holway, Erhard Rostlund, David Blumenstock, and Richard Russell—and two scholars in related fields—Homer Shantz and Herbert Bolton. They had produced their work over a sweep of time between 1800 and 1971. Four of these essays were written for the 1934 *Encyclopedia of the Social Sciences,* five were written to be published as memorials, and the other was written as the foreword to the published Hitchcock Lectures given at Berkeley in 1965.

Three of the encyclopedia articles by Sauer concern the nineteenth-century German geographers Ritter, Peschel, and Ratzel, whose work was already history before Sauer himself got into geography. The three were important among a number of German geographers who put their country far ahead of others in bringing human and physical geography together in a modern scholarly discipline. Sauer's own family heritage was German, and he attended school in Germany during a three-year family residence there. It was not surprising then that he was a leader in introducing German thought to a growing number of American geographers.

The base on which German geography grew was established in the first half of the nineteenth century in the writings of Karl Ritter (1934d, ch. 38 herein) and Alexander von Humboldt. Ritter's great work was the encyclopedic region-by-region accounts of the lands and peoples of Asia and Africa. Humboldt, next to Napoleon the most widely known name of his day, wrote on many aspects of physical geography and other sciences as well as authoring reports on the lands and peoples of his travels. Although the studies of Humboldt and Ritter were very different, each drew on and supported the other in his work. Both were living in Berlin for more than thirty years, and they had opportunity for frequent communication (Hartshorne, 1939:49, 52–53). They were so widely appreciated that their followers were able to establish both human and physical geography in German universities. The dualism so formed in geography has never been completely reconciled or eliminated.

The views of Ritter (and later those of his students who emphasized his religious mission) soon drew criticism (Leighly, 1938:242–248). In 1831, Julius Fröbel, a young student, attacked Ritter's view that the earth was created to be a cradle for mankind, his assumption of environmental influences that had been little considered, and his method of sorting terrestrial information into regions. Information, he believed, should be sorted not by regions, but with other information of similar class within which scientific analysis might lead to process. His criticisms actually effected little change in the practices of geography. Sauer found Ritter's theism, search for environmental influences, and heavy dependence on library sources poor models for his Berkeley students. Hartshorne (1939:237–238) concluded, from sample sections of Ritter's text, that Ritter's emphasis on God's purposes did not deter him significantly from faithful reporting of what his sources offered.

It was apparently in his seventeen years as editor of *Ausland* (Foreign Lands) that Oskar Peschel's interests turned toward geography and then toward its methods (1934b, ch. 36 herein). Thirty-five years after Fröbel's criticism of Ritter, Peschel renewed the battle (Leighly, 1938:244, 249), particularly questioning Ritter's "comparative methods and his assumption of a causal relation between the impression of the Landschaft and the mental expressions of the population" (Hartshorne, 1939:80). Looking ahead, Peschel proposed the scientific study of the morphology of the earth's surface as geography's frame, and soon extended his own work to the geographic distribution of cultural forms also. Peschel's recommendations led to a quick response among geomorphologists, who, for a time, threatened to take over the whole field of geography, as later happened in the United States under William Morris Davis's leadership. Soon, however, Ferdinand von Richthofen, in his regional studies of landforms, was also taking up a variety of human uses in relation to the landforms (Hartshorne, 1939:88). The title of Sauer's "Morphology of Landscape" (1925) and some of its contents echo ideas put into motion by Peschel. Friedrich Ratzel, whose *Anthropogeographie* (1882, 1891) dealt in original ways with both nature and culture, wrote a sympathetic biography of Peschel (1887).

Ratzel's work (1934c, ch. 37 herein) was brought to Sauer's attention by the anthropologist Robert Lowie, at Berkeley. Ratzel is best known in the United States for the first volume of the *Anthropogeographie* (1882), in which he looked at a variety of different environments in terms of the modes of living of the peoples around the world who dwelt in them. The second volume (1891), in which Sauer (1941b:5) found "far more" than in the first, reversed the relationship of land and people (Hartshorne, 1939:91) and was notable for its use of culture distributions and the diffusions that helped explain them. Surprisingly, Sauer suggested in a government report (1934g:246–247) the use of Ratzel's reading of historical process from distributions. Ten years later, he restated Ratzel's principles in his own words as he made his way into "A Geographic Sketch of Early Man in America" (1944a:529–530).

Ellen Churchill Semple (1934e, ch. 39 herein) was the most active of Ratzel's disciples and one among the Chicago teachers of Sauer, who expressed appreciation of her "eloquent and evocative lectures" (Leighly, 1976:338). Her *Influences of Geographic Environment* (1911) started out to be a translation of *Anthropogeographie*'s first volume. Sauer notes her substitution of her own illustrations for Ratzel's, as is witnessed by more than fourteen hundred citations, mostly in English. Beyond this, her version was much more cautious and tentative, without Ratzel's claim of a scientific frame, and compensating for ways in which American readers might be unprepared for German ideas (Semple, 1911:v–viii).

Semple was not out of step with the American geography of her day, focused as it was on relations between the physical environment and its organisms. She has long since been dismissed, however, as an environmental determinist, a position she readily claimed. Her most deterministic statements are general ones, but her treatments of cases are actually laced with historical movement, level of technology, ethnic replacement, and influences of neighboring peoples, evidences that she could understand variables other than physical environment.

Semple's paper on the peoples of the Kentucky mountains (1901) calls on two aspects of the physical environment—isolation and limited arable land— and goes on to describe the local ways of living and attitudes in considerable detail. It would be hard for anyone to give the environment less attention.

Her presidential address to the Association of American Geographers (1922) on livestock in the ancient Mediterranean covers the area in fine detail, with the main premise being that cattle and horses required richer food sources than sheep and goats. Her "Ancient Mediterranean Pleasure Gardens" (1929) treats the subject much as later cultural geographers might have done, taking notice of westward spread of the gardens with the advance of civilization. She did not use the word diffusion, but the fact was explicit.

Sauer's bountiful memories of Homer Shantz leave me little to say (1959d, ch. 41 herein). The very name of Shantz led me to anticipate Sauer's first comment on "the years when geologists and biologists, historians and economists, mingled at geographical meetings." Sauer also mentions Curtis Marbut (soil geologist) and O. E. Baker (agricultural economist). And the list runs on with Association of American Geographers members from government agencies—F. E. Matthes (topographer), H. A. Marmer (tidal mathematician), Roland Harper (botanist, who eventually styled himself a geographer), H. H. Bennett (soil conservationist), Charles H. Birdseye (topographic engineer)—and others who spent more time in colleges and universities—Ernst Bessey (botanist), Edgar N. Transeau (botanist), Frederick Jackson Turner (historian), C. F. Brooks (meteorologist), Henry C. Cowles (botanist), and Robert Cushman Murphy (naturalist, curator).

The other personal sketches included here all grew from Sauer's years in Berkeley, beginning in 1923. Geography had become a department in 1898 to make a place for George Davidson, a distinguished surveyor who had been in charge of major surveys in the West. Ruliff Holway replaced Davidson when he retired in 1905 (1929b, ch. 35 herein). The department's first master's degree was awarded in 1908, and the first of the *University of California Publications in Geography* came out in 1913. Holway's retirement opened up a coveted position for a new chair, and Sauer was selected to fill it. John Leighly came with Sauer from Michigan, and Richard J. Russell, a geology student who had been a teaching assistant in geography for two years, continued as an instructor until his Ph.D. was completed in 1926.

Sauer's memorial of Holway was needed all too soon. It pictures a man who could climb an academic ladder with rungs spaced far beyond the usual reach. Holway's publication topics were among the most attractive available to him within reasonable distance of Berkeley. He studied Lassen Peak through all its 1914–1917 eruption. Sauer doesn't mention an oral paper by Holway on coastal and river terraces noting some evidence of warping. Holway's 1905 study of cold water along the California coast (based on only limited data) pinpoints upwelling as the only possible source of the cold water, but makes no mention of the effect of the earth's rotation in producing it. Unknown to Holway, V. W. Ekman in Sweden had shown in 1902 why a wind pushing a west coast current equatorward on the rotating earth should drive the surface water away from the coast. Ekman's English version (1905) came out in the same year as Holway's article.

Holway had been accepted as a member of the Association of American Geographers. He was also a member of the American Geographical Society Transcontinental Expedition of 1912 on the segment from San Francisco to Salt Lake City (AGS, 1915:38).

Herbert Bolton, a Berkeley historian, parallels Shantz in Sauer's sketches for not being a geographer and being a greatly admired associate about half a generation senior to Sauer (1954b, ch. 40 herein). The interest in historical Latin America that drew Sauer and Bolton together included extensive investigation in both archive and field. Each had developed his Latin American interest *de novo* after moving to a distant university in a border state. With the anthropologist Alfred Kroeber, Sauer and Bolton initiated the long-continuing Ibero-Americana series of research publications in 1932; its fifty-fifth and final item was published in 1992. Although all of Sauer's memorials took note of his subjects' personal qualities, he plumbed Bolton's personality, outer and inner, most deeply.

In 1941, as I started my second year of graduate study at Berkeley, Erhard Rostlund, a stranger to me and probably to all the graduates on hand, showed up in one of my classes (1962a, ch. 42 herein). A decade or two older than the rest of us, he soon became identified as a Swedish sailor. He proved an articulate participant in our classes and seminars before World War II interrupted study for most of us. Back at Berkeley, in 1945, I was appointed as a teaching assistant and, to my surprise, assigned to Rostlund, who would teach the two freshman geography courses over the year. It didn't take long to learn that his lectures were well prepared, his material varied, and his presence in full command of the lecture hall.

The matter of environmental determinism was discussed above in dealing with Ratzel and Semple. The doctrine was waning when Sauer could say in "Morphology of Landscape": "It would, therefore, appear that environmentalism has been shooting neither at cause nor at effect, but rather that it is bagging its own decoys" (1925:52). Although Sauer himself never gave up recognition of the many ways in which environments were important, many other geographers became unduly sensitive about statements that might sound environmentalistic. Rostlund examined that trend with a great deal of clarity, and became one of the first to call for new research toward understanding human responses to natural environments in the light of his conclusion, "Environmentalism was not disproved, only disapproved" (1956:23).

Rostlund's sudden death in 1961 shocked those who knew him. His death was the first among nearly fifty who by then had received the Ph.D. in geography at Berkeley. Two years later David Blumenstock, then also on the faculty, became the third. Warren Thornthwaite, with whom Blumenstock had been fortunate to work as a climatologist in the Soil Conservation Service, had died just two months earlier. (Today, I believe, twelve are still living out of a total of thirty-seven Sauer Ph.Ds.; see appendix.)

Blumenstock had spent twenty varied years in government service and three years at Rutgers University before returning to Berkeley in 1961 (1968b, ch. 44 herein). He wrote an impressive array of government reports connected with his work, some of them classified, hence unavailable. In his spare time during the 1950s, he wrote *Ocean of Air* (1959), a readable, informed, non-technical account of the atmosphere, its behavior, weather, climate, and human significance, with interesting examples of prediction problems and effects on public events and ways of living.

Richard Russell (1967b, ch. 43 herein) and Sauer both had the opportunity in Berkeley of association with William Morris Davis, whose models of land surface features were considered to indicate the conditions that had formed them. Both had doubts about the completeness of Davis's interpretations. When Russell moved to lowland Louisiana, he found himself in an amazingly flat country that was not to be understood by what he had learned in the West. The evenness of the surface, however, did not mean there was a lack of activity to observe, for surface materials were subject to frequent movement by rivers, ocean waves, and wind. Here was an opportunity for Russell and his associates to observe intricate processes that were not in textbooks. He frequently passed on the new findings to others at geography meetings.

Russell attributed his marked success to colleagues, who "seemed to come along at about the right time and place" (McIntire, 1973); to Sauer's intellectual challenges; to Henry Howe, an old classmate and geology head at Louisiana State University; to Harold Fisk, who wrote the monumental report of the Army Engineers on the Mississippi River alluvial plain (1945), and whom Russell had frequently accompanied on field trips; and to Evelyn Pruitt in the Office of Naval Research, who had stimulated Russell's quest for new shores around the world for his Coastal Studies Institute to explore.

The last lines of Mr. Sauer's five memorials bearing warm phrases—such as "avoiding contention and the paths of ambition" (Holway); "his way of life was serene" (Bolton); "great in gentleness, spirit, and vision" (Shantz); "stranger to all guile, rancor, and ambition, he lived, great of heart and perceptive of eye, 'by the side of the road'" (Rostlund); and "unvexed by self-interest and ambition, eager to know, and willing to share" (Blumenstock)—linger in my thoughts.

REFERENCES

AGS. 1915. *Memorial Volume of the Transcontinental Excursion of 1912 of the American Geographical Society.* American Geographical Society, New York.

Blumenstock, David I. 1959. *Ocean of Air.* Rutgers Univ. Press, New Brunswick.

Ekman, V. W. 1905. "Of the Influence of the Earth's Rotation on Ocean Currents." *Arkiv Matematik, Astronomi, och Fysik* 2 (11): 1–55.

Fisk, Harold N. 1945. *Geological Investigation of the Alluvial Valley of the Lower Mississippi River.* Mississippi River Commission, Vicksburg.

Hartshorne, Richard. 1939. *The Nature of Geography.* Association of American Geographers, Lancaster, Pa.

Holway, Ruliff S. 1905. "The Cold Water Belt along the West Coast of the United States." *Bulletin of the Department of Geology of the University of California* 4 (13): 263–286.

Leighly, John. 1938. "Methodologic Controversy in Nineteenth Century German Geography." *Annals of the Association of American Geographers* 28:238–258.

———. 1976. "Carl Sauer, 1989–1975." *Annals of the Association of American Geographers* 66:337–348.

McIntire, William G. 1973. "Richard Joel Russell, 1895–1971." *Geographical Review* 63:276–279.

Ratzel, Friedrich. 1882, 1891. *Anthropogeographie.* 2 vols. J. Englehorn, Stuttgart.

———. 1887. "Oskar Peschel." *Allgemeine deutsches Biographie* 1:416–430.

Rostlund, Erhard. 1956. "Twentieth Century Magic." *Landscape* 5 (3): 23–26.

Sauer, Carl O. For references, see the Sauer bibliography at the end of this volume.

Semple, Ellen Churchill. 1901. "The Anglo-Saxons of the Kentucky Mountains: A Study in Anthropogeography." *Geographical Journal* 17:588–623.

———. 1911. *Influences of Geographical Environment.* Henry Holt, New York.

———. 1922. "Influence of Geographic Conditions upon Ancient Mediterranean Stock-Raising." *Annals of the Association of American Geographers* 12:3–38.

———. 1929. "Ancient Mediterranean Pleasure Gardens." *Geographical Review* 19: 420–443.

35

Ruliff S. Holway, 1857–1927 (1929)

Carl O. Sauer

Ruliff Stephen Holway died in Oakland, California, on December 2, 1927. He had become Professor Emeritus four years previously. A period of uncertain health had been followed by marked improvement, so that we expected many contented years for him with his friends, his garden, and his studies. In less than two years, however, he suffered a stroke, signaling the beginning of his dissolution, to which his body offered lengthy, but finally unavailing resistance.

Born of Quaker ancestry on a farm near Hesper, Iowa, on May 8, 1857, he became a country school teacher in that state and later an instructor in science in high school. In 1884 he was made principal of a California High School, occupying a similar position in San Jose in 1887. This led to appointment as teacher of physical geography at the San Jose Normal School, where he also served later as Vice-President. Here he had the opportunity of completing his college course at Stanford University near by, receiving the bachelor's degree in 1903. In that year he removed to the University of California as Instructor in Education and took his Master's degree the year following. In 1904 he was appointed Assistant Professor of Physical Geography, in 1914 Associate Professor, in 1919 Professor, and retired in 1923.

This bare outline of his career records the quiet steadfastness of purpose, the essential stability of character that marked his nature. His opportunity to enter the academic field came late in life, and he was fully conscious of the difficult task of reorientation that he set for himself in making the change. Always he invested the instruction of youth with a sincerity and devotion that bound to him loyally his pupils, many of whom are now teaching in all parts of the state. Moderate in opinion, punctilious in workmanship, and sane in judgment, he prepared a sound basis for the development of physical geography on the western coast. During his most active period he led the California Physical Geography Club, composed principally of his students, which attracted considerable attention to studies of the physical features of the state. He contributed about two dozen articles to our literature, of which the best remembered are his studies of the cold water belt along the West Coast, the extension of the known area of Pleistocene glaciation to the Coast Ranges, the stream history of the Russian River, the unfinished entrance to San Francisco Bay, and a series of studies on the volcanic activity of Lassen Peak.

His was a gentle spirit, avoiding contention and the paths of ambition. Forbearance, kindliness, and extreme modesty determined his unobtrusive course through life. He lived truly in the light of the humane creed that nurtured him, and he gave the full measure of his strength to his chosen field.

NOTE

Reprinted with permission from the *Annals of the Association of American Geographers* 19:64–65. Copyright 1929, Association of American Geographers.

36

Oskar Peschel, 1826–1875 (1934)

Carl O. Sauer

German geographer. [Oskar] Peschel, who was trained as a jurist, served as newspaper editor in Augsburg from 1848 to 1854 and as editor of the *Ausland* from 1854 to 1871. He was a brilliant publicist, contributing many political and economic essays distinguished by their clarity, their liberal viewpoint, and their insight into historical processes. As a commentator on world events Peschel became increasingly absorbed in geographic studies, and in 1871 he relinquished his editorial chair to become professor of geography at the University of Leipzig. His first book, *Geschichte des Zeitalters der Entdeckungen* (Stuttgart, 1858; 2nd ed. 1877), revealed his views on the interrelation of economic principles, geographic backgrounds, and historical events, and expressed most of the ideas now current regarding the bearing of oriental trade on the progress of discovery. He was then commissioned by the king of Bavaria to write *Geschichte der Erdkunde* (Munich, 1865; 2nd ed. 1877); in executing this task he formulated the definition of the objectives and methods of geography which later found expression in his principal work, *Neue Probleme der vergleichenden Erdkunde* (Leipzig, 1870; 4th ed. 1883). During his academic career Peschel wrote his well known *Völkerkunde* (Leipzig, 1874; 6th ed. 1885; trans. as *The Races of Man,* London, 1876), which discusses general anthropology with emphasis on the geographic distribution of cultural forms. Posthumous publications, compiled from his notes, indicate the wide range of interests to which he was devoting his attention at the time of his premature death (*Abhandlungen zur Erd- und Völkerkunde,* 3 vols., Leipsig, 1877–1879).

Peschel initiated modern academic geography by disposing through vigorous polemic of the historical philosophic orientation of [Carl] Ritter and by introducing the morphologic viewpoint. The half century of Ritter's dominance had resulted in many textbooks but in little productive scholarship. Peschel pointed out a field of investigation and blocked out many problems for study. By their originality, order, and comprehensiveness his lectures and books became programs of inquiry which students still consult. In Germany there sprang up after him a generation of important investigators and teachers, all of whom were indebted to him as their master. The major tradition of modern geography runs in uninterrupted line from Peschel by way of Ferdinand von Richthofen to the currently dominant school of continental geographers.

NOTE

Reprinted from the *Encyclopaedia of the Social Sciences* 13:92. Copyright 1934, Macmillan, New York.

37

Friedrich Ratzel, 1844–1904 (1934)

Carl O. Sauer

German geographer. [Friedrich] Ratzel came into the field of geography from zoology by way of journalism. His first publication (1869) dealt with organic evolution; all his subsequent geographic work was based on biologic views, which re-expressed the tradition of [Carl] Ritter in harmony with later nineteenth-century science. In order to finance his zoological studies, Ratzel wrote travel sketches, which met with immediate success. As the result he was engaged by the *Kölnische Zeitung* to visit and report on various parts of the world, including the United States and Mexico. From *feuilleton* writing Ratzel passed directly to university lectures in geography at Munich and then at Leipsig, where he attracted a steadily growing audience until his death.

Ratzel is remembered most widely for his formulation of anthropogeography as the conditioning of culture by environment, which he called the dynamics of human geography and for which he supplied most of such generalizations as are still current. In the second volume of his *Anthropogeographie* (2 vols., Stuttgart, 1882–1891; 4th ed. 1921–1922) he considered the geographical distribution of man and his culture. From Moritz Wagner, the zoologist, Ratzel adapted the theory of development of new organic forms by migration and isolation as the fundamental thesis of human history. By pioneering studies of the geographical distribution of individual culture elements, by first stating the case for diffusion of culture, and by anticipating the culture area concept Ratzel became parent to the modern cultural history school in anthropology.

Out of the biologic viewpoint proceeded also his *Politische Geographie* (Munich, 1897; rev. ed. 1903; new ed. 1925). The state was considered a quasi-organism, bound to a particular *Lebensraum,* in contest for space with other states and exhibiting distinguishing characteristics of vitality or decadence.

Ratzel directly stimulated, and wrote the philosophy of, the first actual *Weltgeschichte,* carried out by his pupil Hans Helmolt (9 vols., Leipsig, 1899–1907).

With the exception of the culture area concept, Ratzel made relatively slight contribution to present scholarship. The reason perhaps is to be found in his general philosophic view that there are sciences of actuality and sciences of abstraction and that geography belonged among the latter. It has been excessively difficult to establish satisfactory inquiries directly into these relationships. On the other hand, Ratzel's fruitful thesis, that of the diffusion of culture, is applicable inductively in the examination of the arrangement of phenomena and has contributed therefore to a "science of actuality," with which he was not concerned.

NOTE

Reprinted from the *Encyclopaedia of the Social Sciences* 13:120–121. Copyright 1934, Macmillan, New York.

38

Carl Ritter, 1779–1859 (1934)

Carl O. Sauer

German geographer. Carl [Sauer uses "Karl"] Ritter was tutor for nearly twenty years to the family of Bethmann-Hollweg; he thus became acquainted in 1807 with Alexander von Humboldt, who had just returned from his New World travels. Through Humboldt, Ritter found the objective henceforth to occupy him, which is expressed in the title of his major work, *Die Erdkunde im Verhältnis zur Natur und zur Geschichte des Menschen* (2 vols., Berlin, 1817–1818; 2nd rev. ed., 21 vols., 1822–1859). In 1820 he became professor of geography at the University of Berlin, where he remained until his death.

Prior to Ritter geographic treatises had been mainly compendia of data selected because they might be useful. Against this utilitarian political role, Ritter asserted for geography the status of an independent discipline. He thought of the earth as a unitary organism, divinely planned, created for man, and conditioning human life. [Oskar] Peschel has clearly defined Ritter's views as teleologic; Wisotzki has traced the connections between them and Schelling's transcendental idealism and the teachings of Fichte. Ritter was significant for his period as an exponent of theistic environmentalism; he considered the earth as "the preparatory school of the human race."

Two kinds of inquiry were indicated by Ritter's general thesis: that which sought to determine the content of regions and that which noted the recurrence or permanence of environmental influences on human groups. With deep piety he regarded the earth as planned to provide differing habitats for peoples. His definition of natural regions was based not on interest in the genesis of physical land forms but on a desire to discover the potentialities of human progress. From a knowledge of the physical world he proceeded to his ultimate theme, the relation or value of the earth to man. Some of his most familiar conclusions deal with the superior design of the European continent, especially as to its coast line and the smallness of its natural regions, and with the barrier of highland and desert in inner Asia.

Ritter redefined for his time the Aristotelian and Strabonic view of environmental influence as determining the course of history. Among his many followers were [Henry] Buckle in England and [Arnold] Guyot, who for decades set the form in public school instruction in the United States. Ritter's influence on schoolmasters was greater than on scholars, since he made no field observations, devised no new method of gathering and inspecting data, and reformulated an age-old problem without serious analysis of its component elements. His enormous industry in combing literature still lends some bibliographic importance to his volumes on Asia, but he is now rarely read, whereas Humboldt, who ranks with him as a founder of modern geography, remains an important source of contemporary knowledge.

NOTE

Reprinted from the *Encyclopaedia of the Social Sciences* 13:395. Copyright 1934, Macmillan, New York.

39

Ellen Churchill Semple, 1863–1932 (1934)
Carl O. Sauer

American human geographer. Coming under the influence of Ratzel in his early Leipsig days, Ellen Churchill Semple transplanted the authentic anthropogeography of Ratzel to the United States and with undiminished enthusiasm devoted her life to its propagation. In no other country did Ratzel's views find so active and successful an apostle. Through a wide range of stimulating lectures, given for many years at the University of Chicago and later at Clark University, she exerted an influence over virtually the entire living generation of American geographers and recruited to the field a large number of its present representatives. Prior to her time geography in the higher institutions of the United States was physical geography. She pioneered human geography so successfully that many geographers have completely dissociated themselves from physical geography, a result which Miss Semple always deplored.

The whole of her scientific philosophy is derived from Ratzel. She too dealt not with things but with a relationship between things; anthropogeography was a view of life. Her *Influences of Geographic Environment* (New York, 1911) was undertaken originally as a translation of Ratzel's *Anthropogeographie*. The finished study differed from the original not in its concepts but in its illustrations; it was the master speaking through the pupil. Although it is the least original of Miss Semple's contributions, its effect was great because few Americans had read Ratzel.

Miss Semple defined her objective as "the study of the influence of geographical conditions on the development of society." This end is clearly expressed in her *American History and Its Geographic Conditions* (Boston, 1903; rev. ed. 1933). The volume broke new ground in American history and became the model of historical geography in the United States. It develops with spirit and skill the theme that the historical event is to be understood in terms of its physical environment. Miss Semple was never concerned with the origin, content, and succession of culture areas as the *Siedlungsgeographer* have been.

Her final work and most fully documented study, *The Geography of the Mediterranean Region* (New York, 1931), is more strictly anthropogeography than historical geography. It deals with the ancient Mediterranean area without regard to historical sequences. Such themes as barrier boundaries, geographic conditions of ancient tillage, templed promontories and their relation to navigation are developed from a broad knowledge of classical literature. Miss Semple remained true to Ratzel in making no restriction as to kinds of cultural data taken under observation but in being interested in whatever data could be related to environmental conditioning. She did no more than Ratzel in solving the methodological difficulty as to how such a relationship could be evaluated scientifically.

NOTE

Reprinted from the *Encyclopaedia of the Social Sciences* 13:661–662. Copyright 1934, Macmillan, New York.

40

Herbert Eugene Bolton, 1870–1953 (1954)

Carl O. Sauer

Herbert Eugene Bolton died at his home in Berkeley on January 30, 1953 in his eighty-third year as the result of a cerebral hemorrhage six months earlier. Until this last, and only serious illness, he continued in full vigor of body and mind, unravaged by the years and undiminished in his zest for working with his beloved documents on the Spanish Borderlands.

Bolton was born July 20, 1870 in Wilton, Wisconsin, then and still a small village in the hilly section of the Driftless Country to the east of La Crosse. In this belated frontier of the middle of the last century small farms were cleared on creek bottoms and ridge tops, with added employment in the cutting of pine and oak. Indians, especially Menominee, continued to live on the fringes of the white settlements, a nearby Indian school being maintained until a few years ago. Bolton recalled that in his youth flocks of passenger pigeons came to roost in the woods above his native Kickapoo River. He grew up on a frontier and frontier life remained his foremost interest.

His university work was taken at the University of Wisconsin, with an interval of two years as principal of a school in a village near his old home. The course pursued was law, in which he was graduated in 1895. Graduation was followed by marriage, to Gertrude Janes, surviving companion of a long and harmonious life. After one more year of teaching school, he returned to Madison as graduate student in history and assistant to Frederick Jackson Turner, the historian of the frontier. I never heard him refer to his law school days, nor did he show interest later in matters of jurisprudence. In 1897 he went to the University of Pennsylvania on a Harrison Fellowship, receiving his doctorate in 1899 at this then leading center of historical studies. To the impress of Turner at Wisconsin was added that of McMaster and Cheyney at Philadelphia. The Milwaukee Normal School, where a number of scholars got their first academic posts, had him as teacher of history until he was invited to the University of Texas in 1901, mainly to instruct in medieval history.

I do not know how Bolton the medievalist-to-be failed to materialize, but at Texas he found the direction of his life work very promptly. Again he was on a frontier, but one very different from that of Wisconsin and Turner. This was a strange world of which he had no previous knowledge and which lay pretty well outside of the ken of American scholars. His own interpretation was that by chance and good luck he got dropped into a virgin field and that he had sense enough to see that this was opportunity rather than exile from the traditional haunts of historians. He bothered no more with questions of the Middle Ages but started to learn Spanish, of which he knew nothing. On his first Texas vacation, the summer of 1902, he headed for Mexico City to get acquainted with the greatest Spanish archive of the New World, scarcely known this side of the border. Language was still a partial barrier and he was not facile at language. His first inquiry was directed mainly at Texas materials, but [he] soon discovered that here lay historical treasures unimagined. For the first time he knew what he wanted to do and he set about it with the single-mindedness and tenacity that were to characterize his life. Thereafter all his great energy and ceaseless industry were to be directed toward the neglected Spanish frontier.

Careful craftsman that he was, he worked his way slowly into the greater field by doing small pieces concerned with Texas. The Spanish settlements in Texas were imbedded in a diversity of Indian peoples, some hunter warriors, others more primitive and timid collectors. The latter were the objects of mission activities, the more turbulent interior tribes were contained by military expeditions and posts. The Spanish records had meaning as he learned to know the Indian background, then quite unconsidered. Frederick Hodge was at that time undertaking the *Handbook of American Indians North of Mexico* (published in 1907–1909) for the Bureau of Ethnology, and [he] found in Bolton the collaborator who, by a round hundred articles, reconstructed the life of the Indian peoples for the western Gulf Coast and the interior plains behind. Meanwhile also John F. Jameson, for the Carnegie Institution, had begun a series of bibliographies of foreign archives for United States history and committed Bolton to prepare the volume on Mexican archives (published in 1913). This gave [him] the opportunity, realized to the fullest possible extent, to investigate what was accessible in the National Archive of Mexico, and also to uncover a number of collections in religious and provincial centers.

In 1909 he moved on to Stanford University and the Pacific frontier of Spain. Two years later came his final relocation, to the University of California, at Berkeley. This was one of the incidents that resulted in the "gentleman's agreement" between the presidents of the two universities thereafter to refrain from raiding the other campus. Bolton introduced at Berkeley a new beginning course on the History of the Americas which caught on so promptly with the undergraduate body as to shift the whole structure of the department. Graduate students also formed a rapidly growing center of Latin American studies about the new maestro.

The preparatory years elsewhere became at Berkeley the period of fruition. Spain had thrust northward into the interior plains and Southwestern plateaus but also into California and the Atlantic Coast of the Deep South; the Spanish Borderlands from coast to coast henceforth were his range. His output of publication in the [nineteen] tens and twenties was prodigious and supplemented by a host of studies that he set for the apprentice scholars. The Spanish northwest having been richer in events and personalities than the northeastern fringe, California, Arizona, and Sonora were most fully exploited as to documentation of exploration and colonization.

Bolton was a dogged and skillful detective in finding and piecing together the pertinent documentation. This he amassed in his own and the Bancroft Library collections, and it served his students as well as himself and will continue to do so for future generations. He enjoyed translating into English full texts of major journals and relations, for, he said, he wanted others than Latin American scholars to share with him the deeds and thoughts of these his constant companions. He was a model of historiography in the exhaustiveness of his documentary searching and the precision of annotation. A later [Francis] Parkman, [Bolton] delighted in retracing expeditions step by step on the ground, in rediscovering the sites of Indian villages, in identifying Spanish constructions. I cut some of his earlier trails in Mexico, where he was remembered with wonder as the American *sábio* who persisted in following lost trails across ridge and arroyo, not for lost mines as might other Americans, but to find out where some Spanish captain had camped two hundred years ago or where a Jesuit had preached to forgotten Indians. He carried verification all the way and so, when he composed, the land and skies, the desert plants and water holes found their place in his canvases. None other has been as good an historical geographer of the Southwest.

Mainly, he was [a] narrative historian. Most of his work was cast in books about individual pioneers, missionaries such as Kino, Palóu, and Escalante, captains such as Anza and Coronado. Only rarely did he try large synthesis. On the occasion of being the Faculty Research Lecturer (1917) he gave a memorable presentation of the Spanish mission as frontier institution. Unlike Turner, his interest in the frontier was not to test a social or political thesis; he liked the open spaces and the people who lived there; his motive was esthetic, not moral or teleologic.

There was in him awareness of the sweep, the interlocking of all New World history and of its roots in Europe. He believed that Anglo-American history needed to be put into the context of the overseas spread of all European peoples and institutions. Such was the theme of his course on the History of the Americas and the substance of his presidential address to the American Historical Society at Toronto in 1932.

The master, at work almost all of his waking hours, and in great good spirit, set the pattern and pace for the many students who came to him. Perhaps a hundred doctoral dissertations and several hundred master's theses were somewhat too many. It was in Bolton's nature to encourage effort and to share the enthusiasm of the eager student. The apprentice who kindled to the task found his professor's study always open for discussion and counsel. There would be fewer devoted historians in this country today without his cordial invitation, and some of them have surpassed expectations, including their own. The regard and affection in which he is held by his progeny is recorded in two *Festschriften, New Spain and the Anglo-American West* (two volumes, 1932) and *Greater America* (1945). It also speaks in the memorials after his death, such as the one by his successor as Director of the Bancroft Library, George Hammond (in *The Americas,* vol. 9, with record of principal publications, honors, services). He was elected a member of the American Philosophical Society in 1937 and frequently attended its meetings.

The historical characters with whom Bolton dealt were men after his own heart, straightforward, simple, and steadfast. Once I gave him a document showing that one of his captains was a conniving scoundrel. He was a good deal shaken; the man had betrayed his trust; the planned volume was dropped. It was not in Bolton's nature to expose the rascal, but he would have no more to do with him. The sidelight may help to explain why Bolton kept out of diplomatic and political history.

Stratagems were foreign to his nature. He worked on matters unconcerned with devious personalities and concealed motives. He disliked very few people, and these he put out of his mind. Politics did not interest him. If he found himself in a group engaged in political discussion, campus or national, he would drift away to get back to his study. Administrative duties he disposed of quickly and directly. His way of life was serene.

NOTE

Reprinted with permission from *The American Philosophical Society, Yearbook 1953*, 319–323. Copyright 1954, American Philosophical Society, Philadelphia.

41

Homer LeRoy Shantz, 1876–1958 (1959)

Carl O. Sauer

The passing of Homer Shantz on June 23, 1958, closes a volume in American geography, of the years when geologists and biologists, historians and economists, mingled at geographical meetings and helped the rising generation of professional geographers to find footing and direction.

The Shantz family settled in the Grand River country of western Michigan when it was a farming frontier against the north woods, but Homer grew to manhood in Colorado—a Colorado that still was largely mining camp and stock ranch. The years he lived at Colorado Springs and studied at Colorado College exposed him to a community of strong and fine Eastern habits, planted at the foot of the Rampart Range but spiritually still rooted in the Hudson or Connecticut Valley. Here he was well instructed in the traditional liberal arts, but here also the need grew upon him to go out from the campus and its interests and study nature on the high mountain lakes and in the foothills where grassland and pine forest meet. On graduation in 1901 he was named instructor in biology at the college and promptly married Lucia Soper, who was to be his companion for more than half a century.

The next formative step was graduate work at the University of Nebraska, aided by an instructorship in agricultural botany. Professor C. E. Bessey and the Botany Seminar were at the height of their productiveness, and the Nebraska Botanical Survey was doing original work in plant geography under Roscoe Pound, then only partly committed to jurisprudence. [Frederic] Clements was beginning to formulate his system of plant ecology. Shantz prepared a dissertation on the vegetation east of Pikes Peak and received the doctorate in 1905. Grasslands were the center of interest for all.

After three years of teaching botany at the Missouri and Louisiana state universities came a long, highly productive period with the United States Department of Agriculture. The original objective was to develop alkali- and drought-resistant plants for our semiarid lands, and the first publication that resulted therefrom was *Bulletin 201* of the Bureau of Plant Industry, *Natural Vegetation as an Indicator of the Capabilities of Land for Crop Production in the Great Plains Area* (1911). There followed the classic papers on plant physiology, mainly done in collaboration with Lyman Briggs, on wilting coefficient, water requirements, and transpiration rates of plants. These studies represented a happy union of experimental biophysics and field observations.

Through Isaiah Bowman the American Commission to Negotiate Peace invited Shantz at the end of the First World War to advise on natural vegetation and crop potentialities in Africa, and the Smithsonian Expedition of 1919–1920 gave him his first chance to try out his Great Plains background experience in observations from Cape Town to Cairo. "Urundi, Territory and People" was his first publication (1922) in the *Geographical Review* (12:329–357); the following year the Society issued the monograph by Shantz and C. F. Marbut on *The Vegetation and Soils of Africa*. A second trip, to East Africa, followed in 1924, under the auspices of the International Education Board and the Phelps-Stokes Fund.

Meanwhile Shantz had become a valued member of geographic circles. In 1924 he was elected Corresponding Member of the American Geographical Society, and he was a steady contributor to the [*Geographical*] *Review* for many years. He was equally at home in the Association of American Geographers, enlivening its meetings with informed, original, and at times quizzical comment. A major paper presented before the Association was "The Natural Vegetation of the Great Plains" (1923). When Wallace W. Atwood, Sr., became president of Clark University, he invited Shantz, Marbut, and O. E. Baker to become visiting lecturers in his new School of Geography, a connection that was continued until Shantz left Washington in 1926 for Illinois. Geographers everywhere know and use the Shantz and [Raphael] Zon "Natural Vegetation" section of the *Atlas of American Agriculture* (1924) and the nine articles by Shantz on the "Agricultural Regions of Africa" in *Economic Geography* (1940–1943).

From 1928 to 1936 Shantz served as president of the University of Arizona. I had the pleasure of being present at the inauguration, at his invitation. Principal speakers were geologist George Otis Smith, sculptor Lorado Taft, and paleontologist John C. Merriam. Shantz himself spoke movingly of the university that might rise in Arizona, nurturing liberal learning and searching for new knowledge, much of which might be awaiting the seeker within the state. He expressed his faith that "each man must choose for himself which lines of work he can most successfully pursue." In part his hopes were realized, and he was able to bring good men to the staff. He had hoped to continue to teach, and for a while he did give a course on wide horizons in ecology. Also, he was able to make some study of the Arizona desert, of which an extensive collection of photographs, with his notes, are at the university. But he was increasingly regretful that there was less and less left of him as scholar and teacher and was convinced that he did not wish to become wholly absorbed into administration.

In 1936 he was made chief of the Division of Wildlife Management in the United States Forest Service, a position he held until his retirement in 1944. This was only a partial relief from administrative duties, but it did give opportunity for a return to ecology in an inclusive sense—plants, animals, man. What he learned and thought about in those years found its way into print chiefly in reports and hearings for Congressional committees and is brought out best in his answers to questionings on wilderness areas, livestock versus game, proper harvesting of the game crop, the delicate balance of nature and unforeseen effects of its disturbance.

The last remove was to a modest home in Santa Barbara, where he and his wife lived quietly and happily for ten years more. Part of his leisure was spent in going over the thousands of vegetation photographs, with their site notations, which he had taken in his active years. The death of Mrs. Shantz left him desolate, but he found partial relief in working on a study for the Forest Service (with his old associate W. [G.] McGinnies) of changes that had taken place in 45 years between grassland and forest in Colorado. He was back to the scenes and memories of the years when he began his life as a student of nature.

At the Geography Branch of the Office of Naval Research the idea came up in 1956 that he might do the same thing for Africa, revisiting the sites he had studied long ago. When this was presented to him, his immediate response was, "Now I have something to live for." The ONR has put at my disposal the African letters of this man of 80 years, briefly noting that his locomotion was limited, but delighted or dismayed at finding an old observation site with the original photographs and notes for comparison. "We desperately want to finish the job. So far ([in] Tanganyika) we have fared well, only three nights with our boots on." After his return to Santa Barbara he reported himself "in good shape. Can work long hours and am happy while at work."

Before this job was completed (it was in press at the time of his death), he had presented plans for three more comparisons of "vegetation then and now"—the northern Great Plains, the Great Basin, and the Southwest, for which he had more than three thousand of his own early photographs and their site notations. In his opinion, about eighteen months (three field seasons) would be needed, and he was in a hurry to get going. The ONR promptly extended his contract, and he set out for the northern Great Plains with Professor W. S. Phillips of the University of Arizona as his associate. For the rest I shall quote from a letter by Phillips. "While we were in Spearfish we found one of the old photo-points perfectly, and it was while we were setting up the photographic equipment that he became dizzy. He asked me to finish the picture and then we were going into town. We went into town, saw a doctor, and the doctor sent him to Rapid City hospital where he spent a very comfortable night, even talking with the nurses about the work we were doing. At 5 o'clock Monday morning the nurses heard him breathing deeply and discovered his pulse was very weak. He died in 10 minutes. He was contented and happy right up to the last." There could have been no better end to the long life of a man great in gentleness, spirit, and vision.

NOTE

Reprinted with permission from *The Geographical Review* 49:278–280. Copyright 1959, American Geographical Society of New York.

42

Erhard Rostlund, 1900–1961 (1962)

Carl O. Sauer

Erhard Rostlund died suddenly on the evening of July 13, 1961, as he was about to come to the campus to introduce the public speaker of the evening. We had known vaguely that he had been under a doctor's orders for some time, but neither by word nor by mood did he indicate that he was living in the shadow of his end. The memory of his last day, in fact, is of his good spirits.

He was born on January 21, 1900, in the Swedish forest-products port of Norrsundet on the Gulf of Bothnia, where his father worked in a sawmill. Orphaned in boyhood, Erhard helped support the family by jobs in the mill while he finished his common-school education. As a husky youth of sixteen he left home to become a seaman for the next half-dozen years and thus came as an immigrant to San Francisco in 1921. He worked at construction jobs in the Monterey area and soon entered the Pacific Grove High School, from which he went to Los Angeles to graduate at Franklin High School in 1924. The humanitarian Asilomar [Conference Center] influence under which he had come at Pacific Grove turned him to Oberlin College (1926–1928), where he started to study psychology. The record is incomplete; he sailed on Great Lakes vessels and returned to the West Coast to ship out on freighters along and across the Pacific.

In 1932 he entered the University of California at Berkeley, and in 1934 he graduated with highest honors in anthropology. A year of graduate work in that department followed. At this time the relations between Anthropology and Geography at Berkeley were closest and most fruitful; the students in one department were almost equally at home in the other. [Fred] Kniffen and [Peveril] Meigs in Geography had done important field studies contributing to ethnography. [Bob] Richardson, [George] Carter, and [Dan] Stanislawski in Anthropology gradually shifted over to Geography. There was a joint seminar and even some consideration of a joint department. Rostlund thus came to take active part in our classrooms and field work.

But the call of the sea was still insistent. Between 1936 and 1940 he was mostly at sea, earning in this time third and second mate's papers. The sea gave him a sense of physical activity and accomplishment; it also gave him time to read and think and write. From this period, date some published short stories and two unpublished sea novels. By 1940 he had come to the final decision; he would be a teacher of geography. So he came back to the university to study geography and work out a secondary teaching credential. When the United States entered the war, he went as a civilian instructor to the Santa Ana Air Base; at that time he married Esther Gilbert, who survives him. He volunteered to become a G. I., was soon commissioned second lieutenant, later was transferred to the Engineer Corps, and finally was given a discharge from Walter Reed Hospital in the fall of 1944. The College of Marin gave him a year of teaching and rehabilitation work with prisoners at San Quentin State Prison, and then he came back to us for good.

By reason of his teaching experience and known ability he became a lecturer in the department in 1945, before he finished his master's thesis in 1946. According to the prevalent practice of the University of California, he was kept in this semi-academic status until he received his Ph.D. in 1951, though he was soon in full charge of the heaviest teaching commitment of the staff, that of giving the Freshman courses. He might have submitted a creditable doctoral dissertation in much less time [had he not been] divided between it and instruction, but he said that he wished to satisfy himself as well as his committee. The result, "Freshwater Fish and Fishing in Native North America" (*Univ. of California Publs. in Geogr.*, Vol. 9, 1952), was in its first part an original review of the fish resource with respect to the food value of principal species, quantitatively considered as to the productivity of their waters. The second part was a masterly inspection of aboriginal fisheries and fishing methods, fully documented by map distribution for each technique. Of this work Professor [Alfred] Kroeber commented that it was "a monument more lasting than bronze."

Rostlund became associate professor in 1957 and from then to 1960 was assistant dean in the College of Letters and Science.

Rostlund was of that rare and precious breed of teachers to whom the communication and meaning of knowledge come above all else. No one else among us at Berkeley lit the lamp of learning so well for the young student. He gave no humdrum lecture, and none into which he had not put every effort at preparation. With an armful of maps he would stride down the hall saying "Time to preach now," and in the best sense this is what he did.

His *Outline of Cultural Geography,* done in 1955, has influenced markedly the second step in college geography, especially in the Far West. For upperclassmen the course in the regional geography of North America throve under his care: its students were not only directed to the materials students in geography are accustomed to use; they were also given appreciation of some of the best insights into our land and its life as found in accounts of early travel and in American letters. Not the least of his contributions was that students were induced to search out the good literature they otherwise might not have known. The last extension of his curriculum was to take on the course in natural resources and their exploitation, to which he brought his interests of long standing in sea and stream, forest and range.

What Rostlund wanted to do as a scholar is expressed in general terms in a brief article in *Landscape* (1956), in which he gave his affirmation of the stated purpose of that journal as "the Study of the Earth's Surface as Modified by Man." The *Annals of the Association of American Geographers* published two studies that show well the manner in which he asked questions in historical geography and how he sought the answers, the first on the origin of the Alabama Black Belt (1957) and the second on the ranging of the bison into the southeastern states (1960). He was interested first of all in the land as it was under aboriginal occupation and then in the earlier and simpler rural American scene, in farmsteads, barns, fences, and country schools (for example, "The Road to Brown School," *Landscape,* 1959). The forest and its management had his attention in several articles and reviews; he liked to get out into the field where lumbering operations were going on and observe them with an experienced eye. He was engaged at the time of his death on a number of projects that were keeping him happily busy. No running out of ideas or interests as a student was in his prospect. How he thought and sought is shown in various notes and book reviews in the *Geographical Review. Landscape* was his favorite medium; here he felt relaxed among fellow amateurs looking at the true attainments and sad failures of men in their brief lives as inhabitants of earth. Perhaps his deepest satisfaction, though he might have winced at the word, was in experiencing and somehow expressing the esthetics of man in harmony with his place.

A stranger to all guile, rancor, and ambition, he lived, great of heart and perceptive of eye, "by the side of the road." Those who went with him for a piece will remember.

NOTE

Reprinted with permission from *The Geographical Review* 52:133–135. Copyright 1962, American Geographical Society of New York.

43

Richard J. Russell, 1895–1971 (1967)

Carl O. Sauer

The Hitchcock Lectures, begun at the University of California in Berkeley in 1909, have been unrestricted as to their "scientific or practical" range except that they are not to be "for the advantage of any religious sect, nor upon political subjects." They are public, annual, usually a single series by a scholar from other parts. The Earth Sciences had early attention in the lectures of Harry Fielding Reid on earthquakes and in those on vertebrate evolution by Henry Fairfield Osborn. In later years a half dozen series have been concerned with aspects of the history of the earth and the organic and physical processes that are expressed in its changing face. The Russell lectures of 1965, here published, are the first to take features of land and sea as their theme, a subject now known as geomorphology and which in earlier days and with different emphasis was considered a part of physiography or physical geography.

The appointment of Richard Russell as Professor on the Hitchcock Foundation had several good reasons. He has been a principal in revitalizing geomorphology, giving it new directions, new and sharper means of inspection, and linking it to other disciplines. It was hoped that in meeting the obligation of the lectures he would give an overview of the several lines of inquiry he has followed, their interrelation, results, and prospects. He has done so with a synthesis and perspective and simplicity that will reward any reader who is attentive to the features of land, stream, or seacoast. Also, it was proper to bring him back to the place where he spent his formative years and from which he went out to a life of greatly independent discovery.

In the twenties the doctrine of William Morris Davis prevailed in all countries of English speech: that the surfaces of the land were to be explained by cycles of erosion, characterized by stages proceeding from youth to old age. The system had the elegance of attractive models that were proposed as representing the stage to which any given land surface would be assigned. By accepting premises and presumed criteria, attention to event and process in actual (geological) time was excluded. Landforms were thus taken out of the context of earth history.

When Professor Davis retired from Harvard University he moved to California, repeatedly lecturing at Berkeley. Here the young Russell became companion to the old master of physiography, and here also he began to have doubts that the Davisian doctrine was adequate or even valid. His Hitchcock Lectures begin with the influence of Davis and how he found his independence by moving to Louisiana in 1928. This base, with which he has chosen to remain, gave him a great new field of study, beginning with the Mississippi River, its flood plain, and delta. Its terraces led to new insights into the course of glacial and later time in lower latitudes. The Gulf Coast plain with its shores and shallow waters came into his widening range of inquiry, as did alluvial valleys in other continents. Finally his Coastal Studies Institute has engaged in work in the morphology of the borders of sea and land about the world.

This is the record and reading of the forty-year trail of discovery Russell has followed, told in sequence and thus also partly autobiographical. It tells of observations at first casually noted, becoming significant clues, and continued to new understanding of forms and processes. Meanders, levees, terraces that disappear below the flood plain; deltas that are continuing accumulations of sediment and which do not grow as to extent of surface; shapes of lagoons and beaches; beach rock and coral strand and reef—these are some of the items he discusses here. His assurances to the reader that geomorphology is an exciting science in its infancy are substantiated by the new vistas he opens on the nature and origin of lowlands and fringing seas.

NOTE

Reprinted with permission from *River Plains and Sea Coasts,* by Richard J. Russell, v–vi. Copyright 1967, Univ. of California Press, Berkeley.

44

David I. Blumenstock, 1913–1963 (1968)

Carl O. Sauer

David I. Blumenstock first came to Berkeley in the summer of 1935, newly graduated from the University of Chicago, where he had studied geology and mathematics along with geography. He made his own way through college in those depression years, somewhat by aid of scholarships, then of very modest amounts and under sharp competition. That he held three scholarships in his undergraduate days does not mark him as an award hunter but as necessitous and enterprising. His interest in money never extended beyond immediate and minimal needs until his marriage in 1946 when his wife Nancy began to inform him gently of the uses of a household budget. I do not know nor think that he ever took a position for its salary or prestige.

The graduate years at Berkeley turned him to climatology, working under John Leighly, jointly exercising their mathematical bents on the dynamics of weather. David was a member of my weekend field course, at that time directed to the Marin shore north of San Francisco. It was a lively group, sparked in particular by himself and Erhard Rostlund, complementary personalities who formed a lifelong friendship. Surf and tide, cliffed headlands about pocket beaches, marine terraces, coastal chaparral and wind-sheared laurel and oaks, redwoods in the canyons, sheep and cattle pastures, and houses sheltered and designed against the sea wind were the stuff for observation and discussion. The Chicago city lad saw alertly and sharply and at noon lunch and evening camp would spice learning with spontaneous fun.

As teaching assistant, he communicated to undergraduates his and our interests in the wider substances and problems of physical and cultural geography. To our instruction he added work in geology and plant physiology, and courses in anthropology with [Alfred] Kroeber and [Robert] Lowie. Professor Kroeber at the time was drawing up the trait list for California Indians and trying to determine their statistical correlations, to which David contributed a critique of relevance and validity.

From 1938 to 1941 he was Research Climatologist in the Soil Conservation Service [in] Washington under Warren Thornthwaite, completing his dissertation on drought lengths and frequencies in the United States. The USDA Yearbook of Agriculture of 1941 carried their joint memoir, *Climate and the World Pattern*. The Climatic and Physiographic Division was engaged in various studies of good promise as to the processes and incidence of soil erosion until interrupted by the war.

During the war years David served as meteorologist and climatologist in various capacities. He was in the Army Air Force Weather Service [in] Washington until 1944, with terms at UCLA and California Institute of Technology. The studies which he directed, made, or shared concerned possible military objectives and contingencies and were restricted within the Defense Department. In 1944 he was assigned by the U.S. Naval Reserve to Pan American Airways as weather forecaster of the Naval Air Transport Service; he was stationed in New Caledonia, the New Hebrides, and Hawaii. In 1945 he returned to the Army Air Force Weather Service as Consultant and Senior Meteorologist, working in Asheville, N.C.; at Langley Field, Virginia; and in Washington, D.C.

Meanwhile, he had begun to compose his thinking on the mobile atmosphere that was to result after a decade in the *Ocean of Air* (1959). After the war he was attached to the Navy Electronics Laboratory at Point Loma in charge of publications and as a member of the administration. Also, he tried out his climatology at San Diego State College and at San Francisco State College.

In 1951 he returned to the Atlantic coast as consultant to the Army Air Force in Washington and Baltimore, adding a position as lecturer at Rutgers University from 1952 to 1955. He was an active participant in the Wenner-Gren Symposium at Princeton in 1955 and guided an international group across New Jersey as an excursion in human ecology.

In 1956 he came back to the Pacific to be Pacific Area Climatologist for the U.S. Weather Bureau. Travel between Hawaii, Guam, Samoa, and the islands of our Trust Territory provided knowledge of the islands as a backlog for later thinking. Diverse studies of island climates were formal results. In the Ninth Pacific Science Congress (1957) at Bangkok he headed a UNESCO symposium on the humid tropics, [and] at the Tenth (1961) one on Pleistocene and post-Pleistocene climates in the Pacific. His appointment to the Pacific Science Board followed in 1962. Also, he found time to take part in the Graduate Faculty of the University of Hawaii.

In 1961 the wanderer came back to join us at Berkeley and to settle down with the observations and inquiries he had been gathering in many places for more than twenty-five years. The trial period as Lecturer was wholly congenial, and he was made Associate Professor the next year and Professor in 1963. He did not live to enter his third year, dying on August 23, 1963, after a heroically borne illness.

At Berkeley, he had less than two years before he was stricken, but he lived them to fullness and brought new horizons to students, both advanced and beginning. He held a memorable seminar on past climates that was well and actively attended from various departments. He was thinking years ahead as to studies of and about the Pacific Ocean that might be individual and associative explorations of nature and culture. He kept reaching for discovery and understanding until the darkness closed in.

Of his friend Rostlund he said that integrity was his clear mark. Of himself, also the simple epitaph might read *Integer Vitae*. He was unscarred and undemeaned by life, generous in all ways, unvexed by self-interest and ambition, eager to know, and willing to share.

NOTE

Reprinted with permission from the *Yearbook of the Association of Pacific Coast Geographers* 30:9–11. Copyright 1968, Association of Pacific Coast Geographers.

IX

INFORMAL REMARKS

45

Introduction

Philip L. Wagner

> [M]an—the most perilous social animal—is viewed as increasingly the modifier of the environment—perhaps to his greater good and glory, perhaps to his own decline or destruction.
>
> Carl O. Sauer, "The Seminar as Exploration," 1976

Before plowing through what I will now say, the reader would profit greatly from reading Carl Sauer's short evocation of the Cabrillo voyage in "The Quality of Geography," simply in order to sense the enchantment of his personal style (1970a:7–8, ch. 48 herein). And in a letter to the fine frequent standard bearer of the "Berkeley" viewpoint, the magazine *Landscape*, Mr. Sauer exposes in a genial way his personal faith and philosophy (1960, ch. 46 herein).

Mr. Sauer once told me that when he received the invitation to teach at Berkeley, influential faculty members at Michigan urged him to consider remaining there—in the economics department!

The four selections of informal remarks that follow ought to show how alien an environment an economics department would have proved to be for Mr. Sauer. The whole corpus of his later writing corroborates the point. Quite apart from skepticism about quantification and model building, these articles clearly demonstrate that the geography Mr. Sauer advocated was built on a foundation of direct immersion in field observations, interpreted in a broad, rather speculative fashion, informed by solid science, and directed above all toward environmental change over time in a close reciprocal relation with cultural development. Preformed deductive schemes could not compete with alert and insightful firsthand observation in concrete earthly context. Such an inductive strategy necessarily implied some form of mapping and orientation in historical time as well. Moreover, it relied on a sense of local relatedness and mutual influence (holism, *Gestalt*) among the phenomena examined.

These selections do not, however, amount to a set of rigid didactic instructions, by any means. Although Mr. Sauer, as I recklessly ventured in conversation with him once, did tend in his early writings to tell other people what to do (cf. Section III of this volume on "Toward Maturity"), in later years when I knew him he would scarcely tolerate such preaching. For instance, he gave indications of acute suffering when confronted with Richard Hartshorne's *Nature of Geography* (1939). When, while a graduate student, I incautiously brought a copy of the book into the department office, Mr. Sauer harrumphed, picked it up in a gingerly gesture, and asked me scornfully, "You're not reading *that*?"

The geography he advocated and exemplified employed a direct reading of the country for both an insight into how it all fitted together and an understanding of how it all came about. Yet it certainly did not seek simply to delimit regions, in the manner that Hartshorne commended. On the day of my doctoral defense, high in the Berkeley hills at the home of Professor Wolfram Eberhard, whom a broken foot had immobilized, both Mr. Sauer and his faithful "first sergeant," Professor John Kesseli, exposed my naivete by eliciting through dogged grilling my embarrassed admission that, according to regional geography logic, even a single California oak tree might constitute a region.

Contrary to some criticism I have heard, Mr. Sauer did not rest content with speculations based on inspection of the scene. He reached out eagerly for further evidence and judgment both from original archival documents (a practice of his not perhaps sufficiently well represented in this present collection but attested by many of his monograph publications) and from a wide variety of physical and biological (but definitely not behavioral) science literature. He also drew whenever he could from the reports of scientifically inclined travelers, along with what he learned in friendly conversations with the simple folk of each particular place. Thus he read not only the earthly evidence but also "hard" science, Spanish colonial documents, local knowledge, and the observations of idiosyncratic German and other curious wanderers. (He would have loved that latter term!)

I like and frequently use the phrase "Culture and Geography." You can't have one of them without the other. And nothing better describes the enduring theme of Mr. Sauer's quest than this paraphrase of the somewhat less revealing title of one collection of his reprinted writing, *Land and Life* (1963a).

That theme reached far beyond the assignment of given cultures to particular locations, although Mr. Sauer did pay close attention to such basic facts, and investigated and wrote about them early (cf. especially 1932a, 1934a, and 1935a). The phrase "site and culture" in the title of an early study (1927b) offers an indication of what became an enduring focus. An underlying, yet more general commitment to distributional plotting, furthermore, shows up strongly in his paper "The Seminar as Exploration" (1976a, ch. 47 herein).

As he said in his talk "The Quality of Geography," Mr. Sauer held that "the map represents an assemblage of things that you think belong together" (1970a:6, ch. 48 herein). He clearly sensed in a person's childhood enthusiasm for maps a hint of the sort of curiosity that he esteemed as essential for his kind of geography. Even a number of the leading quantitative geographers have also, in their writings or in their personal conversations with me, confessed to a near-obsession with maps early on. How else to explain their alternative description as "spatialists"?

Plainly, however, Mr. Sauer looked for more than simply a map-sense to identify his kind of geographer. He squarely embraced "a kind of understanding that is other than an examination by analytic methods" (1970a:6, ch. 48 herein), and assigned a decidedly limited role, accordingly, to quantification. Even so, he had a high regard for the complicated, even abstruse, mathematical reasoning used by Milutin Milankovich (see Zeuner, 1950:138–141) to account for Pleistocene glacial periodicity. In the same worthy cause, Mr. Sauer animated some graduate students, myself included, to tramp around monotonously making precise measurements of the coastal terraces in northern California and elsewhere.

Precise measurement and mathematical method found their proper place in the realm of physical geography, which itself sometimes—nay, frequently—illuminated problems of human geography. Mr. Sauer steadfastly championed a solid grounding in physical geography, and himself wrote a number of papers entirely on physical topics (cf. 1929a, 1930a, 1932c, 1936c). He had little but scorn for geographers who disclaimed any interest or competence in physical geography.

Thus the map, definitely taking heed of physical features, serves as "the best common ground by which we can identify the convergence of our interests" (1970a:5, ch. 48 herein) and goes right along with "explanatory description" (1976a:79, ch. 47 herein) right on the spot. Combine the two and you arrive at geographic distributions, yet refuse to become bogged down in "regions" for their own sake, because you must take to heart the *explanatory* aspect of your descriptions. Turn away from that, and you turn off curiosity—a prime virtue for Mr. Sauer.

The distributional pattern, as Mr. Sauer wrote in "The Seminar as Exploration" (1976a:80, ch. 47 herein), becomes "an incitement to secure missing evidence, to consider evidence as to its correctness," and to raise "alternative hypotheses of explanation." The step might otherwise lead, as in the case of William Morris Davis or Ellsworth Huntington, to sterile "model building" (cf. 1976b:72, ch. 49 herein), which he accused, perhaps a little cavalierly, of disregarding the time element. After all, both Davis and Huntington created models of imputed temporal progression. Presumably, Mr. Sauer simply wanted to indict an unrealistic disdain on the part of those scholars for circumstantial history in the concrete—as we might say these days, a scorn for detectable "real time" events and processes.

For Mr. Sauer, then, explanatory description emphatically did not mean an appeal to a prefabricated deductive scheme such as those of Davis and Huntington or the pioneering ecologists Henry Cowles and Frederic Clements (1976b, ch. 49 herein). Bereft of such neat preconceptions, explanation had to cast about freely among the "multiple working hypotheses" (1976a:80, ch. 47 herein) about which Mr. Sauer learned from the geologist Thomas Chamberlin at Chicago.

Within the arsenal of such working hypotheses lurked certain inveterate favorites, one of which, cultural diffusion, figured particularly prominently. Now, the concept of cultural diffusion does not imply just a wide scatter of random transmissions of new ideas in all directions from all conceivable sources, without any detectable order. Rather, it tends to expose what the German geographer Alfred Hettner, well known to Mr. Sauer, wrote of as the march of culture over the earth. In the works of the English prehistorians Harold Peak and [Herbert] Fleure, mentioned in "Casual Remarks" (1976b, ch. 49 herein), diffusion through space and over time exhibited very definite paths, pulses, and trends. It traced out of the ancient Middle East, through the Mediterranean basin, up into northwestern Europe, then out overseas. Given the paucity of knowledge of other parts of the world at the time those scholars wrote, such an account left a large part of human history still in the dark. It did, however, itself constitute some sort of model. Mr. Sauer's vigilant awareness of "non-commercial peoples" (1976a:80, ch. 47 herein), and his awe at the very depth of human time, protected him from uncritical acceptance of a plain "Eurocentric" account of cultural diffusion. He entertained, albeit cautiously, the possibilities of cultural comings and goings in other directions as well. He actually welcomed the wide intercontinental speculations of his students in seminars, but always voiced judicious reservations.

Culture—or more specifically and accurately, human activity governed by cultural understandings—works in concert with (other) natural processes to develop particular habitats. Mr. Sauer's pathbreaking 1925 monograph, "The Morphology of Landscape," depicted this interaction as the transformation of the natural into the cultural landscape. Yet I do not recall hearing Mr. Sauer, in my time, trying to evoke "natural" landscapes. In fact, he took satisfaction, as other selections in this volume show, in demonstrating the ancient and pervasive role of human work in extensively modifying the earth's surface elements, such as vegetation. As for "cultural landscape," the term does not appear at all in the four selections in this section, and even "landscape" shows up only once.

Contrary to some rather jejune assertions, he did not regard culture as some mysterious anonymous force. He always spoke specifically about actual people he had observed or, admittedly, conjured up in his nearly febrile imagination, or learned to know through archival documents, behaving in their own distinctive cultural ways.

So if Mr. Sauer in his later years, and those he directly inspired, did not seek to discover either regions or "cultural landscapes" as such, and declined to build grand abstract models, just what did they look for? In his "Casual Remarks," geography focuses on "a culture's environment" (1976b:71, ch. 49 herein). But if "Seminar as Exploration" proposes "selecting students with a bump of curiosity and thereafter . . . giving them the maximum freedom to choose their own direction" (1976a:78, ch. 47 herein), what common commitment emerges? Geographical distribution receives repeated mention in the four present selections, but can obviously serve as the main subject of ordinary fieldwork only exceptionally, for only the net collective results of many people's individual field observations allow for broad distributional conclusions. Perhaps, as stated in "Seminar as Exploration," "our major expectation is to make sense out of changes in the geographical patterns of human populations and activities, or human ecology" (1976a:80–81, ch. 47 herein)—again, an appropriate goal for collaborative enterprise.

Probably many or most of Mr. Sauer's graduate students would find congenial one clear research objective he confessed to: "to present life and nature as a whole in whatever area I had been studying" (1970a:6, ch. 48 herein). Such a formula would not suffice, nevertheless, to characterize fully what many of those students have accomplished. Some of them would uncover unexpected ancient land- and water-use systems in flood-prone tropical lowlands; others tracked down distinctive food preferences and prohibitions; some documented in detail the subsistence and societies of isolated mountain tribes; someone would bring to light unusual cultivated plants, cultivation methods, and uses of wild plants; yet a different group reconstructed provincial ways of life and work in the Spanish colonial domains. Remarkably, even a much more complete list of the topics pursued by Mr. Sauer's students would largely echo "the Old Man's" interests, although not because of any explicit instructions from him, or any pressure at all. Mr. Sauer would seldom go farther than suggesting some puzzling faraway situation, some odd anomaly that might "deserve a look." He would scramble around and somehow dig up some money, then let the investigators chart their own course, and he would savor their findings. Sometimes he would drop around in the field if he fancied the topic or locality.

Mr. Sauer possessed both wisdom and patience. He gave his students not arbitrary assignments but genuinely shared curiosity, lavish enthusiasm, inspiration, and firm support. Myself, I owe deep gratitude to his rather fatherly forbearance. He let me flounder through frustrated fantasies of studying the acorn as a human dietary staple, subsequently sent me off to Costa Rica to cover the belated pioneer colonization of a distant valley, then finally allowed me to sign off from the project, distasteful to me, and go my own way and document a tenuously surviving old-fashioned rural society.

Methodology one definitely did not discuss in the Berkeley department, but research had to measure up to and make proper use of any and all relevant scientific concepts and data, and of any available historical documentation. The possible accusation of acting as "mainly retail distributors of other people's wares" (1976a:78, ch. 47 herein) counted for little as long as the researcher maintained a full and sound command of those wares and put them to fruitful use.

What then did constitute the distinctive character, the common core, of the work of Mr. Sauer and his students? A "large measure of coherence and common interest" (1976b:73, ch. 49 herein) does unmistakably show through in that work. For one thing, a glance through the writings of Herodotus and Humboldt, Ratzel and Reclus, Vidal de la Blache and Varenius, will clearly substantiate that it resolutely maintains "the immemorial tradition of geography," as Mr. Sauer urged (1976b:70, ch. 49 herein). It envisions culture and geography as an inseparable unity, contemplating their linkages and interactions, their mutual evolutions and adjustments, at firsthand in the field and as registered in the surviving historical, archaeological, and genetic record.

Origins and dispersals of peoples and cultural traits, including domestic plants and animals, as well as local history and tradition in each particular place, and of course always the local physical habitat entire, frame the grand perspective, illuminating the autonomy and idiosyncrasy of local cultures and habitats. Economic rationality and calculation do not apparently dominate in most of those cultures; varied sentiments and symbolisms often count for much more. Hence, according to Mr. Sauer, "What makes a people choose and develop a certain mode of life and retain it" becomes a "major objective" of study. A supposed "necessity" then does not fairly account for invention (1976b, 74–75, ch. 49 herein). I should say, in fact, that we invent our necessities, and I think that Mr. Sauer, as a skeptical and downright nostalgic witness of change in the American life of his time, would have said the same (ch. 46 herein).

Then what vision did Mr. Sauer finally communicate? In addition to a somewhat romantic fondness for the traditional past for its own sake, his primary message cautioned us to not just tolerate, but to respect, preserve, and rejoice in diversity, whether of countrysides or folk behavior, as a safeguard against our own follies, heeding the dangers of reckless change. He encouraged alert curiosity, and advised us to focus it on any possible source of greater knowledge and wisdom, no matter how elusive and even recondite. He warned us, thus armed, to proceed with due caution into an uncertain future.

REFERENCES

Hartshorne, Richard. 1939. *The Nature of Geography*. Association of American Geographers, Lancaster, Pa.

Sauer, Carl O. For references, see the Sauer bibliography at the end of this volume.

Zeuner, Frederick E. 1950. *Dating the Past: An Introduction to Geochronology*, 2nd ed. Methuen, London.

46

Letter to *Landscape* [on Past and Present American Culture] (1960)

Carl O. Sauer

My span has covered seven decades that reach back into a greatly different world, one it was very good to have lived in. The Midwest of my youth was unaware of the gathering gloom over Europe, nor did we know that we were of the Gay Nineties or Mauve Decade, as later authors inferred. Mainly we were country-bred of prairie and woodland soil, and kept this knowledge and quality when we went to the cities to live. Main Street had not become a label of derogation for the way of our living, which was rooted in communities our people had founded and formed. Country or city we were still the sons of the Middle Border; it was in Chicago that I knew Hamlin Garland, presiding at the Cliff Dwellers, and Lorado Taft, at the Art Institute. Chicago, brash and bustling, was also the principal seat of art and learning. Its young University attracted from all parts of the Midwest staff and students into an academic community of rarely equaled quality. I was there in its golden age.

We have since become a greatly nomadic people at all levels and occupations, from the unskilled workman to the executives of business and government. The moving van and house trailer travel our highways coast to coast unendingly. Home ownership may be more practical and convenient than rental, but in either case there may be similarly short expectation of staying put. The community ties are greatly loosened or lost, the home a temporary address, not the place where the family has put down its roots. Our once great and sedentary farm population has dwindled to a tenth of the nation's total, is largely overage, and its farms are more and more dependent on ambulant services of machinery and trucks. Farm, orchard, garden, and barnyard are becoming vestigial along with the country school house.

The dominant American way of the present seeks conspicuous and continuous innovation. The automobile industry, most successful in this respect, has called it planned obsolescence. We have made a vast business success of changing styles, even more so than of improved function. Change for the sake of change, discarding what we have and are for something new because it is new, may be less the progress we like to think it to be than loss of moderation and stability, a cutting loose from the moorings we need.

The Restless American has also become our image abroad, somewhat disturbing to our friends, the object of caricature to others. We might do well to think how we appear as seen from other cultures, sensitive as they are to limited impressions. There are some of our tourists who cut loose abroad in appalling fashions as to diversions and demands. The only picture of ourselves that many persons abroad know is by seeing our trashiest movies, widely exported, sensational, violent, stupid, and sybaritic. The more strident and dissonant forms of American night club music have an appalling distribution. We can hardly blame others if we are eyed somewhat askance.

The spirit of *Landscape* is quiet, sensitively observant and often subtle. I find it hard to define beyond its search for humanity living in some sort of state of grace. What the magazine gives to me and others is appreciation of people (communities) living contentedly and with enlightenment (ecologically in balance) in their own fashion (culture) anywhere, now or in the past, or looking ahead to a desired future. You have thus provided a place for those planners who consider what the old Romans called the *genius loci*, the harmonious joining of people and place. You have also invited historically minded people like myself to consider the times and places of such achievement elsewhere and at other times. Classical Greece gave to it the name *kalon*, the dictionary says "with an implication of moral as well as aesthetic beauty." The mechanistic view of life is not enough; man does not live by bread alone. Keep reminding us of the quality of his dwelling on Earth.

NOTE

Reprinted with permission from *Landscape* 10 (1): 6. Copyright 1960, *Landscape*.

47

The Seminar as Exploration (1976 [1948])
Carl O. Sauer

For the purposes of the following remarks it is assumed that the seminar will take place frequently (probably weekly) and in leisurely fashion (evening long?) and will be a prepared conference on research done, doing, or projected. Students and faculty participate, ideally because they are interested in getting at each other's information and views. Perhaps such a seminar does not exist because of the credit system that piles pressures on graduate work; and such a colloquium should depend only on the need felt by those who take part in it. Direction should be by the staff member who himself is concerned as a student of the subject; but other staff members should feel free to join. We have gone too far in setting up seminars as properties of individual professors and in passing the students around from colleague to colleague. Since we are talking of things as they should be, the parenthetic remark is permissible that better seminars require administrative relaxation of required "teaching loads" as well as of "study credits."

The seminar and the field course best meet our needs for intellectual participation, for they are suited to the give and take of observations and alternative hypotheses, for the exchange of opinions to produce new and more valid ideas, pieces of which may have been floating around in individual minds. This is no depreciation of the uses of solitary scrutiny and contemplation, nor is it encouragement to the blithe and gregarious spirit who thrives on conferences in which he may display his talents at academic small talk and dialectics. Understanding may suddenly clarify and change out of patient and thoughtful discussion of a problem, when a flash of insight comes to some participant and the parts of the evidence begin to rearrange themselves into a new form and meaning. When this happens to a young student, he has probably crossed the threshold into a life of scholarship. The expectation of such experience from the meeting of young and mature minds is the ever-present hope and joy of those who profess the academic life.

Above all, the seminar should stimulate and focus the curiosity of its young members. Our classroom system, built around lectures and exercises, may be adequate to hand on learning and perhaps to instruct how to use the tools of our craft. Mainly it is not designed to explore the borders of the known. Indeed, we see undergraduate classes become restive when we leave the better charted areas of knowledge and conjecture about the things we should like to know. For the most part we learn to reserve such excursions to the graduate level. The danger is, however, that graduate work continues to be learning in the undergraduate manner, in which performance is still continuously directed. Of the graduate student we ask not only proficiency but discovery, increasing independence from his teacher, growing ability to chart his own course. Hence, as rapidly as we may, we direct the graduate student to the limits of the known and encourage him to consider how he may proceed beyond them.

There are plenty of people of sound, even good minds, studious learners, clear reasoners in universities who remain quite incurious all their lives, unaware of the fact that they never got beyond being assimilators of things known or accepted. Perhaps in geography we are mainly retail distributors of other peoples' wares, and the dissatisfaction now being shown by some of our younger associates at being middlemen of knowledge instead of originators is healthy self-diagnosis. We need to reduce further our complacency, and we can do so by examining not only the kind of successors to ourselves we are attracting or accepting but what we do with them in those plastic years when they are under our care (too often to become jelled as practitioners). Our present weakness is remediable, first by selecting students with a bump of curiosity and thereafter by giving them the maximum freedom to choose their own direction.

The topical seminar is especially attractive as a means for developing exploratory bent leading to discovery. Its field should be conveniently large and loosely bounded and one in which at least some staff and student members have serious prior interest and experience and for which literature and maps in reasonable amount and quality are available. Topics we have found usable have included coastal terraces, mountain glaciation, season of temperature extremes, grasslands, primitive agriculture, subtropical horticulture, plantation systems, and soil erosion. In contrast to classroom treatment of such a subject, there is no attempt at complete topical coverage.

In the introductory meetings the discussion is largely an *exposé* by the professor of what he considers to be interesting and promising questions, incompleteness of data, contradictory evidence, or blanks in information. The students are encouraged to volunteer topics they wish to explore. They receive counsel as to how the topic might be developed, and how evidence may be sought.

In the next stage, while the individual inquiries are being started, meetings are occupied by reports on journal articles relevant to the overall topic. Some of the reporters stop with abstract of content; others add their critique of presentation; a few try to examine the validity of the data and of the generalizations and find out what addition to knowledge has been made.

Finally comes the attempt at the synoptic reports, in which, with some luck and talent, the reporter brings significant questions rather than answers and some notions of how the inquiry might be continued profitably. If the venture has been worthwhile, it is because there has been . . . not only appreciation of what others have done in the pursuit of knowledge, but because the trial composition presented at the seminar becomes a first draft of a real research problem. It is here that the student may find for himself a thesis subject and an interest that will stay with him in later years.

As to mode of operation, it will be useful to revive that classical geographic phrase, "explanatory description." Of necessity we must begin by a certain amount of identification of the things to be described. If the general topic is "grasslands," there must be some preliminary sorting into subcategories of the general term, most probably only as to dominant plant elements and habits. Genetic implications are best avoided at this stage of incipient inquiry. (In human geography this is the stage of identifying culture traits by sufficient labels, e.g., house types, as to dominant forms, recognized as to functional differences.) As inspection proceeds, judgements will develop as to the merits of the labels being used and revisions will be proposed. Caution is indicated that these questions of classification must not become ends in themselves and that the job does not get hung up on the matter of name-giving.

The next and perhaps most illuminating step may be inspection of the geographical distribution of the elements being studied. Grasslands of each type, for instance, are charted on maps as exactly as possible as to position and limits, as to the sharpness of their boundaries, and as to enclaves of other vegetation and exclaves of the grassland type beyond its continuous area. Possibly correlated distributions are charted, e.g., climatic elements and terrain. The topical maps, as well executed as the data permit, constitute arrangements of data for comparative study. Presence or absence of the trait, continuity or disjunct occurrence, and absence of information have been presented cartographically. The descriptive tasks of recognition, extension, and arrangement are completed.

These descriptive activities are carried out at the same time that thinking goes on about the meaning of the labels and of the distributions. Morphography becomes morphology. Description becomes explanation. *Science is always genetic* and asks about origins, development, and disappearance. Perhaps the most serious weakness of present-day geography is the extent to which it has lost sight of processes proper for it to study and has become static and hence sterile description.

The abandonment of physiographic studies in particular has been a loss for which we have made no adequate replacement. When geographers began by being physiographers, they learned classification of forms by close field observation, acquired facility in recognizing the diagnostic features in the physical landscape, and an understanding of the processes that fashioned them. We are sadly lacking in dynamic or processes approaches at present. I am quite at loss to understand the prevalent resistance of ourselves, now usually professed human geographers, to anthropogeography (defined as the geography of noncommercial peoples), advantageous for basic discoveries because of concern with simple organisms and organizations, and to historical geography generally.

To all the learned world the "geographic method" is known as the study of the localization and extension of terrestrial phenomena, capable of giving insight into centers of origin, processes of spread (diffusion), and time differentiations (as in the age-and-area concept), yet we work much less with this appropriate and distinctive mode of inspection than others who have borrowed it from us. This later torpor which has settled upon us seems to me to be due to substitution of applied ends for scientific interest.

The preparation of the topical distributional pattern, if it has been worth doing, has gone well beyond a compilation or an editorial job. It has become an incitement to secure missing evidence, to consider evidence as to its correctness, and it should have raised alternative hypotheses of explanation. One of the things that killed off physiography was that the explanation became merely a confirmation or minor variation of the cycle of erosion hypothesis, and that sort of thing became pretty dull. Multiple working hypotheses, each of which has some possibility of validity, become most helpful as the inspection of distributional data really gets under way. In our hypothetical grassland seminar, for instance, the claims for the existence of climatic grasslands will get attention, as will the hypotheses of their correlation with the low relief and the thesis of their cultural origin. At the least, the elements of the problem should become recognized out of comparative distributional study. Perhaps some participant has seen that he must take his knowledge into the field to collect evidence that his erstwhile preceptors, the authors on the subject, have missed. He has then ceased to rely on authority and is on the way to becoming a contributor to knowledge. Even the preliminary short-term search afforded by such a seminar may well be the requisite focusing from which will come later sustained inquiry, specialized knowledge, and independent conclusion, adding up to significant discovery.

Since mostly we profess to be human geographers, our major expectation is to make sense out of changes in the geographical patterns of human populations and activities, or human ecology. The trouble with the use of that phrase has been its unilateral and static interpretation. If it is understood as a dynamic man-environment relation in which man's activity has brought change or disturbance of the ecologic balance, it becomes the most critical approach to human history, past, present, and future. Then the external changes of the environment, though not excluded, become secondary, and man—the most perilous social animal—is viewed as increasingly the modifier of the environment—perhaps to his greater good and glory, perhaps to his own decline or destruction. Thus human "progress" becomes dissociated from the shortsighted optimism of our present-day social sciences; ecologic equilibrium or its lack is placed in the long view of trends and end results; and social values are judged as to whether they are compatible with enduring and meliorative human occupation. In these terms, it must be repeated, no part of human time, no human society is to be put aside as unworthy of scrutiny for comparative cultural understanding and evaluation. Any geographic study of human activity must be pervaded by this awareness of the kind and quality of cultural action as affecting, perhaps irreversibly, the ecologic balance, and thus by the necessity of recognizing qualitative ecologic differences in the behavior of cultures.

NOTE

Reprinted with permission from the *Journal of Geography* 75:77–81. Copyright 1976, National Council for Geographic Education. Originally delivered at a symposium on graduate study at the 1948 meeting of the Association of American Geographers in Madison, and thereafter as a Sauer seminar handout. There is minor editing by Dan Stanislawski, who says that Sauer probably "would have wanted to change some of the phraseology, but I cannot know that and it seemed best to alter it as little as possible, which is what I have done."

48

The Quality of Geography (1970)

Carl O. Sauer

As I look over this vast concourse, the thing that impresses me most is the ruddiness of complexion of the participants. Thereby, I have the feeling that I am among my kind of geographers. This raises the question as to how you came to be geographers? When I was a young fellow, there were testimonial meetings, especially in wintertime, when people would gather together and testify as to how they came to undertake a Christian living. The same thing is of interest with regard to how you came to be geographers.

I start with the premise that you became geographers because you like geography. This is an important premise that does not apply to all people in all professions. The reasons are undoubtedly various, but they probably go back with most of you into quite early years. Perhaps the thing that you would find most common among yourselves is that you liked maps. I think this is important: all geographers who have been any good, in my judgment, have been people who have liked maps, or the conversion of maps into language.

THE MAP AS GEOGRAPHY'S SYMBOL

I think that there may be one person here—maybe only one—who remembers Miss Ellen Semple and her lectures. When Ellen Semple lectured, you could see the people trailing across the Appalachians and into the Bluegrass Region; you could follow them mile by mile. Or when she was dealing with the ancient Mediterranean, you could watch the ships turning the promontories on which the temples stood. You went with her! I think this is an important quality. With some it is a God-given gift; she had it. With others it is a learning that follows upon an inclination in that direction. The map is the common language.

There is an old saying, largely true, that "if it is geography, you can show it on a map." Maps speak an international language that, in large measure, is dissociated or quickly dissociated from particular training. The map, of course, is a wandering by the mind's eye; the feeling of wandering depends largely upon one's particular desire and yearning to wander.

Kids like maps! So far as I know, they genuinely *like* them. It may be that education gets in their way so that they no longer wonder about the world, taking pleasure in recognizing again on the map what they have known visually, adding to it some sort of insight into what they have not seen and may see or may not see. I think that the map is perhaps the best common ground by which we can identify the convergence of our interests.

I suspect that the map has been with us ever since there has been geography, and that is almost ever since man existed. I am thinking of the first human who scratched a line on the sand and, with a stammering speech, said "Now you go this way and then you go that way and you get to that point and that's where the fat oysters are." Therein lies the beginning of geographic nomenclature. That place became known as "Fat Oyster Point," a proper name that had to be learned by the people who grew up in that community. I look on geographic thought as having that antiquity, supported by fundamental interests.

CONTINUITY IN GEOGRAPHY'S BREADTH OF INTEREST

The map represents an assemblage of things that you think belong together. This "belonging together" business is one of the most rewarding and one of the most difficult things about trying to be a geographer.

I am reminded of a good parallel in psychology. Prominent some years ago was a form of psychology called *gestalt*, of which Professor Köhler was the principal exponent. Gestalt psychology has always been respected but not followed very much because it operates on the idea that the person or group is more than components, the parts. Not lending itself readily to analysis or experimentation, gestalt psychology is regarded as more of an intuition than a discipline. Well, I accept that same sort of thing for geography. One of the things that I have always hoped to do was to present life and nature as a whole in whatever area I had been studying. I know that this is not a matter that is satisfactorily subject to analysis; I have been trying for a kind of understanding that is other than an examination by analytic methods.

In addition to the inclusive map, from which one starts with information and extends his learning process, there is the topical map, which also is very old. A topical map is one on which there is no attempt to show the *gestalt*—the everythingness—but which establishes the distribution or ranges of some particular thing. This concern about the description and distribution of whatever you are working at and the concern as to the meaning of its range is the only thing that is recognized in the rest of the world as "the geographic method."

I have implied that geography is a broad subject, that it has always been broad, that it should remain broad, and that it has no lesser task than the one that it has always had. I have no sympathy with these people who say, "Well, let's not have anything to do with physical geography because it is not significant to the kind of human geography in which we are interested." I think that whenever a group who call themselves geographers try to reduce the field, to make it only the kind of field in which they are most interested, they are taking a step in the wrong direction.

We are now in a time when innovation is tremendously "in," when the only thing that is recommended for me to buy as being old is distilled liquor (and one beer, I believe). For the rest, innovation, the newness, is the thing that recommends it. In geography, we are having rather a time with innovators. I do not object to their playing their particular kind of a game, but I do object to these folks coming and saying, "*We* are the geographers." I do not object to a person who really likes statistics doing quantifying work; there are things that he probably can do that are worthwhile doing. I should even admit that a person is entitled to be a simulator and a model builder; if that is the game that interests him, that is all right. The one thing that I am worried about is when geographers offer themselves as decision makers. This seems to me to be going a bit far, that those who have a title in geography can then set themselves up as decision makers.

I think that if they take over in geography, my kind of geography is gone. If they take over in geography, I also think that public school geography is gone. Can you imagine getting grade school kids interested in regression analysis in order to study geography?

I think that ever since primitive man scratched the route to the oyster bed on the sand, the basic continuity of direction of geography has been set. We redefine geography in terms of our needs and our interests—but are these reformulations so very different? I do not object to one of the definitions: "Geography is the organization of space." But why be so esoteric about it? We all know, or ought to know, what the essential interests of geography are: the diversity of the earth, the patterns of resemblance and repetition, and (this is my personal addition which I find necessary) how things came to be.

By and large, geography has been historical. It has been historical in the physical sense, also. Professor William Morris Davis made the most gallant attempt ever to make physical geography non-historical, and he failed. No one may ever do as well as that again! When man is introduced into the geographic scene and into the geographic process, explanations can only be in terms of origins and changes. The Bible, as you know, begins with *Genesis*, and the second book is *Exodus*. I think that is just about the way geography has its main problems presented to it. This is true, you know, of primitive geography (I am not apologizing for being interested in primitives). Primitive peoples have their creation myths and they have their migration myths, the equivalents of the Jewish *Genesis* and *Exodus*. There is a need in mankind to look back and see how he came to be and how he changed and how he went from one place to another. There is thus a greatly enduring reason why the historical approach has its basic place in geography.

THE CALIFORNIA EXPERIENCE

You are all well aware that I am not a California specialist, sometimes to my regret. But many of you are primarily interested in California, which is good. California geography, in my view, is a greatly under-cultivated field.

If I were to undertake its study, I would start with the first known human beings in California. This would take me to the Channel Islands off the coast of Santa Barbara, where an interesting story of human antiquity raises some very intriguing geographic questions. In establishing a reasonable view of where the natives lived and what they did, geographers have contributed very little. The notable study by the late Erhard Rostlund on Indian fishing in the streams of California is first-class geography, but there is not much along that line. There is, then, a field of Indian geography which is intellectually interesting and on which there is material available.

On the coming of the white man, I have done a little reading in the past few days. I want to share with you some discoveries of mine of new knowledge and new meaning.

For example, I think all Californians know the name of [Juan Rodríguez] Cabrillo, since his voyage of discovery for Spain is taught in all the histories, but always his name is mispronounced. Cabrillo was not a Spaniard, but a Portuguese, which is rather interesting; his second in command was a Christian Levantine, which is also interesting; and the third was a Corsican. This expedition of two ships, one of which had no deck to it, started off from the known—which at that time was Cedros Island, halfway north along the coast of peninsular California. That far these people knew where they were; beyond was untrodden ground to civilized people. As they came north, their descriptions of the coast throughout are quite interesting. Although all the names they used have been forgotten with one exception, the whole route can be established quite well. The coast was pretty bleak until they approached the present-day Mexico-California border. Their interest picked up markedly when they sailed into the Bay of Ensenada, to use the current place-name. Ashore, they saw groves of trees and savannas of grass intermingled, and repeatedly herds of animals ranging from 100 to 150 in number. Even without their description of these animals, the mere designation of the savannas and the herds would establish prong-horn antelopes.

Farther north at San Diego, the Cabrillo expedition encountered unusually stalwart Indians. Nothing much happened in San Diego (sorry!), but then they went on to the Port of Smokes. Because this occurred in September or the beginning of October, fires were burning on the hills behind this Port of Smokes—San Pedro. Continuing to a point northwest (they were quite good in directions, not good on latitude), they came upon villagers living in houses, whereupon the story becomes quite interesting in human terms. From this point, which later is identified in the text as Mugu, they were then going through the Santa Barbara Channel for weeks, naming and counting the villages, 50 or more. They were impressed by the quality and the size of the canoes, made and used by the best native navigators anywhere south of Puget Sound. They were very freely served sardines, fresh and in great diversity. The houses interested them and are fairly well described.

They came to areas in the villages where there were poles. Spaniards are always kind of casual about the size of timber (for example, using the phrase *palo colorado*, the "red pole," as their name for the redwood). But these poles were not redwoods, being described as painted and having figures on them. This is the first evidence that these Santa Barbara Channel people were kinfolk of the people of the Pacific Northwest, a suggestion that has had a good deal of reinforcement since then.

The Cabrillo expedition had a marvelous time in the Santa Barbara Channel country. They provide some very appreciative descriptions of how nice this country was and how nice the people were, but in so fooling around they consumed several weeks of precious time. About the first of November, when they decided to go on, they learned a significant lesson about California's climate. They were just beyond Point Concepción when they were hit by their first storm, which drove them back south to the town of the sardines. They started out again, and the next storm took them in the other direction, chasing them north until they entered a bay where the pine trees came down to the water's edge. This was Monterey, and from then on they were in continual trouble. They were experiencing a season in which the fall and winter storms had come early and unusually hard. When they were being driven along the coast south of Monterey they recorded how the waves broke unceasingly against a cliffshore and how the mountains were so close and so high and covered with snow (in November, which is possible) that they feared the mountains might fall down and crush the ships. From their description, they did not know where they were a good deal of the time, because of being battered back and forth by storms. The question whether they reached as far north as Cape Mendocino or not, is academic and no longer of interest, really, to the geographer.

Here is a remarkably good first account of the presence of white men on the coast of California, from which I have taken some of the highlights. After this introduction to the historical geography of California, for 60 years there is nothing more. It is an interesting matter that there is no more concern with California for a while. But finally the colonial period begins with the settlement of the whites that leads on to all that you know so well about California in the nineteenth century.

Then in 1923, C. O. Sauer came to California, and that was some California! I am sorry that I never studied it; I just experienced it. To use the current term, I have had only some "environmental perception." We came out for the first time on the Santa Fe Railroad, all the way from Kansas City. During this trip, the train stopped for every meal and we put on our coats before entering the Harvey dining room, an interesting cultural note. Then we got to

Southern California, to Pasadena, the goal of the old folks who had a moderate income. What a wonderful place these towns were down there: Pasadena, Sierra Madre (still a bit on the youngish side), and Laguna Beach!

After I had been in California for not too long a time, I was invited to a very august place right in the middle of the best of Southern California to make a geographical address. In reassuring my audience about California, I said, "You have a wonderful state here. You have people who enjoy it, who appreciate it, people who have the means and the good sense to come out here and choose it as a place to live, and you won't need to worry about the future. This is so because California has climate, but it does not have the resources that will sustain industrialization. You will enjoy your citrus fruits, your palm trees, your living, and you're not going to be overwhelmed by industrial and urban growth." I am afraid that shortly after this I quit predicting the future!

AN APPRECIATION OF DIVERSITY

What a state this California was! And now we have a magazine called *Cry California*. Just this past week I was sent a copy of a new book by Raymond Dasmann, who was at Humboldt State College at Arcata for a good long time. He should have been brought into a geography faculty, but he did not have a Ph.D. in geography and, as you know, this is required now. Dasmann is now one of the members of the Conservation Foundation in Washington, and a very good observer; a wildlife man, among other things. His book is called *A Different Kind of Country* [1968]. It has California in it although it is not a book about California, but about what is happening to the United States at a dreadful rate. I refer to the loss of diversity, a thing that I have felt so very, very much in a particular way in respect to the farms.

When I left the Middle West, nobody would have thought that the then American farmer was on the way to extinction. By the American farmer I mean a person who grew a diversity of crops, who rotated his crop system, who had animals on the farm, who produced some of the food for himself, the pigs, the farm orchard, and that sort of thing. This type of person is more nearly extinct in the American scene today than the Indian is. Nobody would have dreamed of that sort of thing a half century ago. The agricultural colleges, the experimental stations, were telling the farmer to keep on being a farmer of diversified crops, that the family farm was the good thing toward which American rural and small town life was looking.

I cannot describe this change for California as I can for the Middle West, which is my home country. Clover there was almost the sign of farming decency. Now, over the Mississippi Valley one must go a long way to find a field of red clover. I happen to be a farmer, in absentia, by inheritance. The farm used to be a farm; it is now just a piece of a larger operation in the middle of the best country in Illinois. It has yielded up to 210 bushels of corn an acre, which is something that no self-respecting piece of land can do by itself. This is just feeding it by the bottle. In that township I do not think there is a single living domestic animal; only a few birds and a very few bumblebees.

Now this is the sort of thing that has happened. The barn, over a large part of the United States, is as much a relic as the hand pump in the yard. The farm houses are coming down. The land may even bring a somewhat better price if it does not have a house on it. The change between the world of my youth and the present rural world is fantastic!

There are other illustrations of this process in California, but here more strongly compounded with actual urban and industrial expansion. I am talking about the greatest granary of the world and the manner in which the American Midwest has become completely unrecognizable as to the way of life.

The thing that scares me is how fast things can come in this country. In the early 1930s one rarely saw soybeans; if so, one was likely to stop and ask what they were. The American farmer had grown corn by the Indian method for more than 300 years, "laying by the corn." This was the final cultivation by which the earth was heaped up or "hilled" around the corn. The American farmer did all of his cultivation in the Middle West by the Indian method of mounding the earth around almost whatever he grew; corn, potatoes, and everything he planted. It was not necessary to do it, but a tradition that lasted and lasted, although people no longer knew how it started. This hilling cultivation of the Midwestern American farm was knocked out, beginning in the 1930s. The cornfields of today are an entirely different sort of thing from the cornfields on which America lived for hundreds of years.

Not liking a lot of change, certainly not too much change, I come back to an old creed and delight, as a geographer, of enjoying the uses of diversity, which I think is one of the most attractive features of human living. For diversity we currently have substituted the word "development." As you may have gathered, I am very, very cool on development. Our old friend, Lewis Mumford, who I think is one of the most interesting people in this country, recently wrote a book on *The City in History* [1961]. Long before that he had introduced the term "megalopolis." Concerned about the life of cities (before there were riots or anything like that), Mumford thought that large cities represented a great possibly insoluble problem, an impasse for civilization. Then along comes a geographer, Jean Gottman, who writes a book called *Megalopolis* [1961], in which all is sweetness and light.

I should like to call to your attention the fact that geographers do not worry enough. I worry a lot, but I cannot get enough people to worry along with me. I am terribly worried about megalopolis. I am very worried about underdeveloped countries, mainly because I am afraid of how they are being developed.

This is a finite world. It is a terrible truth, but what a truth it is! We are running along as though everything was infinitely expandable and we are not concerned about the expansion. I believe that more geographers should become recorders of where we are. A very, very serious question is whether we are going to have a world that is kept tolerable by the restraint of its civilized people who realize the importance of diversity in making the world attractive.

In closing, I have just one question to ask. It was asked of me by a youngster, and I think it is the biggest question of all: "Do you wish you were a teenager now?"

NOTE

Reprinted with permission from *The California Geographer* 10:5–10. Copyright 1970, California Council of Geography Teachers. Banquet address, annual meeting of the California Council of Geography Teachers, May 4, 1968, at California State College, Hayward. Edited by William L. Thomas Jr.

49

Casual Remarks (1976)

Carl O. Sauer

FIRST SESSION

You know, it is an interesting thing to watch and see what the boys have been into. Some of them are things that I would not dream of myself. I had the notion while I was sitting here that what we are doing, without being particularly aware of it, is what might be called a Gestalt Geography, by which I mean that the parts are different from the whole. In various individual ways, we are trying to put analytic work into synoptic purviews. Take Brig Arnold for instance on the philosophy of geography; now, I would trust Brig Arnold probably better than anybody I know to be able to distinguish between accidental fracture of rock and man-determined percussion of rock by instruments, but Brig in this capacity is something new to me. Here is also Al Urquhart who goes into Angola and does a good, rugged job of field work and now he gives us a talk on the distribution of scientific societies and their communication in Europe.

All these presentations rather move me as to the way in which thoughts are going out in new directions. Now I wouldn't wish to say, indeed I can't say, that we have wanted to be innovators in geography. But I do think that what we should try to do is to maintain the immemorial tradition of geography, the geography that goes back beyond the written word. I still think that Herodotus is an awful good person for one to read as a cultural interest of man that may go back almost as far as culture goes. The where, what, and why checklists in the imagination of human reference points are intrinsic, or at least very early in cultural development.

Lately, I have been somewhat interested in going through French literature of the seventeenth century in North America, and note repeatedly how a person, in one case [Samuel de] Champlain, gets information from an Indian who takes a stick and draws a map of waterways, portages, places where you get beaver, places where you get buffalo, and so forth, and puts this before the Frenchman so that the French explorations are following a geographic orientation presented by the natives. I think this is probably quite characteristic. Simple people need to know for survival, or for well-being, what there is in the land within their knowledge that is of interest to them. I wouldn't want to put this down in terms simply of economics—I am getting to be less and less of an economic geographer the older I get—but of symbolic interest, of being something that is good for you, good for your spirit, or bad for your spirit. I will turn it over to George [Carter]; he can work on that theme.

Now, however, I have to make some flat declarations. One is that to me geography is an earth science and I mean that literally. It is not a social science, it is not a behavioral science, it is an earth science! It is concerned with the physical environment; it is concerned with a culture's environment; everyone knows that unless they have gone to certain universities.

When the University [of California] was building new quarters and we had a choice as to where we wanted to go, there was no doubt in our minds that we would like to go into the Earth Sciences Building. They housed paleontology on the bottom, with geology in the middle, geophysics above that, and the geographers on the top floor—which nobody wanted because it wouldn't carry heavy machinery. This is the way we of the oldest vintage of American geographers got started. We all got started in earth science. Most of us started out to be geologists, and we came in at the time when the older specialization in field geology was going out of existence. Geologists no longer concerned themselves with the surface of the land, with its landforms, with its soil, with its vegetation, or with the people who lived on it. The earlier geologist did.

In actuality, of course, it was geologists who established geography in the universities of this country. It was Thomas Chamberlin, the greatest geologist of his time, who was the head of the committee of natural sciences that drew up the original proposals for the establishment of geography as a university addition. In the first years of the meetings with the Association of American Geographers, the liveliest participants were geologists. I think [those] were better years than those you have got now. These old boys, like Alfred Brooks, had been out in Alaska and done field work in other parts of the United States and beyond. They were the backbone of the meetings of the early [geographers].

Then, along about 1919, came the declaration of independence by the then-president [of the AAG] [Charles R. Dryer], geography now having nothing to do with geology and being independent of other earth sciences. I think this was a very grave mistake that we got into. Geology has a strong historical face to it, but we got away from it for awhile.

The geophysicists have put time, as specific time, in order as it's never been before. The geophysicists have become historians again. Yet, we have the rumor adrift now in certain quarters of American geography that the geographer has nothing to do with the realities of time, the statements of time, or the orientation of his knowledge in terms of when it takes place in time. This is something to which I cannot in the least subscribe. In this little retrospective glance that I'm taking with you this evening, if I summon my courage, I should say to you that our little group has done its darnedest to keep geography from getting stranded in the timeless shoals on which it has a tendency to get out of action every once in awhile. This has been going on for quite awhile.

The first such stranding that I recall was model building. The first and greatest model builder that we had was an old gentleman by the name of William Morris Davis. Davis did not describe landforms and landscapes, he construed them. He was beautiful at this. His block diagrams were lovely things to look at, but they weren't based on observation—they weren't based on reality. He was a wonderful old gentleman, but he was wrong, that's all. He also invented for geography stages and cycles in which, of course, he was joined by Ellsworth Huntington. Huntington was a nice fellow and very intelligent in some ways. Here we have two people with a tremendous influence, saying "time is of no account."

Mr. Davis was out on one of his trips to the coast, and I was getting a little interested in terraces at that time. On one field trip I mentioned to Mr. Davis, "I think this is a late terrace; it might be a Wisconsin [period] terrace." He replied, "That's a term for the geologist; the geographers do not use time terms."

Mr. Davis' statement was important, for we had two distinguished people at the University at the same time. One was Professor Davis, and the other was Professor Albert Penck. Penck is the man who really worked out the sequence of events of the Alpine Pleistocene in Europe and largely the basic sequence of ice-age events; Penck did a lot of other things too. Penck was visiting with us, and Davis was visitor in the Department of Geology. One day Davis shook his head and said, "It is a strange business here when a geologist is found in the Geography Department and a geographer in the Geology Department." Here's what his difference was: If you were interested in actual time, you were a geologist. If you were interested in stages and cycles, then you were a geographer. Now, this thing still goes on.

I think, however, it's a handicap to ecology. Ecology—I've been around for a long time—was started in this country by two men. One of them we [knew] very well, Henry Cowles of Chicago, and the other was Frederic Clements at the University of Nebraska. Both of them based their ecology on Davis' notion of stages and cycles—I'm inferring this about Clements only from what he's written—but I knew Cowles quite well. For that reason both of them eliminated actual succession, which was not a change or advance in stage but the intrusion of something at a particular time. Also, both believed there was a climax which was then changeless, unless something happened to the external environment that started the "war" over again.

I think this has been a rather major weakness in ecologic studies. I'm all for ecology, I wish we practiced a lot more of it. But ecology is not vegetation, landforms, or the works of man. What we've tried to do is the sort of thing we've been hearing about this evening, getting at distributions. You noticed how informed some of these guys are on what they are describing.

Now before you get around to distribution and mapping outriggers and things of that sort, you've got to know what the things are, and you've got to know them in valid detail. We have at least a modest start in this direction because we all got started learning about landforms. We were not duds when it came to glaciated areas. We had the patience to learn to put names on forms of physical land; others learned to give recognizable names to phenomena of weather and their seasonable distribution.

We also had a partial introduction, not a large introduction or a sufficient introduction, to vegetation. But I was surprised a matter of some 25 years ago when a few of us older fellows, including the ranking professors of economic geography at a school of finance, were in the field in Michigan with a group of younger geographers. All these old chaps, even though they were teaching economic geography, could tell one tree from another and could tell kinds of marsh vegetation, and that sort of thing. The young fellows just stood there and listened to us, and thought it queer that we would be interested in that sort of thing.

There was a generation gap that was established then; I'll tell you how I think it got started. It was probably started by identification, naming, and then getting interested in distribution. Distribution, of course, brings up the question raised a half a dozen times here this evening: was it something that diffused—carried by man—and serves to trace his coming and his action?

My time is up. I just want to say I think the next trouble the youngsters got into was World War I, when some of the bright young geographers were invited to Washington to serve on shipping boards that allocated tonnage and directions of cargo ships. Geographers were accustomed to making maps and they gathered statistics and put them on maps. These young geographers spent several years on the shipping board (I knew the head of the shipping board very well—I wasn't invited in on this) and they became economic geographers; that is, they became geographers of statistics. They needed statistics as something to deal with. You had economic geographers who dealt with agriculture and didn't know a thing about how farmers plant, or why they plant. If you concentrated on statistics you are likely to miss the most important things, because very often it is the small things that don't total up in bills of exchange and enumerations.

Too often we become overconfident about statistics. This has been a thing that has not been sufficiently evaluated; censuses are not as exact as they are supposed to be. There's this infatuation with numbers, quantities, and values; some people should have it, but you should only have it if you are forced to have it. With this you get the curious absorption of the interest of the geographer in growth. This is one of the strangest phenomena, and the most disturbing one that I know of in contemporary geography. So maybe we'll talk about growth tomorrow.

SECOND SESSION

I cannot tell you how greatly I appreciate the way you men have rallied around to Joe Spencer's invitation, and have given us examples of the work that you are in. Somehow I am reminded of an old line of Kipling's: ". . . each in his separate star, so draw the things as he sees it." Yet, there is a large measure of coherence and common interest that shows through this. I can sort of recapture some of those years that we talked about certain things in seminar, and then you found something to love and work on. Somehow as [John] Kesseli used to say, "You don't give them any training, but they go out and do something."

Now, an item or two: one of the things, a minor but important thing, that has repeatedly come up in these presentations is the question of what we call diffusion, dispersal at the hand of man. This is obviously not a thing that falls within the scope of what is known as theoretical geography, which talks about diffusion of the human race. There has been this interesting illustration of Lauren Post about the movement of the bee ahead of man—it's a very good field, and it is very well presented here. The question of the spread of the banana has also been brought up as being slow, obviously slow, because the banana must have been carried as a plant. The taking of plants, the planting of plants, and then taking shoots from them and so on, is not going to work fast under very many circumstances. I recall, on the other hand, the terrific rate at which the watermelon spread. In the course of something like fifty years it was taken from northern Mexico as far as Illinois, grown by Indian tribe after Indian tribe. It took only one season of eating watermelon to put them in business.

Last night I took a poke at the economic geographers, and I think I should elaborate a little bit. My objection, which I didn't really express, was twofold. One was that it [statistics] was a substitute for observation, and to me this is a serious objection to any geographic work that deals, by preference, with other people's observation. You often have to use other people's observations, that are summarized in terms of numerical series gathered for their purpose and used for your purpose. This, I think, has gone too far and too strongly. We had not an economic geography coming out of it, we had a commercial geography coming out of it.

Now rather interesting to me was an old work by George Chisholm in England, which was called Commercial Geography, and was a much better job of economic geography than the things that have prevailed in this country as economic geography. He talked, not simply about quantities and values, but about why people grew certain things and how they happened to grow them. He placed them into a cultural context. I think a job that is really worth doing shouldn't have to be revised every two or three years to bring it up to date. This is one of the major weaknesses in economic geography that I see, that if a book is 10 years old it isn't worth anything. Now, I don't think it's worthwhile to spend your time writing a book that's going to be out of date in two or three years.

We need a timeless quality introduced into economic geography. Of course, this is again a sort of a submission to economics; economics has become too narrow a subject. Economics is, or was, or should be, *economia*, portraying the way of making a living, the way of living with your resources, with your talents. I hope that economic geography can reach back into that, but this will mean—I don't know quite how to say this—well, let me try this. Consider economic geography as a subject only incidentally dealing with quantities, to simplify everything, and that the main concern is with the quality of living. Behind economic interests are interests that are not economic, for which we don't have a proper sort of a name. Yet, I think this sort of thing, as to what makes a people choose and develop a certain mode of life and retain it, is pretty well the major objective to which we should like to contribute.

I was interested in Carl Johannessen's reference to the very slow evolution of maize, and the reason that he suggested for it. I thought that this was a very striking example, and probably a very important one. Here you have maize grown in a pretty rudimentary and certainly economically very unsatisfactory form for a long, long time. Then something comes to it, comes into it, by which it changes its character and its importance. If I remember Carl correctly, he said he didn't think this was an economic start, but that there was something else that kept maize under the attention of people for a long time. Somebody, I think, suggested that maybe it was making beer for ceremonies that gave it a place before it had a food producing utility.

So then, I am sort of edging into this vague boundary line between cultural geography and other kinds of geography. But as you know, or may surmise, from some of the things that have been said here, the economic motivation of invention doesn't impress me very much. I don't think that the desire to have more food, the desire to possess something, the need of something, has been operative in this interesting and diverse and long history of mankind. Necessity! I don't believe it has been the mother of invention. The objective of invention, I think, is only secondarily to satisfy material wants. We who are obsessed by material wants, or think we have these material wants, are perhaps most likely to disregard the ways and means and ends of life of other people from other times. I would go so far as to say that the American perhaps has more to learn from these less materially-oriented ways of other people in other times than almost any other people.

Because these lads have worked with me, there is a good deal of emphasis on domestication that shows up in their work and this, of course, is a matter of major importance. This might be something that is really significant to the understanding of mankind; that animals were taken under the care and management of man, not to have more food, not to have better hides, not to have wool to keep them warm; these are secondary results. This thesis, of course, was worked out a long time ago and has been greatly substantiated since the early work of [Eduard] Hahn. The older I get the more I am inclined to assign similar, non-utilitarian non-want satisfying objectives to the selection and promulgation of plants.

The point Carl Johannessen raised, I think, is one of the most important questions of all. Why fool around with this odd, miserable thing that was a primitive maize? It had an extraordinarily queer inflorescence. It had a strange way of bearing its seeds, and it didn't amount to much of anything. Yet, we have had, for thousands of years, a concern with this plant, when there were much better food-producing plants available. I think Carl also mentioned the fact that curiously maize is associated with religious ceremonies and with divinities. Now don't sell the divinities short; they have helped the people a lot, in a lot of ways that we have difficulty understanding because we are rational people. We think we are rational people—just watch the stock market to see how rational we are!

This interest in growth habit, in color, in domesticated plants, especially in seed plants, is again a very interesting thing. In this whole matter of color we have a great diversity. In the case of the American bean, color didn't represent a particular food pattern that carried a different taste pattern, but there was something in a red color, or a calico color, that interested these people and kept on interesting them. Of course, it runs through the whole business of animal domestication. White animals, black animals, and pied animals are selected for something that has nothing to do with meat or milk production or anything of the sort.

We reach back into time as far as it seems to tell us something. This is one of those things for which I am very grateful, because when I came into geography, and the American situation that I knew, the geographer was concerned with the things of today and tomorrow. It always bothered me because I couldn't shake loose from an interest in what was and became something else, and I followed my way back slowly, and hesitantly. It was the larger tradition of the European geographer that really freed me from the worry about today and tomorrow. An important step here was not only the continental geographers but the work of two Englishmen, Harold Peak and [Herbert] John Fleure, who did a series of volumes that attempted correlations of time in which they reached all the way back to what was thought then to be the beginning of mankind. In other words, it was very misstated but the perspective was very, very good. I am always glad when somebody can reach back farther into time, and make some sense out of it, and link us into a long past.

We know now that this past is terribly long; we have known this only for a short time. The actual huge jump in human time is really a matter of about twenty years, and, of course, the Leakey family comes in very, very strongly in this. I would say to you not to apologize if you are interested in what man was like and was doing here and there a hundred thousand years ago, a million years ago, or more than that if you have insight into this sort of thing.

I don't want to end up with a sermon here, but I think it may be especially significant to us at present that time is accelerating at such a fearful rate. Forty years ago—when I was already a geographer—seems closer to the time of the American Revolution than it is to the present time. In the thirties, in this country, an extraordinary revolution in agriculture took place. In the thirties—some of you can remember the thirties—before we had an entirely different farming system, we had a quite different farming population. Aldous Huxley, who lived in Southern California as you know for quite a few years, told us a number of very interesting things. Aldous should be read by the geographer beginning, of course, with the *Brave New World* which was a pioneering job in the twenties [1932]. Another important book he wrote in Los Angeles, one called *Time Must Have a Stop* [1944], has an awful lot to say.

There is an awful lot to time and the rate at which we are spinning along. I think that for me the only sanity is to look backward, instead of looking forward to the year 2000. Isn't that awful that here am I, who have had 83 years of good life, and in a sense am sort of glad that I am not going to be around 20 years from now, and you are facing it. At any rate, please God, don't you geographers whoop it up for growth and development!

NOTES

Reprinted with permission from the *Historical Geography Newsletter* 6 (1):70–76. Copyright 1976, *Historical Geography Newsletter*. Offered at the "Special Session in Honor of Carl O. Sauer: Fifty Years at Berkeley," 36th Annual Meeting, Association of Pacific Coast Geographers, San Diego, June 14–15, 1973.

Note by Ralph Vicero, editor of the 1976 *Historical Geography Newsletter*: "These informal comments were delivered in two parts as Mr. Sauer's response to the papers offered in his honor by his former students.... Mr. Sauer's remarks were taped and transcribed by Dr. David Hornbeck, California State University, Northridge. Light editing was carried out [by Vicero] so that the text would flow more smoothly."

Volume editors' note: The papers in the two sessions, in sequence, were by John Leighly,

John Vann, Brigham Arnold, Clinton Edwards, Alvin Urquhart, Edwin Doran Jr., George Carter, Andrew Clark, Dan Stanislawski, Carl Johannessen, Donald Brand, Henry Bruman, Lauren Post, Levi Burcham, John Street, Donald Innes, and William Denevan, with Joseph Spencer as Chair. See the *Yearbook of the Association of Pacific Coast Geographers* 36 (1974): 129–183, for abstracts.

Appendix

Doctoral Dissertations Supervised by Carl Sauer

1927 Leighly, John B. (1895–1986). "A Study in Urban Morphology: The Towns of Mälardalen in Sweden."

1930 Kniffen, Fred B. (1900–1993). "The Colorado Delta."

1930 Thornthwaite, C. Warren (1899–1963). "Louisville, Kentucky: A Study in Urban Geography."

1931 Dicken, Samuel N. (1901–1989). "The Big Barrens: A Study in the Kentucky Karst."

1932 Meigs, Peveril, III (1903–1979). "Historical Geography of the Dominican Mission Frontier, Baja California."

1933 Brand, Donald D. (1905–1984). "The Historical Geography of Northwestern Chihuahua."

1935 Raup, Hallock F. (1901–1985). "The Pennsylvania Dutch at the Forks of the Delaware, Northampton County, Pennsylvania."

1936 Spencer, Joseph E. (1907–1984). "The Middle Virgin River, Utah: A Study in Culture Growth and Change."

1937 Post, Lauren C. (1899–1976). "Cultural Geography of the Prairies of Southwest Louisiana."

1938 Kesseli, John E. (1895–1980). "Pleistocene Glaciation in the Valleys between Lundy Canyon and Rock Creek, Eastern Slope of the Sierra Nevada, California."

1940 Bruman, Henry J. (1913–2005). "Aboriginal Drink Areas in New Spain."

1940 Hewes, Leslie (1906–1999). "Geography of the Cherokee Country of Oklahoma."

1940 McBryde, Felix W. (1908–1995). "Native Economy of Southwest Guatemala and Its Natural Background."

1941 Bowman, Robert G. (1912–2007). "Soil Erosion in Puerto Rico."

1942 Carter, George F. (1912–2005). "Plant Geography and Cultural History in the American Southwest."

1944 Clark, Andrew H. (1911–1975). "The South Island of New Zealand: A Modification of British Rural Patterns and Practices Associated with the Exotic Plants and Animals of the Island."

1944 Stanislawski, Dan (1903–1997). "Historical Geography of Michoacán."

1946 West, Robert C. (1913–2001). "The Mining Community in Northern New Spain: The Parral Mining District."

1948 Parsons, James J. (1915–1997). "Antioqueño Colonization in Western Colombia."

1950 Price, Edward T. (1915–). "Mixed Blood Racial Islands of Eastern United States as to Origin, Localization, and Persistence."

1951 Rostlund, Erhard (1900–1961). "Freshwater Fish and Fishing in Native North America."

1953 Doran, Edwin, Jr. (1918–1993). "A Physical and Cultural Geography of the Cayman Islands."

1953 Wagner, Philip L. (1921–). "Nicoya: Historical Geography of a Central American Lowland Community."

1953 Zelinsky, Wilbur (1921–). "Settlement Patterns of Georgia."

1954 Arnold, Brigham A. (1917–). "Land Forms and Early Human Occupation of the Laguna Seca Chapala Area, Baja California, Mexico."

1954 Aschmann, H. Homer (1920–1993). "The Ecology, Demography, and Fate of the Indians of the Central Desert of Baja California."

1954 Gordon, Burton L. (1920–). "Human Geography and Ecology in the Sinú Country of Colombia."

1954 Sopher, David E. (1923–1984). "The Sea Nomads of Southeast Asia."

1956 Simoons, Frederick J. (1922–). "The Peoples and Economy of Begemden and Semyen, Ethiopia."

1957   Kramer, Fritz L. (1918–). "Distribution of Primitive Tillage."

1957   Merrill, Gordon C. (1919–2004). "The Historical Geography of St. Kitts and Nevis, British West Indies."

1958   Innis, Donald Q. (1924–1988). "Human Ecology in Jamaica."

1959   Johannessen, Carl L. (1924–). "The Geography of the Savannas of Interior Honduras."

1959   Mikesell, Marvin W. (1930–). "The Northern Zone of Morocco: A Study of Rural Settlement and Its Effect on the Land."

1962   Edwards, Clinton R. (1926–). "Aboriginal Watercraft of Western South America: Distribution, History, and Problems of Origin."

1963   Talbot, Lee M. (1930–). "Ecology of the Serengeti-Mara Savanna of Kenya and Tanzania, East Africa."

1967   Sawatsky, Harry L. (1931–). "Mennonite Colonization in Mexico: A Study in the Survival of a Traditionalist Society."

Bibliography of Publications by Carl Ortwin Sauer

Compiled by William M. Denevan

This bibliography of writings by Carl Sauer is primarily based on "Works in Print" by Sauer assembled by Bob Callahan (Sauer, 1981a:367–374), the most complete Sauer bibliography to date. To that we have added thirty-eight other items either not listed by Callahan (twenty-seven) or published since 1981 (eleven). These consist of:

- Thirteen book reviews, 1914a, b, c, d, e; 1915d, e; 1916c, 1919c, 1931d, 1933, 1934h, 1935d.
- Sauer's doctoral dissertation, 1915a.
- A note on "Man's Influence Upon the Earth," 1916b.
- A programmatic meeting abstract on "Geography as Regional Economics," 1921b.
- Syllabus, *An Introduction to Geography*, 1924b.
- A short commentary on "Soil Conservation," 1936b.
- *Handbook for Geomorphologists* for the Soil Conservation Service, 1936c.
- Syllabus, *Culture Regions of the World*, 1940.
- A note on the "Man's Role" Symposium, 1954d.
- A conference session introduction, 1959b.
- A response to a review of *Land and Life*, 1966b.
- "Foreword to the Second Edition," *Agricultural Origins and Dispersals*, 1969.
- "Casual Remarks" at a session in his honor, 1976b.
- "Regional Reality in Economy," 1984, from a 1936 manuscript.
- Published correspondence, 1979, 1982, 1985c, 1996, 1999a, b.
- Published class lectures, 1985a, b; seminars, 1963d, 1987a; and public lectures, 1981b, 1987b.

Reprinted articles and later editions of books and monographs are indicated; however, these are incomplete, especially those in foreign languages.

BIBLIOGRAPHY

1911–1912. "Educational Opportunities in Chicago." Council for Library and Museum Extension, Chicago.

1914a. Review of *Die Niederelbe*, by Richard Linde. *Bulletin of the American Geographical Society* 46:692–693.

1914b. Review of *Das Russische Reich in Europa und Asien*, by A. von Boustedt and Davis Trietsch. *Bulletin of the American Geographical Society* 46:851.

1914c. Review of *Mountains: Their Origin, Growth and Decay*, by J. Geikie. *Bulletin of the American Geographical Society* 46:852–853.

1914d. Review of *Der Ozean*, by O. Krümmel. *Bulletin of the American Geographical Society* 46:854.

1914e. Review of *Das Meer als Quelle der Völkergrösse*, by Friedrich Ratzel. *Bulletin of the American Geographical Society* 46:855–856.

1915a. "The Geography of the Ozark Highland of Missouri." Ph.D. diss., University of Chicago, Chicago.

1915b. "Exploration of the Kaiserin Augusta River in New Guinea." *Bulletin of the American Geographical Society* 47:342–345. Chapter 6 herein.

1915c. "Outline for Field Work in Geography," 2nd. author with Wellington D. Jones. *Bulletin of the American Geographical Society* 47:520–526.

1915d. Review of *Island: Das Land und das Volk*, by Paul Herrmann. *Bulletin of the American Geographical Society* 47:789–790.

1915e. Review of *Der erdlundliche Unterricht an höheren Lehranstalten*, by Richard Lehmann. *Bulletin of the American Geographical Society* 47:796–797.

1916a. *Geography of the Upper Illinois Valley and History of Development*. Illinois State Geological Survey, Bulletin No. 27.

1916b. "Man's Influence Upon the Earth." *Geographical Review* 1:462. Chapter 8 herein.

1916c. Review of *Länderkunde der osterreichischen Alpen*, by Norbert Krebs. *Geographical Review* 2:317–319.

1917a. "The Condition of Geography in the High School and Its Opportunity." *Journal of Geography* 16:143–148. Originally in the *Journal of the Michigan Schoolmaster's Club, 51st. Annual Meeting, 1916*, 125–129.

1917b. "Proposal of an Agricultural Survey on a Geographic Basis." *Michigan Academy of Science, 19th Annual Report*, 79–86.

1918a. "Geography and the Gerrymander." *American Political Science Review* 12:403–426.

1918b. "A Soil Classification for Michigan." *Michigan Academy of Science, 20th Annual Report*, 83–91.

1918c. "Part I, Geography." In *Starved Rock State Park and Its Environs*, by Carl O.

Sauer, Gilbert H. Cady, and Henry C. Cowles, 1–83. Geographic Society of Chicago, Bulletin No. 6.

1919a. "Mapping the Utilization of the Land." *Geographical Review* 8:47–54.

1919b. "The Role of Niagara Falls in History." *Historical Outlook* 10:57–65.

1919c. Review of *Verkehrsgeschichte der Alpen,* vol. 1, by P. H. Scheffel. *Geographical Review* 7:190–191.

1920a. "The Economic Problem of the Ozark Highland." *Scientific Monthly* 11:215–227.

1920b [1915]. *The Geography of the Ozark Highland of Missouri.* Geographic Society of Chicago, Bulletin No. 7. Reprinted by Greenwood Press, New York, 1968. "Preface," chapter 7 herein.

1921a. "The Problem of Land Classification." *Annals of the Association of American Geographers* 11:3–16.

1921b. Abstract, "Geography as Regional Economics." *Annals of the Association of American Geographers* 11:130–131. Chapter 15 herein. Also in *The Association of American Geographers: The First Seventy-Five Years,* by Preston E. James and Geoffrey J. Martin, 71–72 (Association of American Geographers, Washington, D.C., 1978).

1922. "Notes on the Geographic Significance of Soils: A Neglected Side of Geography." *Journal of Geography* 21:187–190. Chapter 9 herein.

1924a. "The Survey Method in Geography and Its Objectives." *Annals of the Association of American Geographers* 14:17–33.

1924b. *An Introduction to Geography,* with John B. Leighly. Edwards Brothers, Ann Arbor. Later revised editions to 1932.

1925. "The Morphology of Landscape." *University of California Publications in Geography* 2 (2): 19–53. Reprinted 1938. Also in Sauer, 1963a:315–350. Also in *Introduction to Geography: Selected Readings,* ed. Fred E. Dohrs and Lawrence M. Sommers, 90–120 (Thomas Crowell, New York, 1967). Also in *Treballs de la Societat Catalana de Geografia* 43:155–186, 1997. Also in *Paisagem, Tempo e Cultura,* ed. Roberto Lobato Corrêa and Zeny Rosendahl, 12–74 (EdUERJ, Rio de Janeiro, 1998). Also in *Polis: Revista de la Universidad Bolivariana* (Santiago, Chile) 5 (015): 1–21, 2006. Long excerpt in *A Question of Place: The Development of Geographic Thought,* ed. Eric Fischer, Robert D. Campbell, and Eldon S. Miller, 424–435 (R. W. Beatty, Arlington, Va., 1967).

1927a. *Geography of the Pennyroyal.* Kentucky Geological Survey, Series 6, vol. 25. Extracts in Sauer, 1963a, 23–31.

1927b. "Lower California Studies. I, Site and Culture at San Fernando de Velicatá," with Peveril Meigs. *University of California Publications in Geography* 2 (9): 271–302.

1927c. "Recent Developments in Cultural Geography." In *Recent Developments in the Social Sciences,* ed. Edward C. Hayes, 154–212 (Lippincott, Philadelphia). Also in *Geografia Cultural: Um Século,* ed. Roberto Lobato Corrêa and Zeny Rosendahl, 15–98 (EdUERJ, Rio de Janeiro, 2000).

1927d, ed. "Vereinigten Staaten." In *Stieler's Handatlas*, 10th. ed., sheets 95–100. Justus Perthes, Gotha.

1927e. "The Field of Geography," with John B. Leighly. In *An Introduction to Geography: I. Elements*, 4th. ed., with John B. Leighly, 1–13. Edwards Brothers, Ann Arbor. Chapter 11 herein.

1929a. "Land Forms in the Peninsular Range of California as Developed about Warner's Hot Springs and Mesa Grande." *University of California Publications in Geography* 3 (4): 199–290.

1929b. "Memorial of Ruliff S. Holway." *Annals of the Association of American Geographers* 19:64–65. Chapter 35 herein.

1930a. "Basin and Range Forms in the Chiricahua Area." *University of California Publications in Geography* 3 (6): 339–414.

1930b. "Historical Geography and the Western Frontier." In *The Trans-Mississippi West: Papers Presented at a Conference Held at the University of Colorado June 18–June 21, 1929*, ed. James F. Willard and Colin B. Goodykoontz, 267–289. University of Colorado, Boulder.

1930c. "Pueblo Sites in Southeastern Arizona," with Donald Brand. *University of California Publications in Geography* 3 (7): 415–459. Also in *Field Study in American Geography*, ed. Robert S. Platt, 141–160 (Research Paper No. 61, Department of Geography, University of Chicago, 1959).

1930d. "Thirty-Two Ancient Sites on Mexican West Coast." *El Palacio* 29:335–336.

1931a. "Geography: Cultural." *Encyclopaedia of the Social Sciences* 6:621–624. Macmillan, New York. Also as "Cultural Geography," in *Readings in Cultural Geography*, ed. Philip L. Wagner and Marvin W. Mikesell, 30–34 (Univ. of Chicago Press, Chicago, 1962). Also in *Geografia Cultural: Um Século*, ed. Roberto Lobato Corrêa and Zeny Rosendahl, 12–14 (EdUERJ, Rio de Janeiro, 2000). Chapter 12 herein.

1931b. "Prehistoric Settlements of Sonora, with Special Reference to Cerros de Trincheras," with Donald Brand. *University of California Publications in Geography* 5 (3): 67–148.

1931c. Review of *Anza's California Expeditions*, by H. E. Bolton. *Geographical Review* 21:503–504.

1931d. Review of *The Ancient Mimbreños Based on Investigations at the Mattocks Ruin, Mimbres Valley, New Mexico*, by Paul H. Nesbitt. *American Anthropologist* 33:636–637.

1932a. *Aztatlán: Prehistoric Mexican Frontier on the Pacific Coast*, with Donald Brand. Ibero-Americana, No. 1. Univ. of California Press, Berkeley.

1932b. Correspondence [on Physical Geography in Regional Works]. *Geographical Review* 22:527–528. Chapter 13 herein.

1932c. "Land Forms in the Peninsula Range." *Zeitschrift für Geomorphologie* 7:246–248.

1932d. *The Road to Cibola*. Ibero-Americana, No. 3. Univ. of California Press, Berkeley. Also in Sauer, 1963a:53–103.

1933. Review of *Koreanische Landwirtschaft*, by M. Heydrich. *American Anthropologist* 35:529.

1934a. *The Distribution of Aboriginal Tribes and Languages in Northwestern Mexico*. Ibero-Americana, No. 5. Univ. of California Press, Berkeley.

1934b. "Peschel, Oskar (1826–1875)." *Encyclopaedia of the Social Sciences* 13:92. Macmillan, New York. Chapter 36 herein.

1934c. "Ratzel, Friedrich (1844–1904)." *Encyclopaedia of the Social Sciences* 13:120–121. Macmillan, New York. Chapter 37 herein.

1934d. "Ritter, Karl [Carl] (1779–1859)." *Encyclopaedia of the Social Sciences* 13:395. Macmillan, New York. Chapter 38 herein.

1934e. "Semple, Ellen Churchill (1863–1932)." *Encyclopaedia of the Social Sciences* 13:661–662. Macmillan, New York. Chapter 39 herein.

1934f. "Preliminary Recommendations of the Land-Use Committee Relating to Soil Erosion and Critical Land Margins," with C. K. Leith, J. C. Merriam, and Isaiah Bowman. In *Report of the Science Advisory Board, July 31, 1933 to September 1, 1934*, 137–161. Washington, D.C.

1934g. "Preliminary Report to the Land-Use Committee on Land Resource and Land Use in Relation to Public Policy." In *Report of the Science Advisory Board, July 31, 1933 to September 1, 1934*, 165–260. Washington, D.C.

1934h. Review of *Agriculture of the American Indians: A Classified List of Annotated Historical References*, by Everett E. Edwards. *American Anthropologist* 36:129.

1935a. *Aboriginal Population of Northwestern Mexico*. Ibero-Americana, No. 10. Univ. of California Press, Berkeley.

1935b. Review of *The Peninsula of Yucatan*, by G. C. Shattuck et al. *Geographical Review* 25:346–347.

1935c. "Spanish Expeditions into the Arizona Apacheria." *Arizona Historical Review* 6:3–13.

1935d. Review of *Macht und Erde*, by Karl Haushofer. *Political Science Quarterly* 50:449–452.

1936a. "American Agricultural Origins: A Consideration of Nature and Culture." In *Essays in Anthropology Presented to A. L. Kroeber in Celebration of His Sixtieth Birthday, June 11, 1936*, 278–297. Univ. of California Press, Berkeley. Reprinted 1968. Also in Sauer, 1963a:121–144.

1936b. "Soil Conservation." *Geographical Error*. Department of Geography Newsletter, University of California, Berkeley, 2–3. Also in *Process and Form in Geomorphology*, ed. D. R. Stoddart, 370–372 (Routledge, London, 1997). Chapter 22 herein.

1936c. *Handbook for Geomorphologists: The Inauguration of Geomorphological Re-

search in the Southwest. Division of Climatic and Physiographic Research, Soil Conservation Service, Department of Agriculture, Washington, D.C.

1937a. Communication [in reply to Ronald L. Ives regarding Melchior Díaz]. *Hispanic American Historical Review* 17:146–149.

1937b. "The Discovery of New Mexico Reconsidered." *New Mexico Historical Review* 12:270–287.

1937c. Discussion [of Isaiah Bowman, "Influence of Vegetation on Land-Water Relationships"]. In *Headwaters Control and Use: Papers Presented at the Upstream Engineering Conference Held at Washington, D.C., September 22 and 23, 1936,* 104–105. Government Printing Office, Washington, D.C.

1937d. "The Prospect for Redistribution of Population." In *Limits of Land Settlement: A Report on Present-Day Possibilities,* Isaiah Bowman, Director, 7–24. Council on Foreign Relations, New York. Chapter 27 herein.

1938a. "Destructive Exploitation in Modern Colonial Expansion." *Comptes Rendus du Congrès International de Géographie, Amsterdam, 1938* 2:494–499. E. J. Brill, Leiden. Chapter 23 herein.

1938b. "Theme of Plant and Animal Destruction in Economic History." *Journal of Farm Economics* 20:765–775. Also in Sauer, 1963a:145–154. Also in *CoEvolution Quarterly* 10 (1976): 48–51. Also in *Environmental Essays on the Planet as a Home,* ed. Paul Shepherd and Daniel McKinley, 52–60 (Houghton Mifflin, Boston, 1971).

1939. *Man in Nature: America before the Days of the White Man: A First Book in Geography.* Scribner's, New York. Reprinted by Turtle Island Foundation, Berkeley, 1975, 1980. Excerpt, chapter 28 herein.

1940. *Culture Regions of the World: Outline of Lectures.* Department of Geography, University of California, Berkeley. Various other versions, editions, and authors.

1941a. "The Credibility of the Fray Marcos Account." *New Mexico Historical Review* 16:233–243.

1941b. "Foreword to Historical Geography." *Annals of the Association of American Geographers* 31:1–24. Also in Sauer, 1963a:351–379. Also in *Revista de Geografía* (Bogotá) 1 (1980): 36–56. Also in *Serie Traducciones* 3 (1971): 9–30 (Instituto de Geografía, Universidad Nacional del Nordeste, Resistencia [Argentina]). Also in *Geografía histórica,* ed. C. Cortéz, 35–52 (México, 1991). Also in *Half-Way to the Present,* comp. H. J. Walker, 29–52 (Department of Geography, Louisiana State University, Baton Rouge, 1977).

1941c. "The Personality of Mexico." *Geographical Review* 31:353–364. Also in Sauer, 1963a:104–117.

1942a. "The March of Agriculture across the Western World." *Proceedings of the Eighth American Scientific Congress Held in Washington, May 10–18, 1940* 5:63–65. Also in Sauer, 1981a:45–56.

1942b. "The Settlement of the Humid East." In *Climate and Man, Yearbook of Agri-*

culture, 1941, 157–166. Government Printing Office, Washington, D.C. Also in Sauer, 1981a:3–15.

1943. "A Section of 'Notes and Queries.'" *Acta Americana* 1:134.

1944a. "A Geographic Sketch of Early Man in America." *Geographical Review* 34:529–573. Also in Sauer, 1963a:197–245.

1944b. Review of *Les origines de l'homme américain,* by Paul Rivet. *Geographical Review* 34:680–681.

1945. "The Relation of Man to Nature in the Southwest." *Huntington Library Quarterly* 8 (2): 116–125; discussion, 125–149. Also in *New World Journal* 1 (1): 37–47, 1975. Chapter 24 herein.

1947. "Early Relations of Man to Plants." *Geographical Review* 37:1–25. Also in Sauer, 1963a:155–181.

1948a. *Colima of New Spain in the 16th Century.* Ibero-Americana, No. 29. Univ. of California Press, Berkeley. Reprinted by Greenwood Press, New York, 1976. Also in Sauer, 1981a:181–238. Spanish editions were published in 1976 and 1990.

1948b. "Environment and Culture during the Last Deglaciation." *Proceedings of the American Philosophical Society* 92:65–77. Also in Sauer, 1963a:246–270.

1950a [1943]. "Cultivated Plants of South and Central America." In *Handbook of South American Indians,* ed. Julian H. Steward, 6:487–543. Bureau of American Ethnology, Bulletin 143. Smithsonian Institution, Washington, D.C.

1950b [1943]. "Geography of South America." In *Handbook of South American Indians,* ed. Julian H. Steward 6:319–344. Bureau of American Ethnology, Bulletin 143. Smithsonian Institution, Washington, D.C.

1950c. "Grassland Climax, Fire, and Man." *Journal of Range Management* 3:16–21. Chapter 25 herein.

1952a. *Agricultural Origins and Dispersals.* Bowman Memorial Lectures, Series 2. American Geographical Society, New York. Second edition, with additions, retitled *Agricultural Origins and Dispersals: The Domestication of Animals and Foodstuffs* (MIT Press, Cambridge, Mass., 1969). Third edition retitled *Seeds, Spades, Hearths, and Herds* (MIT Press, Cambridge, Mass., 1972).

1952b. "Folkways of Social Science." In *The Social Sciences at Mid-Century: Papers Delivered at the Dedication of Ford Hall, April 19–21, 1951,* 100–109. Univ. of Minnesota Press, Minneapolis. Also in Sauer, 1963a:380–388.

1954a. Comments [on Paul Kirchhoff, "Gatherers and Farmers in the Greater Southwest"]. *American Anthropologist* 56:563–566.

1954b. "Herbert Eugene Bolton (1870–1953)." *The American Philosophical Society, Yearbook 1953,* 319–323. Philadelphia. Chapter 40 herein.

1954c. "Economic Prospects of the Caribbean." In *The Caribbean: Its Economy,* ed. A. Curtis Wilgus, 15–27. Univ. of Florida Press, Gainesville. Chapter 17 herein.

1954d. Note. "An International Symposium: Man's Role in Changing the Face of the Earth," with Marston Bates and Lewis Mumford. *Man* 54:139.

1956a. *Man's Role in Changing the Face of the Earth,* collaborator with Marston Bates and Lewis Mumford. Ed. William L. Thomas Jr. Univ. of Chicago Press, Chicago.

1956b. "The Agency of Man on the Earth." In *Man's Role in Changing the Face of the Earth,* ed. William L. Thomas Jr., 49–69. Univ. of Chicago Press, Chicago. Also in Sauer, 1981a:330–363. Also in *Readings in Cultural Geography,* ed. Philip L. Wagner and Marvin W. Mikesell, 539–557 (Univ. of Chicago Press, Chicago, 1962). Also in *Cultural Geography: Selected Readings,* ed. Fred E. Dohrs and Lawrence M. Sommers, 2–27 (Crowell, New York, 1967).

1956c. "Summary Remarks: Retrospect." In *Man's Role in Changing the Face of the Earth,* ed. William L. Thomas Jr., 1131–1135. Univ. of Chicago Press, Chicago.

1956d. "The Education of a Geographer." *Annals of the Association of American Geographers* 46:287–299. Also in Sauer, 1963a:389–404. Also published as *La Educación de un Geógrafo,* trans. and introduction by Hector F. Rucinque, 1–29 (GEOFUN, Bogotá, 1987). Also in *Geographie* 2 (4): 137–150.

1956e. "Time and Place in Ancient America." *Landscape* 6 (2): 8–13.

1957. "The End of the Ice Age and Its Witnesses." *Geographical Review* 47:29–43. Also in Sauer, 1963a:271–287.

1958a. "Man in the Ecology of Tropical America." *Proceedings of the Ninth Pacific Science Congress, 1959,* 20:104–110. Bangkok. Also in Sauer, 1963a:182–193. Also in *The Cultural Landscape,* ed. Christopher L. Salter, 136–142 (Duxbury Press, Belmont, Calif., 1971).

1958b. "A Note on Jericho and Composite Sickles." *Antiquity* 32:187–189.

1958–1959. Review of *History of the Ancient Southwest,* by Harold Gladwin. *Landscape* 8 (2): 31.

1959a. "Middle America as Culture Historical Location." *Actas del XXXIII Congreso Internacional de Americanistas,* San José, Costa Rica, 1958, 1:115–122. Lehmann, San José. Also in *Readings in Cultural Geography,* ed. Philip L. Wagner and Marvin W. Mikesell, 195–201 (Univ. of Chicago Press, Chicago, 1962). Chapter 29 herein.

1959b. "Introductory Statement for Session on Aboriginal Plant Diffusion." *Actas del XXXIII Congreso Internacional de Americanistas,* San José, Costa Rica, 1958, 1:213–214. Lehmann, San José.

1959c. "Age and Area of American Cultivated Plants." *Actas del XXXIII Congreso Internacional de Americanistas,* San José, Costa Rica, 1958, 1:215–229. Lehmann, San José. Also in *Agricultural Origins and Dispersals: The Domestication of Animals and Foodstuffs,* 2nd ed., by Carl O. Sauer, 113–134 (MIT Press, Cambridge, Mass., 1969). Chapter 19 herein.

1959d. "Homer LeRoy Shantz." *Geographical Review* 49:278–280. Chapter 41 herein.

1959e. Communication: "The Scope of Geography." *Landscape* 9 (1): 13.

1960. Letter to *Landscape* [on Past and Present American Culture]. *Landscape* 10 (1): 6. Chapter 46 herein.

1961a. "Sedentary and Mobile Bents in Early Man." In *Social Life of Early Man,* ed.

Sherwood L. Washburn, 258–266. Viking Fund Publications in Anthropology, No. 31. Washington, D.C. Also in Sauer, 1981a:114–128.

1961b. "Fire and Early Man." *Paideuma* 7:399–407. Also in Sauer, 1963a:288–299.

1962a. "Erhard Rostlund." *Geographical Review* 52:133–135. Chapter 42 herein.

1962b. "Homestead and Community on the Middle Border" (abridged). *Landscape* 12 (1): 3–7. Also in Sauer, 1963a:32–41.

1962c. "Maize into Europe." *Akten des 34. Internationalen Amerikanisten-Kongresses, Wien, 1960,* 777–788. Also in *Agricultural Origins and Dispersals: The Domestication of Animals and Foodstuffs,* 2nd ed., by Carl O. Sauer, 147–167 (MIT Press, Cambridge, Mass., 1969). Chapter 20 herein.

1962d. "Seashore—Primitive Home of Man?" *Proceedings of the American Philosophical Society* 106:41–47. Also in Sauer, 1963a:300–312.

1962e. *"Terra Firma: Orbis Novus."* In *Hermann von Wissmann-Festschrift,* ed. Adolf Leidlmair, 258–270. Geographisches Institut der Universität, Tübingen. Chapter 30 herein.

1963a. *Land and Life: A Selection from the Writings of Carl Ortwin Sauer,* ed. John Leighly. Univ. of California Press, Berkeley. Reprinted 1967, 1983, 1992.

1963b. "Homestead and Community on the Middle Border." In *Land Use Policy in the United States,* ed. Howard W. Ottoson, 65–85. Univ. of Nebraska Press, Lincoln. Also in Sauer, 1981a:57–77.

1963c. "Status and Change in the Rural Midwest—A Retrospect." *Mitteilungen der Oesterreichischen Geographischen Gesellschaft,* 105:357–365. Also in Sauer, 1981a: 78–91. Also in *The Evolution of Geographic Thought in America: A Kentucky Root,* ed. Wilfred A. Bladen and P. P. Karan, 115–122 (Kendall/Hunt, Dubuque, Iowa, 1983).

1963d. *Plant and Animal Exchanges between the Old and the New Worlds: Notes from a [1961] Seminar Presented by Carl Ortwin Sauer,* ed. Robert M. Newcomb. Los Angeles State College, Los Angeles. (Mostly notes on Sauer's comments, only partially verbatim; includes remarks by Newcomb and others.) Excerpts in *Historical Geography Newsletter* 6 (1) (1976): 22–30.

1964. "Concerning Primeval Habitat and Habit." In *Festschrift für Ad. E. Jensen,* 513–524. Klaus Renner Verlag, Munich. Also in Sauer, 1981a:95–113.

1965. "Cultural Factors in Plant Domestication in the New World." *Euphytica* 14:301–306. Also in *Agricultural Origins and Dispersals: The Domestication of Animals and Foodstuffs,* 2nd ed., by Carl O. Sauer, 135–146. (MIT Press, Cambridge, Mass., 1969).

1966a. *The Early Spanish Main.* Univ. of California Press, Berkeley. 4th printing, with new foreword by Anthony Pagden, 1992. Spanish edition, Fondo de Cultura Económico, Mexico City, 1985 (?).

1966b. Commentary on a review of "Land and Life." *Geographical Review* 132:448.

1967a. "On the Background of Geography in the United States." Festschrift für Gottfried Pfeifer, *Heidelberger Geographische Arbeiten* 15:59–71. Also in Sauer, 1981a: 241–259.

1967b. "Foreword." In *River Plains and Sea Coasts*, by Richard J. Russell, v–vi. Univ. of California Press, Berkeley. Chapter 43 herein.

1968a. *Northern Mists.* Univ. of California Press, Berkeley. Reprinted by Turtle Island Foundation, Berkeley, 1973.

1968b. "David I. Blumenstock, 1913–1963." *Yearbook of the Association of Pacific Coast Geographers* 30:9–11. Chapter 44 herein.

1968c. "Human Ecology and Population." In *Population Economics*, ed. Paul Deprez, 207–214. Univ. of Manitoba Press, Winnipeg. Also in Sauer, 1981a:319–329.

1969. "Foreword to the Second Edition." In *Agricultural Origins and Dispersals: The Domestication of Animals and Foodstuffs*, by Carl O. Sauer, vii–x. MIT Press, Cambridge, Mass.

1970a. "The Quality of Geography." *California Geographer* 10:5–10. Chapter 48 herein.

1970b. "Plants, Animals and Man." In *Man and His Habitat: Essays Presented to Emyr Estyn Evans*, ed. R. E. Buchanan, Emrys Jones, and Desmond McCourt, 34–61. Routledge and Kegan Paul, London. Also in Sauer, 1981a:289–318.

1971a. "The Formative Years of Ratzel in the United States." *Annals of the Association of American Geographers* 61:245–254. Also in Sauer, 1981a:260–278.

1971b. *Sixteenth Century North America: The Land and the People as Seen by the Europeans.* Univ. of California Press, Berkeley.

1974. "The Fourth Dimension of Geography." *Annals of the Association of American Geographers* 64:189–192. Also in Sauer, 1981a:279–286.

1975. "Man's Dominance by Use of Fire." In *Grasslands Ecology*, ed. Richard H. Kesel, 1–13. Geoscience and Man, vol. 10. Geoscience Publications, Department of Geography and Anthropology, Louisiana State University, Baton Rouge. Also in Sauer, 1981a:129–156.

1976a [1948]. "The Seminar as Exploration." *Journal of Geography* 75:77–81. Also in *Historical Geography Newsletter* 6 (1) (1976): 31–34. Also in Sauer, 1963d:v–viii. Chapter 47 herein.

1976b. "Casual Remarks," offered at the "Special Session in Honor of Carl O. Sauer: Fifty Years at Berkeley," Association of Pacific Coast Geographers, 1973, ed. David Hornbeck. *Historical Geography Newsletter* 6 (1): 70–76. Chapter 49 herein.

1976c [ca. 1940]. "European Backgrounds [of American Agricultural Settlement]." *Historical Geography Newsletter* 6 (1): 35–57. Also in Sauer, 1981a:16–44.

1979. "The Correspondence: Charles Olson and Carl Sauer" (1949–1960). *New World Journal* 1 (4):140–167.

1980. *Seventeenth Century North America.* Turtle Island Foundation, Berkeley. Excerpts, chapters 31, 32, and 33 herein.

1981a. *Selected Essays 1963–1975: Carl O. Sauer,* ed. Bob Callahan. Turtle Island Foundation, Berkeley.

1981b. "Indian Food Production in the Caribbean." Lecture at the University of Wisconsin-Madison, May 12, 1965, ed. William M. Denevan. *Geographical Review* 71:272–280.

1982. *Andean Reflections: Letters from Carl O. Sauer while on a South American Trip under a Grant from the Rockefeller Foundation, 1942.* Ed. Robert C. West. Dellplain Latin American Studies, No. 11. Westview Press, Boulder, Colo.

1983. "Geographers on Film: The First Interview: Carl O. Sauer Interviewed by Preston E. James." August 24, 1970. *History of Geography Newsletter* 3:8–12.

1984 [1936]. "Regional Reality in Economy." Edited with a "Commentary" by Martin S. Kenzer. *Yearbook of the Association of Pacific Coast Geographers* 46:35–49. Chapter 16 herein.

1985a. "North America: Notes on Lectures by Professor Carl O. Sauer at the University of California, Berkeley, 1936," taken by Robert G. Bowman. Occasional Paper Number 1. Department of Geography, California State University, Northridge.

1985b. "South America: Notes on Lectures by Professor Carl O. Sauer at the University of California, Berkeley, 1936," taken by Robert G. Bowman. Occasional Paper Number 2. Department of Geography, California State University, Northridge.

1985c. Letter to Professor John F. Frick, Central Wesleyan College, October 26, 1908. In "Carl O. Sauer: Nascent Human Geographer at Northwestern," by Martin S. Kenzer, 7–8. *California Geographer* 25:1–11. Originally published in the *Central Wesleyan Starr* 26 (2) (November 1908): 19–20.

1987a. "'Now This Matter of Cultural Geography': Notes from Carl Sauer's Last Seminar at Berkeley," March 1946 (verbatim), ed. James J. Parsons. In *Carl O. Sauer: A Tribute*, ed. Martin S. Kenzer, 159–163. Oregon State University Press, Corrallis.

1987b. "Observations on Trade and Gold in the Early Spanish Main." Lecture at the University of Wisconsin-Madison, May 1965, ed. William M. Denevan. In *Carl O. Sauer: A Tribute*, ed. Martin S. Kenzer, 164–174. Oregon State Univ. Press, Corvallis.

1996. Excerpts from letters to Joseph Willits, Rockefeller Foundation, February 4, February 5, March 12, 1941; February 22, March 15, 1942; February 12, 1945. In "Farmers, Seedsmen, and Scientists: Systems of Agriculture and Systems of Knowledge," by Stephen A. Marglin. In *Decolonizing Knowledge: From Development to Dialogue*, ed. Frédérique Apffel-Marglin and S. A. Marglin, 185–248, esp. 212–217, 221. Clarendon Press, Oxford.

1999a. Letter to William W. Speth, June 30, 1971. In *How It Came To Be: Carl O. Sauer, Franz Boas and the Meanings of Anthropogeography*, by W. W. Speth, 199. Ephemera Press, Ellensburg, Wash.

1999b. Letter to William W. Speth, March 3, 1972. In *How It Came To Be: Carl O. Sauer, Franz Boas and the Meanings of Anthropogeography*, by W. W. Speth, 200. Ephemera Press, Ellensburg, Wash.

Contributors

WILLIAM M. DENEVAN is Carl O. Sauer Professor Emeritus of Geography at the University of Wisconsin-Madison. His Ph.D. in geography (1963) is from the University of California, Berkeley, chaired by James J. Parsons. A Latin Americanist, his fieldwork has taken him throughout the American tropics for research on cultural and historical ecology, aboriginal demography, traditional and ancient agrosystems, and cultural biogeography—all falling within the Sauerian purview. His books include *The Native Population of the Americas in 1492* (1976) and *Cultivated Landscapes of Native Amazonia and the Andes* (2001).

DANIEL W. GADE is Professor Emeritus of Geography at the University of Vermont. His Ph.D. in geography (1967) is from the University of Wisconsin-Madison, cochaired by Henry S. Sterling and W. M. Denevan. Human interactions with plants and animals occupy a central place in his attentions, with extensive fieldwork in both the New and Old Worlds. His approach is unmistakably Sauerian in its cultural-historical orientation. He is the author of numerous publications, including *Plants, Man and the Land in the Vilcanota Valley, Peru* (1975) and *Nature and Culture in the Andes* (1999).

MARTIN S. KENZER is Professor Emeritus of Geography at Florida Atlantic University. His doctoral dissertation in geography from McMaster University (1986) is "The Making of Carl O. Sauer and the Berkeley School of (Historical) Geography." He is the author of articles and editor of collections on the methods and history of geography, including *Carl O. Sauer: A Tribute* (1987) and *Culture, Land, and Legacy: Perspectives on Carl O. Sauer and Berkeley School Geography* (2003).

W. GEORGE LOVELL is Professor of Geography at Queen's University in Canada, and Visiting Professor of Latin American History at the Universidad Pablo de Olavide in Seville, Spain. His Ph.D. in geography (1980) is from the University of Alberta. His Latin American research was inspired early on by Sauer's cultural-historical perspectives. He is the author of articles, books, and edited collections on indigenous historical geography, demography, and contemporary consequences in Central America, especially in Guatemala. His books include *Conquest and Survival in Colonial Guatemala* (1985) and *A Beauty that Hurts: Life and Death in Guatemala* (1995).

GEOFFREY J. MARTIN is "Distinguished" Connecticut State University Professor Emeritus and Professor Emeritus of Geography at Southern Connecticut State University. His Ph.D. in geography (1987) is from the University of London. He is the author of articles and books on the history of geography, including biographies of Isaiah Bowman, Ellsworth Huntington, and Mark Jefferson, coauthor of *The Association of American Geographers: The First Seventy-Five Years* (1978), and author of *All Possible Worlds: A History of Geographical Ideas* (2005). He has been the Archivist of the Association of American Geographers since 1986.

KENT MATHEWSON is Associate Professor of Geography at Louisiana State University in the Department of Geography and Anthropology. His Ph.D. in geography (1987) from the University of Wisconsin-Madison, was written under the direction of W. M. Denevan. He is the author of articles, books, and edited collections on the history of geography and ancient and traditional agriculture in Latin America. These include *Irrigation Horticulture in Highland Guatemala* (1984) and *Culture, Land, and Legacy: Perspectives on Carl O. Sauer and Berkeley School Geography* (2003).

JAMES J. PARSONS (1915–1997) was Professor Emeritus of Geography at the University of California, Berkeley. His Ph.D. in geography (1948) from the University of California, Berkeley, was supervised by Carl Sauer. He taught geography at Berkeley from 1947 until 1986. He is the author of articles, books, and edited collections on peoples and landscapes in Latin America, Spain, and California, including *Antioqueño Colonization in Western Colombia* (1948) and *The Green Turtle and Man* (1962).

EDWARD T. PRICE is Professor Emeritus of Geography at the University of Oregon. His Ph.D. in geography (1950) from the University of California, Berkeley, was written under the direction of Carl Sauer. He is the author of articles on the cultural and historical geography of North America, as well as Barbados and Italy, and the book *Dividing the Land: Early American Beginnings of Our Private Property Mosaic* (1995).

WILLIAM W. SPETH is an independent scholar based in Ellensburg, Washington. His doctoral dissertation in geography (1972), "Historicist Anthropogeography: Environment and Culture in American Anthropological Thought from 1890 to 1950," at the University of Oregon, was directed by E. T. Price. He is the author of articles on the history of geography and anthropology, and he edited a collection of his own writings, *How It Came To Be: Carl O. Sauer, Franz Boas and the Meanings of Anthropogeography* (1999).

PHILIP L. WAGNER is Professor Emeritus at Simon Fraser University in Canada. His Ph.D. dissertation on the historical geography of the Nicoya Peninsula of Costa Rica (1953) was written under the direction of Carl Sauer. He is the author of numerous articles and books, including *The Human Use of the Earth* (1960) and *Showing Off: The Geltung Hypothesis* (1996). He is co-editor with Marvin Mikesell of the sub-field-defining *Readings in Cultural Geography* (1962).

MICHAEL WILLIAMS is Professor Emeritus of Geography at the University of Oxford. His Ph.D. in geography (1960) is from the University of Wales. He is the author of many articles and books and editor of collections on the history of geography, human agency, and landscape change, with special emphasis on wetlands and forests. These include *Americans and Their Forests: A Historical Geography* (1989), *A Century of British Geography* (2003), and the highly acclaimed *Deforesting the Earth* (2003).

Index

AAG. *See* Association of American Geographers (AAG)
Abhandlungen zur Erd- und Völkerkunde (Peschel), 355
Aboriginal Population of Northwest Mexico (Sauer), 54
"About Nature and Indians" (Sauer), 273, 292–295
Adams, Henry Brooks, 37, 39
Adelug, Johann Christoph, 31
Afghanistan, 21
Africa: Europeanization of, 246; grain sorghums in, 221; grasslands of South Africa, 284; and Leo Africanus, 225–226; maize in, 194, 195, 220; population of South Africa, 285; Shantz's research on, 368, 369–370; and slavery, 223–225; vegetation and soils of, 111, 368
"Age and Area of American Cultivated Plants" (Sauer), xi–xii, 189–190, 195, 198–212
Agency. *See* Human agency
"The Agency of Man on the Earth" (Sauer), xi, xviii
Agricultural hearths, 188, 200–202, 207

Agricultural Origins and Dispersals (Sauer), xvii, 17–19, 39–40, 55, 185, 188, 190, 195
"Agricultural Regions of Africa" (Shantz), 368
Agriculture: in Africa, 194, 195, 220, 368; and agricultural hearths, 188, 200–202, 207; American cultivated plants, 198–212; in Caribbean, 154–155, 157, 177–182, 191, 201, 207; and *chinampas*, 199–200; *conuco* planting system, 154–155, 157, 177–178, 182, 203, 210, 212; Corn Belt farming, 286; decline of American farmer, 404–405; and economic geography, 141; and fertilizers, 258; and hoe culture, 141, 251, 289; of Indians, 198, 202; and irrigation, 256–257, 284; in Mexico, 32, 186, 190, 199–200, 201, 207, 208, 210, 211, 256; *milpa* system of, 208, 210, 248; and monoculture, 250–251, 258; mound cultivation or "hilling," 210, 282, 405; in New Guinea, 90; New World crop diversity, 18–19, 185, 188, 194; Old World plant and animal domestication, 18, 20; origins, 45–46, 186–190; and rotation of fields, 248; Sauer on, 17–18; seed agriculture, 189, 190, 198–199, 202, 210–212;

Agriculture (*continued*)
  slope cultivation, 92; and soil erosion, 243, 250–251; in South America, 18–19, 187–189, 193–194, 199, 206–207, 212, 256; technologic advance in, 289–290; vegetative planting, 189, 190, 198–199, 202–208; and vernalization, 290; and weeds, 39–40, 190, 211, 304. *See also* Crops; Domestication; Livestock; and specific crops
Agrogeography, 160–161
Aguilera, Francisco, 214
Alabama, 243, 266, 286, 373
Alcohol, 191, 209–210, 211, 333, 413
Alexander, Charles S., 41, 155
All Possible Worlds (Martin), 23*n*1
Alvarado, Pedro, 330–331
Amaranth, 210, 211
America: naming of, 275, 321. *See also* United States
"American Agricultural Origins" (Sauer), 186–187
American Council of Learned Scholars, 7
American Geographical Society, 7, 188, 237, 273, 349, 368
American Geographical Society *Bulletin*, 89
American Historical Society, 365
American History and Its Geographic Conditions (Semple), 361
American Indians. *See* Indians; and specific tribes
American Nation at Bicentennial, 5
American Philosophical Society, 7, 365
"Ancient Mediterranean Pleasure Gardens" (Semple), 348
Andean Reflections (Sauer), 56
Anderson, Edgar, 187, 192, 193, 216, 227*n*8, 256
Andes, 189, 201, 205, 207, 211, 212
Animals. *See* Livestock; and specific animals
Annals of the Association of American Geographers, 12, 15, 35, 44, 49*n*7, 137, 373
Anschauung (holistic vision), 185
Antevs, Ernst, 111, 112
Anthropogeographie (Ratzel), 137, 165–166, 347, 356–357, 360
Anthropogeography, 127, 245, 249–250, 356–357, 360
Anthropology: and Boas, 17, 23*n*1; comparative study of primitive cultures, 167; and cultural geography, 140; as important for geographers, 12–13; and material and spiritual elements of culture, 170; Peschel on, 354; professional associations in, 168–169; and Ratzel's diffusion of culture, 137, 153, 356–357; Sauer's assessment of, 12–13, 164–165; and Sauer's domestication theories, 17–18; Sauer's references to, 37; at University of California, Berkeley, 372; and zoology, 233
Antigua, 155
Antipode, 15
Apaches, 332
Arabs, 122, 123
Arawaks, 300, 308, 310, 314, 318, 326
"Arbeitsweise der Siedlungsgeographie" (Gradmann), 140
Archaeology, 18, 38, 40, 46, 166, 189–190, 193, 207, 211, 259
Arciniegas, Germán, 322*n*19
Area studies, 156
Areal differentiation, 131, 141, 169
Argentina, 212, 283, 284
Aristotle, 121, 359
Arizona, 19–20, 255, 257, 258, 364, 369
Arnold, Brigham A., 407, 418
Aschmann, H. Homer, 418
Association of American Geographers (AAG), 7, 12–13, 43–44, 111, 114, 116, 150, 348, 349, 368, 408–409
Atlas of American Agriculture, 368
Atwood, Wallace W., 90, 368
Aurousseau, Marcel, 112
Ausland, 346
Australasia, 18
Australia, 246, 247, 284, 285
Aztatlán (Sauer), 54

Baker, Alan R. H., 46, 49*n*5
Baker, Oliver Edwin, 150, 348, 368
Balboa, Vasco Núñez de, 298
Ballesteros y Beretta, Antonio, 316, 318, 322*n*14, 322*n*19
Bananas, 182
Barrera-Bassols, Narciso, 24*n*9
Barrett, Ward, 155

Barrows, Harlan, xiv, 38, 90, 91, 101, 113, 115, 137, 252
Bastidas, Rodrigo de, 312, 316–317, 322*n*14
Bataillon, Lionel, 138
Bates, Marston, 40
Beans, 199, 202, 210–213, 256, 327, 414
Behrmann, Walter, 90, 95
Bellwood, Peter, 46
Bennett, Hugh H., 242, 348
Benton, Thomas Hart, 153
Berkeley school: adherents of, 42; and domestication process, 18; Duncan on, 16; Hartshorne's critique of, 11; Mitchell on, 21; overarching ideas of, 32; and Parsons, 155; references to, 23*n*1, 24*n*10; and Sauer and his students generally, 4–5, 9, 22. *See also* University of California, Berkeley
Berlin Geographical Society, 7, 95
Berry, Brian J. L., 41
Bertand, Claude and Georges, 42
Bessey, Charles Edwin, 367
Bessey, Ernst, 348
Biard, Father Pierre, 333, 334
Bibliographic citations. *See* Citations
"Bibliography of Commentaries on the Life and Work of Carl O. Sauer" (Denevan), xii, xix, 53–86
Bicentennial of America, 5
Bildung (self-education), 35–39, 48
Biodiversity, 19. *See also* Diversity
Biogeography, 129
Biosphere, 132, 232
Biotic regions, 253
Bird, Junius, 212
Birdseye, Charles H., 348
Bisons, 254, 263, 299–300, 373
Black civil rights movement, 13–14
Black Mountain College, 7*n*1
Blanchard, R., 112, 115
Blaut, James M., 14–15, 24*n*5
Blumenstock, David I., 350, 351, 377–379
Blumler, Mark A., 195
Board, C. F., 273
Boas, Franz, 17, 23*n*1
Bock, Hieronymus, 213–214, 219
Bodman, Andrew R., 41, 49*n*4
Body Parts (Lee), xiii

Bolivia, xvii, 207
Bolton, Herbert Eugene, 349, 351, 362–366
Borah, Woodrow, 331
Boulding, Kenneth E., 149
Bowman, Isaiah, 20, 90, 111, 271, 273, 368
Bowman, Robert G., 155, 418
Brand, Donald D., 24*n*9, 44, 417
Braun, Gustav, 112, 237
Brave New World (Huxley), 415
Brazil, 194, 207, 222–225, 275, 280, 283, 299, 316, 321
Britain. *See* England; English scientists and geographers
British Honduras, 176
Broek, Jan O. M., 155
Brooks, Alfred, 409
Brooks, Charles F., 348
Brown, Scott S., 22, 41
Brown, Terry, 196*n*1
Brücher, Heinz, 187
Bruhnes, Jean, 22, 38, 113, 115, 139, 237, 238, 240
Bruman, Henry J., 20, 418
Brunfels, Otto, 213
Bryan, Kirk, 19–20
Bryan, Patrick Walker, 112
Buckle, Henry, 127, 137, 359
Buckman, H. O., 108
Bulletin of the American Geographical Society, 89
Bureau of Ethnology, 363
Burgess, Ernest W., 38
Burkill, Isaac Henry, 221
Burning. *See* Fires
Byrne, Roger, 195

Cabot, John and Sebastian, 306
Cabral, Pedro Álvares, 319
Cabrillo, Juan Rodríguez, 383, 402–403
Cady, Gilbert H., 44
Calabashes, 205
California: agriculture in, 181, 212, 258–259; Blumenstock in, 377–378; Cabrillo expedition in, 383, 402–403; fishing in, 401; geography of, 401–404; Holway's study of cold water along coast of, 348–349; homeless in, 247; Indians in, 378, 401; mining

California (*continued*)
in, 279; Peninsular Range of, 19; religious sects in, 260; route to, 256; Sauer's travel to, 403–404; and Spanish explorers generally, 338, 364
California Institute of Technology, 254, 378
Callahan, Robert, xviii, 8*n*1, 54
Canada, 284, 328, 339–340
Capra, Frank, 153
Carbohydrate food, 204, 205–207
Caribbean, 154–158, 158–59*n*3, 173–182, 201, 205, 207
The Caribbean: Its Economy (Wilgus), 156
Caribs, 308, 310, 314, 316, 318
"Carl Ritter, 1779–1859" (Sauer), 358–359
"Carl Sauer (The Migrations)" (Callahan), 8*n*1
Carnegie Institution, 363–364
Carter, George F., 40, 256, 372, 408, 418
Cartier, Jacques, 195–196*n*1, 334
Cartography, 129, 166. *See also* Maps and mapping
Castro H., Guillermo, 24*n*9
"Casual Remarks" (Sauer), xii, 387, 407–416
Cattle, 181, 283, 327, 348
Cayman Islands, 158, 175
Central America. *See* Middle America
Central Wesleyan College, xiv, 4, 232
Chamberlin, Thomas C., 5, 23, 103, 234, 252, 386, 408
Champlain, Samuel de, 328, 333–335, 340
Charles V, German emperor and Spanish king, 216
Chart of my course, 3, 275, 324–329
"Chart of My Course" (Sauer), 275, 324–329
Chibchan, 301, 326
Chile, 187, 190, 199, 207, 212
Chiloé, 187
China, 123, 226, 251*n*1, 280, 307, 311
Chinampas, 199–200
Chisholm, George G., 412
Chorology, 130–132, 138, 153–154, 231, 233, 237, 240. *See also* Regional geography
"Chorology and Landscape" (Penn and Lukermann), 23–24*n*2
Ciboney, 300
Citations: changes in citation practice, 43–48; and coauthorship, 44, 49*n*7; to Denevan's "Pristine Myth," 34–35; and *Geltung*, 29–30; general thoughts about citation practice, 33–35; in geography, 29–49; Merton on, 29; Sauer's use of, 32, 35–41, 43, 47; to Sauer's writings, 41–43, 46, 47; self-citation, 32, 43–44, 49*n*7
The City as Text (Duncan), 14
The City in History (Mumford), 405–406
City of Knowledge Foundation, 24*n*9
Civil rights movement, 13–14
Clark, Andrew H., 418
Clark University, 360, 368
Cleland, Robert G., 252
Clements, Frederic E., 367–368, 386, 410
Climate: in agricultural hearth north of equatorial area, 201; of Caribbean, 154, 182; classification of, 262; climate regions, 253, 261–262; climatic warming, 262–263; and empty lands, 288; Früh on, in Switzerland, 144; glaciers and glaciation, 262–264, 385; Humboldt on, 126; influences of, on humans, 129; and morphology of landscape, 133, 134; number of climates, 133; in Sauer and Leighly's *Introduction to Geography*, 113; savanna climates, 200; and starch plants, 206–207
Climate and the World Pattern (Blumenstock and Thornthwaite), 378
Climatic grassland climax, 239, 261–268
Climatology, 129
Climax vegetation, 261–268
Cluverius, Philippus, 38
Coal, 280
"Coconuts on a Lava Flow in the Chiricahua Mountains" (Symanski), 20
Coffee, 281
Colbert, Jean-Baptiste, 339
Colima of New Spain (Sauer), 55
Colombia, 24*n*9, 181, 187, 189, 207, 209, 256, 297–298, 302, 303, 317
Colonization: destructive exploitation of, 22, 24*n*8, 238, 245–251; as objective of Columbus, 310–315. *See also* French explorations and colonies; Spanish explorations and colonies
Colorado, 369
Colton, Harold Sellers, 38

Columbus, Christopher: on calabashes, 205; colonial legacy of, 313–315; colonization as objective of, 310–315; discovery and exploration of New World by, 123, 298, 299, 306–309, 320, 326; and maize, 193, 194, 215, 217–220, 222; Sauer's critique of, 40, 271, 274–275, 313, 315; scholarship on, 41; and slaves, 314–315; and Vespucci, 319

Commercial exploitation of land, 242–243, 247–251. *See also* Destructive exploitation

Commercial geography. *See* Economic geography

Commodity geography. *See* Economic geography

"The Condition of Geography in the High School and Its Opportunity" (Sauer), 91–92

Conger, Charles, 114

Conservation Foundation, 7, 404

Conservation movement, 233–236, 242–244, 285

Conuco planting system, 154–155, 157, 177–178, 182, 203, 210, 212

Cook, Sherburne F., 331

Cooking. *See* Food and cooking

Cooper, Father John, 300

Copeland, Aaron, 153

The Coral Reef Problem (Davis), 111

Cordius, Valerius, 215

Corn. *See* Maize

Cornish, Vaughn, 38

Corominas, Joan, 219, 221, 222

Coronado, Francisco Vázquez de, 327, 338, 365

"Correspondence [on Physical Geography in Regional Works]" (Sauer), 144–146

Cortés, Hernando, 330

Cosa, Juan de la, 312, 315–318, 319, 322*n*19

Cosgrove, Denis, 16, 21, 24*n*6

Cosmographiae Introductio, 275, 321

Cosmologic School, 120–121, 125

Costa Rica, 158, 188, 303, 388

Cotton, 188, 208, 251, 256, 283, 286

Cowles, Henry C., 44, 348, 410

Crescenzi, Pietro de, 221

Croce, Benedetto, 37

Crops: age and area of American cultivated plants, 189–190, 195, 198–212; agricultural hearth north of equatorial area, 200–202; in Caribbean, 177–182, 191; cultural factors in plant domestication in the New World, 190–191, 195; diffusion of generally, 188–189, 195, 195–196*n*1; diffusion of New World crops, 282–283; dispersability of, 189; diversity of, 18–19, 185, 188, 194; molecular techniques and genome sequencing for study of, 195, 195–196*n*1; Sauer's interest in generally, 185–188; seed agriculture, 189, 190, 198–199, 202, 210–212; of South and Central America, 18–19; spread of Mesoamerican seed plants, 211–212; spread of nontropical Old World crops, 283–284; spread of Old World garden crops, 280–282; and starch foods, 204, 205–207, 212; tropical origin of American cultivated plants, 198–200; Vavilov on centers of origin and centers of diversity of, 186, 187, 188; vegetative planting, 189, 190, 198–199, 202–208; and weeds, 39–40, 190, 211, 304. *See also* Agriculture; and specific crops

Cross, Hugh B., 195

Cuba, 154, 175, 178, 180, 212, 298–300, 307–311, 313, 326

Cultivated plants. *See* Crops

"Cultivated Plants of South and Central America" (Gade), 18–19

Cultivated Races of Sorghum (Snowden), 220

Cultural diversity, 31

Cultural evolutionism, 17

"Cultural Factors in Plant Domestication in the New World" (Sauer), 190–191, 195

Cultural geography: critiques of, 15, 16, 21; definition of, 115; and Gradmann, 116; major problems of, 141–142; "new cultural" geography, 10, 16–17, 42–43; Price and Lewis on, 16–17; Rostlund on, 373; and Sauer generally, 21, 112; Sauer's articles on, xx, 114, 115, 136–143. *See also* Human geography

"Cultural Geography" (Sauer), xx, 115, 136–143

Cultural-historical geography, 47, 150, 155–

Cultural-historical geography (*continued*) 156, 158, 189. *See also* Cultural geography; Historical geography
Culture: comparative study of cultures, 167; and culture traits, 167, 170–171; degeneration of, 153, 171; diffusion of, 153, 166, 170–171, 356–357; and environmental adaptation, 167–168; Germans on cultural landscape, 139; interconnections between landscape and generally, xx–xxi, 36, 131, 388–389; limitation of concept of culture areas and stages, 302–4; past and present American culture, 390–391; and plant domestication in the New World, 190–191, 195; Sauer on, 15, 167; and site, 385. *See also* other headings beginning with Culture
Culture hearths, 18, 153, 166, 171, 188, 265, 274, 304
Culture history, 17, 31, 45–46, 167–168, 185, 190, 195, 245, 304
Culture traits, 167, 170–171
Curiosity, 32–33
Cushing, Sumner W., xiv
Cycle of erosion, 19, 116, 376

Daly, Herman, 149
Dalziel, John McEwen, 225
Dana, James Dwight, 103
Daniels, Stephen, 16, 21
Darwin, Charles, 254
Dasmann, Raymond F., 404
"David I. Blumenstock, 1913–1963" (Sauer), 377–379
Davidson, George, 348
Davis, William Morris: in California following retirement of, 111, 376; citations by Sauer to, 38; on cycle of erosion, 19, 116, 376; and German language proficiency, 44–45; influence of, on Sauer, 234; and land surface features, 350; leadership of, in geography, 10, 57, 115–116, 146, 346; model building by, 350, 386, 409–410; and physical geography as non-historical, 401; and Russell, 376; Sauer's critique of, 386
De Candolle, Alphonse, 186, 214
De Geer, Sten, 38, 112, 115
De Geer, Gerard, 112

De Martonne, Emmanuel, 116, 144, 146
Decades (Peter Martyr), 218–220, 314, 321
"Decline of Indian Population" (Sauer), 276, 330–337
The Decline of the West (Spengler), 36
Deforestation, 92, 157, 176–177, 235, 248
Delisle, Claude, 336
Demography, 166
Denevan, William: bibliography of commentaries on Sauer by, xii, xix, 53–86; bibliography of Sauer's publications by, 421–431; biographical information on, 433; on Caribbean publications by Sauer, 158n3; on collections of Sauer's writings, xvii–xxi; on early publications by Sauer, 89–94; fieldwork by, 35; as student of Sauer and Parsons, xvii, 24n9
—work: "Pristine Myth," 34–35
Denys, Nicolas, 333, 340
Derrida, Jacques, 45
Description of Africa (Leo Africanus), 226
Desert Trails of Atacama (Bowman), 111
Deserts, 256
Destruction of landscapes, 22, 23, 139
The Destruction of the Indies (Las Casas), 287
Destructive economies, 153, 171, 240
Destructive exploitation, 22, 24n8, 238, 245–251, 251n1
"Destructive Exploitation in Modern Colonial Expansion" (Sauer), xii, 238, 245–251
Determinism. *See* Environmental determinism; Geographical determinism
Diamond, Jared, 24n3
Dicken, Samuel N., 417
Dickson, D. Bruce, 17–18
A Dictionary of the Economic Products of the Malay Peninsula (Burkill), 221
A Different Kind of Country (Dasmann), 404
Diffusion: of crops, 188, 195, 195–196n1, 280–283; of culture traits, 153, 166, 170–171, 356–357; definition of, 412; of maize, 191–195, 195, 195–196n1, 213–228; Ratzel on diffusion of culture, 137, 153, 166, 356–357; Sauer on generally, 195
Dioscorides, Pedanio, 191, 214
Disciplinary world view, 231, 232
Diseases, 276, 287, 327, 330–331, 334, 335, 336

Dissertations: Sauer's dissertation, xviii, 4, 22, 89, 90–91, 114, 236, 275; supervised by Sauer, 350, 417–419
Distribution, 410
Distribution of Aboriginal Tribes (Sauer), 54
Diversity: appreciation of, 404–406; biodiversity, 19; of crops, 18–19, 185, 188, 194; importance of, 389; molecular analysis of genetic diversity, 195, 195–196n1; of plants, 201; of starch food, 205–207
Dobzhansky, Theodosius, 254
Doctoral dissertations. *See* Dissertations
Domestication: cultural factors in plant domestication in the New World, 190–191, 195; Dickson's critique of Sauer's theories of, 17–18; Gade on, 18–19; Harris on, 18; New World crop diversity, 18–19, 185, 188, 194; Old World plant and animal domestication, 18, 20; preconditions and initiators of, 201–202; Sauer on, 17–18, 188. *See also* Crops; Livestock
Dominican Republic, 155, 174, 178, 180, 181, 326
Donkin, Robin A., 40
Doran, Edwin, Jr., 418
Drainage, 249–250, 284
Dryer, Charles R., 11, 24n4, 115, 409
Dunbar, Gary S., 39
Duncan, James S., xviii, 10, 14, 15–16, 24n6, 42–43
Dust Bowl, 236, 272

Early humans, 264–265, 297–298, 347, 415
The Early Spanish Main (Sauer), xviii, 40–41, 54, 55–56, 158, 191, 274, 326
Eastern Europe, 191–192, 215, 216
Eberhard, Wolfram, 384
Ecology: and balance, 396; beginning of, 410; coinage of term by Haeckel, 233; and conservation movement, 233–236; development of Sauer's ecological perspective, 231–240; historical ecology, 240; human ecology, 396; plant ecology, 34, 170; Sauer's first use of term "ecological," 92, 104; Sauer on, 149, 231; use of term for economy, 170
Economia, 149–150, 412–413
Economic development, 156, 158, 246
Economic Geography, 368

Economic geography: of Caribbean, 154–158, 173–182; of crops, 194; and cultural geography, 141; geography as regional economics, 150, 160–161; and hoe culture, 141, 251, 289; land use and commercial geography, 150–151; of Ozark Highlands, 101; Sauer's critique of, 411, 412–413; and statistics, 411; in university curricula, 112. *See also* Crops; Economy and economics
"Economic Prospects of the Caribbean" (Sauer), 156–158, 173–182
Economy and economics: of Caribbean, 154–158, 173–182; destructive economies, 153, 171; geography as regional economics, 150, 160–161; and regional development, 153–156; regional reality in economy, 151–153, 162–172; Sauer on, 149–150. *See also* Economic geography
Ecuador, 201, 207, 209
"The Education of a Geographer" (Sauer), xii, 38–39
Edwards, Clinton R., 419
Effland, Anne B. W., 94
Effland, William R., 94
Eichel, Marijean, 328n1
Ekholm, Gordon F., 302
Ekman, Vagn Walfrid, 349
Elements of Geography (Finch and Trewartha), 113
The Elements of Geography (Salisbury, Barrows, and Tower), 91, 113
"Ellen Churchill Semple, 1863–1932" (Sauer), 360–361
Emory, William H., 258
Empty lands, 288–289
Encyclopedia of Social Sciences, 345
"The End of the Century" (Sauer), xx, 276, 338–341
Energy resources, 156, 174
Engelbrecht, Theodor, 39
England, 16, 123, 126, 137, 207, 327
English scientists and geographers, 107–108, 115, 237
Entrikin, J. Nicholas, 152
Environmental adaptation, 167–168
Environmental adjustment, 137–138
Environmental causality, 126–127, 137

Environmental change over time, 383–384
Environmental determinism, 5, 11, 17, 24*n*3, 111, 347, 349–350
Environmental influences: doctrine of, 127; of humans upon the earth, 5, 23, 92, 103–104, 139; in Ozark Highlands, 100–101
Environmentalism: in geography from 1900 to 1955, 10; and human geography, 231; Ratzel on, 137; rejection of, by geographers, 137–138, 238; Sauer on, 3, 16–17
Eratosthenes, 121–122
Der Erdkunde im Verhältnis zur Natur und zur Geschichte des Menschen (Ritter), 358
"Erhard Rostlund, 1900–1961" (Sauer), 371–374
Erosion: causes of accelerated erosion, 92; of crop diversity, 194; "cycle of erosion" concept, 19, 116, 376; human factors in, 92, 139, 235–236, 242–244; research on, 20; water erosion, 250; wind erosion, 250. *See also* Soil erosion
Escalante, Father Silvestre Velez de, 258
Ethnology, 166
Euphytica, 190–191, 195
Eurocentrism, 21, 32, 34, 386
Europe: destructive exploitation by, in modern colonial expansion, 238, 245–251; destructive exploitation in Old World, 251*n*1; influences from, on geography generally, 44–45, 237–238, 240; maize in, 191–194, 195, 195–196*n*1, 213–228. *See also* Colonization; and specific countries
Europe Centrale (de Martonne), 116, 144
European Conquest, 40–41
Europeanization, 246
Evolution, 254, 259, 375
Examen critique (Humboldt), 309
Explanatory description, 386, 394–395
Exploration, Age of, 122–124, 306–321. *See also* Spanish explorations and colonies
"Exploration of the Kaiserin Augusta River in New Guinea, 1912–1913" (Sauer), 90, 95–98
Extinction, 246, 254

Fairgrieve, James, 112
Fats, 204, 211

Febvre, Lucien, 37, 138
Fenneman, Nevin M., 115
Fernandes, Valentim, 224
"The Field of Geography" (Sauer), xviii, xx, 114, 119–135, 231
Fieldwork: in Caribbean, 154–158, 173–182; by Denevan, 35; in geography generally, 6, 32; in local region, 92; outline for, 90, 113; by Parsons, 154, 155–156; by Sauer, 19–20, 89, 93, 101, 151, 154–155, 325–326; Sauer on importance of field observers, 112
Finan, John J., 191, 213, 215
Finch, Vernor C., 93, 113, 150
Fippin, Elmer, 108
Fires, 265–268
Fischer, Ernst, 103
Fish and fishing, 204, 372, 401
Fisk, Harold N., 350
Five Nations, 335–336
Fleure, Herbert John, 38, 386, 414
Flora of West Tropical Africa (Dalziel), 225
Florida, 173, 298–299, 330, 333, 338–339
Flurformen (field forms), 141
"Folkways of Social Science" (Sauer), 152
Fonesca, Juan Rodríguez, 317
Food and cooking, 204–205, 210–211, 413. *See also* Agriculture; Crops
Foote, Kenneth E., 46
Ford, James A., 38
Forest Service, U.S., 369
Forests, 92, 157, 176–177, 236, 248, 285, 286, 288–289, 298, 373
"Foreword to Historical Geography" (Sauer), xii, xvii, 12–13, 38, 152
Forschungen zur Deutschen Landes- und Volkskunde, 140, 145
Foucault, Michel, 45
The Four Books of Meteorology (Aristotle), 121
Fox, Cyril, 39
Fox, Stephen, 233
France, 123, 141, 195–196*n*1. *See also* French explorations and colonies; French geographers
Franciscans, 257
French explorations and colonies, 276, 327, 328, 333–335, 339–341, 408

French geographers, 10, 45, 113, 115, 237, 240. *See also* specific geographers
"Freshwaer Fish and Fishing in Native North America" (Rostlund), 372
Friederici, Georg, 246
Friedrich, Ernst, 237, 238
"Friedrich Ratzel, 1844–1904" (Sauer), 356–357
Fröbel, Carl Ferdinand Julius, 137, 346
Frobenius, Leo V., 37
Frontiers, 278
Früh, Jacob, 116, 144
Fruits, 177, 191, 198, 207–208, 209
Fuchs, Leonhart, 214, 219, 221, 224
Fur trade, 340

Gade, Daniel W., xix, xx, 18–19, 24n7, 29–52, 46, 57, 185–197, 433
Gaile, Gary L., 46
Gallegos, Hernán, 331
Garland, Hamlin, 390
Geer, Sven de, 38
Geltung (competitive communicative display), 29–30
Gemeinschaft (professional solidarity), 43, 232
General Land Survey, 113
Genetics, 253–254
Genetics and the Origin of the Species (Dobzhansky), 254
Genome sequencing, 195
Geodesy, 129
Geographia Generalis (Varenius), 125
Geographic distribution analysis, 165
Geographic Information Science (GIS), 10
Geographic method, 395–396
"A Geographic Sketch of Early Man in America" (Sauer), 347
Geographical determinism, 10, 116
Geographical Error, 236
Geographical Journal, 91
The Geographical Lore of the Time of the Crusades (Wright), 111
Geographical Review, 11, 89, 92, 111, 116, 144–146, 368, 373
Geographical Society of Chicago, 90, 101–102

Geographie der Schweiz (Früh), 116, 144
Geographische Jahrbücher, 136
Geography: American geography generally, 43–45, 115–116; Augean Period of, 13; changes in, 10–23; chorologic position in, 130–132; citations in literature of, 29–49; continuity in breadth of interest of, 399–401; cultural-historical geography, 47, 150, 155–156, 158, 189; curiosity underlying research in, 32–33; data of, as forms of landscape, 132–33; definitions of, 112, 114, 115, 117n1, 119, 127; diversity of geographers in U.S., 43–44; as earth science, 408–409; European influences on, 44–45; general versus special geography, 127–128; geology's relation to, 116, 126, 128–129, 324, 408–409; history of, 12, 114, 115, 120–128, 141; home geography, 92, 101; maps and mapping in, 91, 93; "new cultural" geography, 10, 16–17, 42–43; and physical forms of landscape, 133–135; quality of, 389–406; quantification in, 13–14, 34; "relevancy orientation" of, 33; Sauer on three-point underpinning for, 12–13; seminars in, 392–396; as social science, 129–130; and spatial positivists, 13–14; subfields of, 92; survey method in, 112–113; theoretical geography, 412. *See also* Berkeley school; Cultural geography; Economic geography; Fieldwork; Human geography; Landscape; Physical geography; Regional geography; Sauer, Carl
Geography and Geographers (Johnston and Sidaway), 23n1
"Geography as Regional Economics" (Sauer), xx, 160–161
"Geography 1957–1977: The Augean Period" (Gould), 13
The Geography of the Mediterranean Region (Semple), 361
The Geography of the Ozark Highland of Missouri (Sauer), 4, 54, 93, 99–102, 236, 275, 325
Geography of the Pennyroyal (Sauer), 54, 89, 93, 325
Geological Survey, xiii–xiv
Geological surveys, 113, 115–116

Geology, 116, 126, 128–129, 324, 408–409
Geology (Chamberlin and Salisbury), 103, 234–235
Geomorphology: and cultural geography, 140; as field of study, 128, 129, 346; and geography texts, 113; and Grandmann, 145; humans as geomorphologic agents, 234; and Peschel, 346; and Salisbury, 324; and Sauer's study of California's Peninsular Range, 19; and slope studies, 249; and soil erosion, 139
Geophysics, 128, 409
Gerland, Georg, 130
German ancestry, 3–4, 31
German geographers: and anti-German wartime sentiment, 49n5; biographies of, 345–347, 354–359; citations by Sauer to, 40; and cultural geography, 115; exploration of Kaiserin Augusta River in New Guinea by, 90, 95–98; and German language proficiency of American geographers, 44–45; influence of, on Sauer, 10–11, 19, 40, 112, 237; and *Landschaftskunde* school, 240; and Leighly, 12, 19; on natural versus cultural landscape, 139; and *Raubbau, Raubwirtschaft* (destructive exploitation), 238, 239, 242, 247–248; and *Siedlungskunde*, 140; at University of California, Berkeley, 11. See also specific geographers
Germany, 4, 37, 144–145, 213–216, 221, 272
Gerrymander, 89
Gershwin, George, 153
Geschichte der Erdkunde (Peschel), 354
Geschichte des Zeitalters der Entdeckungen (Peschel), 354
Gesellschaft (rational self-interest), 43, 232
Gestalt (integrated whole, "everythingness"), 274, 384, 399–400, 407
GIS. See Geographic Information Science (GIS)
Glaciers and glaciation, 262–264, 385
Goats, 348
Goethe, Johann Wolfgang von, 31, 36, 235, 240
Gold, 279, 308, 309, 311–314, 317
Goode, J. Paul, 101, 114
Gordon, Burton L., 41, 418
Gottmann, Jean, 406

Gould, Peter, 10, 13–14
Gourds, 208
Gradmann, Robert, 115, 116, 140, 144–145
Grandeza de Bahia de Todos os Santos (Soares de Souza), 223
Grasses and grassland, 239, 264–267, 283–284, 324–325, 368, 395
"Grassland Climax, Fire, and Man" (Sauer), xviii, 239, 261–268
Great Britain. See England; English scientists and geographers
Great Geological Surveys, 113, 115–116
Great Lakes, 286
Great Plains, 266, 278, 285, 368, 370
Greater America, 365
Greek philosophers, 120–122, 125, 130, 391
Greenland, 327
Gregory, Derek, 16
Guanin, 309, 312, 315
Guatemala, 211, 256, 330–331
Guggenheim Foundation, 7, 151, 187
Guns, Germs, and Steel (Diamond), 24n3
Guyot, Arnold, 359

Habitat, 231, 245, 261, 358–359, 387, 389
Haeckel, Ernst, 233
Hagiography, 20–21
Hahn, Eduard, 38, 40, 141, 414
Haiti, 155, 176, 178, 203, 300, 303, 308, 309, 313–314, 316
Hall, A. Daniel, 107–108
Hamilton, Earl J., 158n2
Hammond, George P., 365
Handbook of American Indians North of Mexico (Hodge), 363
Handbook of Cultural Geography, 24n10
Handbook of South American Indians, 187–188
Harlan, Jack R., 186, 187
Harper, Roland, 348
Harris, David, 18
Harris, David R., 195
Harrisse, Henry, 320
Hartshorne, Richard: citations by Sauer to, 38; critique of Sauer by, 10–13, 91; and German language proficiency, 44; on Peschel, 346; on positivistic social science, 12; on

Ratzel, 347; on Ritter, 346; and Sauer's practice of geography, 237; and Sauer's writings, 115, 117*n*1
—work: *Nature of Geography*, 11–12, 23*n*1, 24*n*4, 384
Harvard University, 19–20, 111, 376
Haudricourt, André, 40
Haushofer, Albrecht, 112
Hawkes, John Gregory, 190, 207
Headland, Thomas, 34
Hearths. *See* Agricultural hearths; Culture hearths
Hédin, Louis, 40
Helmolt, Hans F., 357
Henry of Portugal, Prince, 123, 306
Herbals, 191, 213–215
Herbert, David T., 46
"Herbert Eugene Bolton, 1870–1953" (Sauer), 362–366
Herbertson, A. J., 113
Herder, Johann Gottfried, 137
Hernández, Francisco, 216–217
Herodotus, 121, 124, 388
Herrera, Gabriel Alonzo de, 221
Hettner, Alfred, xiv, 36, 115, 136, 386
Hewes, Leslie, 418
High school geography, 89, 91–92
Hilgard, Eugene Woldermar, 107, 242
Hispaniola, 40–41, 155, 181
Historia (Joâo de Barros), 223
Historia de las Indias (Las Casas), 217
Historia natural y general de las Indias (Oviedo), 214
Historical ecology, 240
Historical geography: on Columbus, 274–275, 306–315; and cultural geography, 141; on Indians, 273, 292–295; on Middle America as a culture historical location, 274, 296–305; and physical geography, 401; redistribution of population, 271–272, 277–291; Sauer's courses on, 114; scholars on, 114; on seventeenth-century North America, 275–276, 324–341; on sixteenth-century North America, 56, 275, 327; and soil erosion, 248–249; survey of, 120–128; as underpinning of geography, 12. *See also* History of geography

Historicism, 36, 195, 233
History of Civilization (Buckle), 127
History of geography, 12, 114, 115, 120–128, 141. *See also* Historical geography
Hitchcock Foundation, 375–376
Hobbs, William H., 49*n*5, 90, 114
Hodge, Frederick, 363
Hoe culture, 141, 251, 289
Hoffpauir, Robert, 1, 57
Hogs, 181, 283
Hohokam, 301–302
Hojeda, Alonso de, 314, 315–316
Hologeographic approach, 249
Holt-Jensen, Arild, 23*n*1
Holway, Ruliff S., 114, 348–49, 351, 352–353
Home geography, 92, 101
Homer, 120
"Homer LeRoy Shantz, 1876–1958" (Sauer), 367–370
Honduras, 299, 301, 312
Hooson, David, 10
Hopi, 332
Horses, 257, 348
Howe, Henry V., 350
Hugill, Peter J., 46
Human agency, 5, 23, 234, 240, 264–265
Human geography: Bruhnes on, 240; and causal relations between humans and their natural environment, 126–127, 137; definition of, 240; destructive exploitation in modern colonial expansion, 238, 245–251; destructive exploitation in Old World, 251*n*1; environmentalist definition of, 231; Gradmann on, 116, 145; and man as geomorphologic agent, 234; in 1970s, 10, 14–15; Sauer on, 231–232, 240; Semple on, 360–361; and soil erosion and soil conservation, 139, 235–236, 242–244; of Southwest in U.S., 237, 252–260. *See also* Cultural geography
Human Geography (Bruhnes), 238
Humboldt, Alexander von, 38, 39, 125–126, 138, 309, 314, 345–346, 358
Humboldt State College, 404
Huntington, Ellsworth, 111, 386, 409
Hurons, 335
Huxley, Aldous, 415

Huxley, Thomas H., 37
Hybridization, 246
Hyland, Ken, 33
Hypothoses, 396

Icarus Complex, 234
Ice Age, 3, 254, 262–263, 324, 409–410
Iceland, 327
Illinois River Valley, xiii–xiv, 89, 92, 114
Illinois State Geological Survey, 324–325
India, 123, 192–193, 307
"Indian Food Production in the Caribbean" (Sauer), 158n3
Indians: agriculture of, 198, 202; in California, 378; decline of population of, 276, 287–288, 327, 330–337; and diseases, 276, 287, 327, 334, 335, 336; on East Coast, 333–336; elementary education textbook on, 235, 292–295; impact of, on environment, 236, 239; in Mexico, 37; and New France, 340–341; and New Spain, 287, 330–333; in Ozark Highland of Missouri, 4, 236; Sauer's *Man in Nature* on, 273, 292–295; in Texas, 363; warfare between, 334–336. *See also* "About Nature and Indians" (Sauer); and specific tribes
Industrial and Commercial Geography (Smith), 150
Industrial geography. *See* Economic geography
Industrial Revolution, 246
Influences of Geographic Environment (Semple), 347, 360
Innis, Donald Q., 419
Intellectual climate, 153, 164, 167–170
International Botanical Congress, Tenth, 190
International Congress of Americanists *Proceedings*, xii
International Congress of Geographers, xii
International Symposium on Man's Role in Changing the Face of the Earth (1955), 23n1, 24n8
An Introduction to Geography: Elements (Sauer and Leighly), 113–114, 117n1, 119–135
Iraq, 22

Iroquois, 334–336, 340–341
Irrigation, 256–257, 284
Isabella and Ferdinand, 310, 321
Italy, 191–195, 215–221, 306

Jackson, Peter, 16, 21, 24n6
Jacobs, Jane, 24n10
Jamaica, 155, 175, 298, 303, 308, 309, 313, 317, 326
James, Preston E., 11, 113, 150
Jameson, John Franklin, 363–364
Jefferson, Mark, 90, 111
Jefferson, Thomas, 242
Jeffreys, Mervyn David Waldegrave, 194, 223, 225
Jesuits, 257, 326, 333, 339, 340, 364
Johannessen, Carl L., 155, 193, 413, 414, 419
Johnson, Douglas W., 90
Johnston, Ron, 23n1
Joliet, Louis, 339
Jones, Martin, 196n1
Jones, Wellington D., xiv, 44, 90, 113
Journal of Farm Economics, 194
Journal of Geography, 91, 93
Journal of the Michigan Schoolmaster's Club, 91
Juana, Queen, 216, 321

Kaiserin Augusta River, 90, 95–98
Kalon (moral and aesthetic beauty), 391
Kentucky, 114, 266, 325, 347–348
Kentucky Geological Survey, 325
Kenzer, Martin S., xviii–xx, 53, 149–159, 158n1, 162–172, 232, 433–434
Kesseli, John E., 19, 384, 411, 417
Keyserling, Herman Graf von, 36
King, Franklin Hiram, 107
Kipling, Rudyard, 411
Kniffen, Fred B., 13, 15, 38, 112, 372, 417
Köhler, Wolfgang, 399
Kräftezentren (hearths), 171
Kramer, Fritz L., 419
Krebs, Norbert, 237
Kroeber, Alfred L., 15–16, 37, 187, 349, 372, 378
Kulturgeschichte (culture history), 31

Kulturkreis (culture-sphere), 37
Kulturlandschaft (landscape as visible cultural imprint), 36, 112

La Salle, René-Robert Cavelier, Sieur de, 339, 340
Lahontan, Louis Armand de Lom d'Arce, 276, 340–341
Land and Life (Sauer), xviii, 5, 54, 55, 384
Land Economic Survey of Michigan, 4, 24*n*8, 30, 93, 237
Land Forms in the Peninsula Range (Sauer), 54
"Land Resources and Land Use" (Sauer), 237
Land surveys, 92–93, 113, 237
Land use, 150–151, 237
Landform studies, 10
Landscape, xix, 383, 390–391
Landscape: and climates, 133, 134; cultural versus natural landscape, 139, 387; data of geography as forms of, 132–133; Davis on, 111; design of, 131; of destruction, 22, 23; Germans on cultural landscape, 139; Hartshorne's critique of Sauer's concept of, 12; interconnections between culture and generally, xx–xxi, 36, 131, 388–389; Mitchell's construction and deconstruction of, 21–22; natural landscape, 133–135; and oceans, 133, 134; patterns and processes of changes in, 17; physical forms of, 133–135; Sauer and description of, 31; and space relationships, 133; Spengler on history of, 37; and vegetation, 133–134. *See also* Geography
Landscape-as-history, 36–37
Landscape journal, 373
Landschaftskunde (landscape science), 240
Lange, Dorothea, 153
Las Casas, Bartholome de, 40, 217, 287, 314, 322*n*14
The Last Glaciation (Antevs), 111
Latin America, 22, 30, 42–43, 187, 192–193, 279, 349. *See also* Caribbean; Middle America; South America; and specific countries
Laufer, Berthold, 226
Lawson, Andrew Cowper, 19

Lebensraum, 272, 357
Lee, Hermione, xiii
Leighly, John Barger: biographical information on, 417; commentaries on Sauer by generally, 53; as editor of *Land and Life*, xviii, 5, 48*n*2, 54; on "Morphology of Landscape," 117*n*1; and natural region, 113; and Penck's analytical method, 19; on Ritter, 346; Sauer on intellectual curiosity of, 114; Sauer's citations of, 40; on Sauer's *Ozark Highland*, 91; on Schlüter, 36–37; on Semple, 347; at University of California, Berkeley, 114, 348
—works: "The Field of Geography," 119–135, 231; *An Introduction to Geography*, 113–114, 119–135; "Methodologic Controversy in Nineteenth Century German Geography," 12; "Some Comments on Contemporary Geographic Methods," 11–12
Leith, Charles Kenneth, 44
Leo Africanus, 225–226
Leo X, Pope, 225
Leonardo da Vinci, 227*n*8
Lepe, Diego de, 319
Lescarbot, Marc, 333, 340
"Letter to *Landscape* [on Past and Present American Culture]" (Sauer), 390–391
Levillier, Roberto, 322*n*19
Lewis, Martin W., 16–17, 24*n*6, 43
Ley, David, 16
Life-writing, xiii
Livestock, 181, 185, 198, 199, 258, 283, 348, 413–414
Livingstone, David N., 23*n*1
Lobato, Roberto, 24*n*9
L'Obel, 215
Louderback, George D., 19
Louisiana, 328, 350
Louisiana State University, 12–13, 15, 350, 368
Lovell, W. George, xix, 91, 271–276, 434
Lowie, Robert H., 347, 378
LSU. *See* Louisiana State University
Lucas, C. P., 103
Lukermann, Fred, 10–11, 23–24*n*2, 43, 117*n*1
Lumbering, 285, 286. *See also* Deforestation; Forests

Lyell, Charles, 103, 234–235
Lyon, Thomas Lyttleton, 108

Macpherson, Anne, 115
Magellan, Ferdinand, 221, 222
Magnaghi, Alberto, 322*n*19
Maize: bread from, 218, 219–220, 222, 225; diffusion of, into Europe, 191–194, 195, 195–196*n*1, 213–228; drought resistance of, 200; early descriptions of, 213–216; evolution of, 413, 414; flood plain cultivation of, 199; and Italy, 191–195, 215–220; Leo Africanus on, 225–226; Leonardo da Vinci on, 227*n*8; for livestock feed, 283; in Mexico, 186, 210, 256; *milho* and *zaburro* as, 222–225; Peter Martyr on, 217–220, 225, 226; and sorghum and millet, 220–222; in South America, 187, 211; in Southwest U.S., 256; and Spain, 216–217, 222; tortillas from, 257; and Turkish corn, 213–216
Maize and the Great Herbals (Finan), 213
"Maize into Europe" (Sauer), xii, 191–195, 213–228
Mamey, 191
Man and Nature (Marsh), 232
Man in Nature (Sauer), 8*n*1, 55, 91, 235, 273, 292–295
Mangelsdorf, Paul C., 193
Manioc (yuca), 199, 200, 202, 203, 206–207, 282
"Man's Influence Upon the Earth" (Sauer), xii, xx, 89, 92, 103–104, 234
Man's Role in Changing the Face of the Earth (Sauer), xii, 5, 55, 92
Man's Role in Changing the Face of the Earth symposium (1955), 23*n*1, 24*n*8
Maps and mapping, 91, 93, 124, 125, 129, 167, 169, 385, 398–401, 411
Marbut, Curtis F., 111, 348, 368
Margarita Island, 41, 155
Margry, Pierre, 328
Marmer, Harry Aaron, 348
Marsh, George Perkins, 22, 38, 232, 240
Martin, Geoffrey J.: on anti-German wartime sentiment, 49*n*5; biographical information on, 434; critical commentary on Sauer by, xx, 111–118; on Davis, 57; on economy/economics, 149–159; and history of geography, xix, 23*n*1; on Michigan Land Economic Survey, 93; on Sauer's correspondence with Hartshorne, 117*n*1; on Sauer's knowledge of Schlüter's work, 36–37

Marxism, 16, 21
Massachusetts State Normal School, xiv–xv, 4, 91
Mathewson, Kent: biographical information on, 434; on citations to Sauer, 41; commentaries on Sauer by, 30, 53, 57; on cultural geography, 43, 46; dissertation by, xix; on economy/economics, 149–159; as editor of Sauer's collected essays, xviii; on Latin America, 42; on Sauer's critics, 9–23; on Sauer's view of culture as superorganic, 43
Matthes, François E., 348
Matthews, John A., 46
Mattioli (Matthioli), Pier Andrea, 192, 213
Maury, Matthew Fontaine, 298
McBryde, Felix Webster, 418
McCann, James C., 194
McGee, William James, 242
McGinnies, William G., 369
McIntire, William G., 350
McLeod, Bentley H., 302
Mechanism, 231
Megalopolis, 405–406
Megalopolis (Gottman), 406
Meigs, Peveril, III, 44, 372, 417
Meinig, Donald W., 8*n*1
Meitzen, August, 38, 140–141
Mercator, Gerhard, 124
Merriam, John C., 369
Merrill, Gordon C., 155, 419
Merton, Robert K., 29
Mesoamerica: alcohol in, 210; Andean culture area versus, 296; decline of native population in, 330–333; and Mexican-Guatemalan complex, 303; seed farming in, 190, 202, 208, 211–212, 304; spread of Mesoamerican seed plants, 211–212; Vavilov's crop theories applied to, 186–187
Messedaglia, Luigi, 191, 192, 195
Metallurgy, 301, 302, 318

Meteorology, 129
"Methodologic Controversy in Nineteenth Century German Geography" (Leighly), 12
Mexico: agriculture in, 32, 186, 190, 199–200, 201, 207, 208, 210, 211, 256; alcoholic beverages in, 209; cotton in, 209; Indian-dominated rural way of life in, 37; Sauer on personality of, 39; Sauer's fieldwork in, 19, 93, 154; silver in, 272; tobacco in, 209; translation of geographical classics on Michoacán in, 24n9
Michigan Academy of Science, 92–93
Michigan Land Economic Survey, 4, 24n8, 30, 93, 237
Michotte, Paul L., 142
Middle America: corridor and crossroads of, 297–299; as culture historical location, 274, 296–305; and early migrations, 299–301; limitation of concept of culture areas and stages, 302–304; position and configuration of land and sea in, 296–297; South American influences in, 301–302
"Middle America as a Culture Historical Location" (Sauer), xii, 274, 296–305
Mikesell, Marvin, xii–xiii, xx, 21, 113, 419
Milankovich, Milutin, 385
Milho, 222–225. *See also* Maize
Mill Springs Field Station, Kentucky, 114
Miller, Hugh, 126
Millet, 220–222
Millikan, Robert A., 252
Milpa system, 208, 210, 248
Mineralogy, 128
Mining and minerals, 257, 258, 279–280, 314
Miracle, Marvin, 194
Missionaries and missions, 257, 315, 326, 333, 335, 339, 340, 363, 365
Missouri, 3–4, 90–91, 99–102, 236, 275
Missouri State University, 368
Mitchell, Don, 21–22, 23, 33
Model building, 386, 409–410
Modern history, end of, 272, 278–279
Modernism, 154, 156
Moe, Henry Allen, 7
Monoculture, 250–251, 258
Montesquieu, Charles de Secondat, baron de, 137

Montfort study, 93
Morality: and biodiversity, 19; Sauer as environmental moralist, 3, 48, 234, 239
Morgan, Adrienne, 328n1
Morgan, Arthur E., 153
Morison, Samuel Eliot, 41
Mormons, 259
Morphologie der Westgeschichte (Morphology of World History), 36
Morphology, 36, 395
"The Morphology of Landscape" (Sauer), xii, xviii–xx, 5, 10–11, 15, 23–24n2, 35–38, 42, 54, 114, 115, 117n1, 152, 233, 237, 346–347, 349, 387
Mosto, Cà da, 306
Motley, Timothy, 195
"'Mr. Sauer' and the Writers" (Parsons), 8n1
Mumford, Lewis, 153, 405–406
Mundus Novus (Vespucci), 320
Murphy, Robert Cushman, 348
Muskhogean tribes, 198

Nansen, Fridtjof, 326–327
Native Americans. *See* Indians; and specific tribes
Natural regions, 112–113, 358–359
Natural resources, 27–80, 237, 252–260
Natural selection, 259, 263
Natural Vegetation as an Indicator . . . (Shantz), 368
"The Natural Vegetation of the Great Plains" (Shantz), 368
Nature and society, 91–92
The Nature of Geography (Hartshorne), 11–12, 23n1, 24n4, 384
Naturlandschaft (natural landscape), 112
Naturphilosophie, 240
Navagero, Andrea, 191
Nazism, 272
Near East, 18
Nemoianu, Virgil P., 232
Nevis, 155
"New cultural" geography, 10, 16–17, 42–43
New France. *See* French explorations and colonies
New Guinea, xx, 89, 90, 95–98
New Mexico, 20, 211, 257, 258–259, 333, 338

New Spain. *See* Spanish explorations and colonies
New Spain and the Anglo-American West, 365
New Zealand, 246
Newfoundland, 327
Niagara Falls, 89
Nicaragua, 176, 300, 301
9/11 terrorist attack, 21
Norse settlements, 327
North America (Smith), 150
Northern Mists (Sauer), 8*n*1, 56, 326–327
Northwestern University, xiii, 4, 19
"Notes on the Geographic Significance of Soils: A Neglected Side of Geography" (Sauer), 93, 105–108

Ocean of Air (Blumenstock), 350, 378
Oceans, 133, 134, 156–157, 175, 204
Odum, Howard W., 38, 153, 252
Office of Naval Research (ONR), 7, 154–156, 350, 369–370
Ogilvie, Alan G., 112
Olson, Charles, 7–8*n*1
Olwig, Kenneth, 21
Oñate, Juan de, 332
ONR. *See* Office of Naval Research (ONR)
Opus Epistolarum (Peter Martyr), 309, 311
Oregon, 247
Organic analogy, 233
Organicism, 231
Ortelius, Abraham, 124
Osborn, Henry Fairfield, 375
"Oskar Peschel, 1826–1875" (Sauer), 354–355
"Outline for Field Work in Geography" (Jones and Sauer), 90, 113
Outline of Cultural Geography (Rostlund), 373
Overgrazing, 258
Oviedo, Gonzalo Fernández de, 40, 178, 203, 214, 219
Ozark Highlands, 3–4, 89, 93, 99–102, 211, 236, 275, 325

Pacific Coast Economic Association, 151
Pacific Railroad Surveys, 258
Pacific Sociological Society, 151
Paleontology, 128
Palestinian towns, 21–22
Panama, 24*n*9, 173, 299
Paraguay, 283
Pareto, Vilfredo, 38
Park, Robert E., 38
Parkman, Francis, 364
Parsons, James J.: and Berkeley school, 155; biographical information on, 434–435; dissertation by, 418; as faculty member at University of California, Berkeley, 155; fieldwork by, 154, 155–156; on Sauer's influence on writers, 7–8*n*1; on Sauer's life, scholarship, and significance, xviii–xix, 3–8; as Sauer's student, 32, 53, 418; students of, 24*n*9
Partsch, Joseph, 237
Passarge, Siegfried, 36, 115, 237
Patterns and processes, 17
Peak, Harold J. E., 386, 414
Peanuts, 203–204, 207
Pearsall, Deborah M., 190
Peattie, Roderick, 116, 144
Penck, Albrecht, 36, 112, 146, 237, 358–359, 409–410
Penck, Walther, 19, 20, 249
Peninsular Range of California, 19
Penn, Misha, 10–11, 23–24*n*2, 43, 117*n*1
Pennyroyal monograph (Sauer), 54, 89, 93, 325
Peopling the Argentine Pampa (Jefferson), 111
"The Personality of Mexico" (Sauer), 39
Peru, 205, 207, 209, 212, 256, 272
Peschel, Oskar F., 137, 138, 345–347, 354–355, 358
Peter Martyr d'Anghiera: and Age of Discovery generally, 306; citations to, by Sauer, 40; on Columbus, 311–312, 315; on maize, 193, 217–220, 225, 226; on millet, 221–222; Sauer's view of, 275
—works: *Decades*, 218–220, 314, 321; *Opus Epistolarum*, 309, 311
Petrie, William Matthew Flinders, 37
Petrography, 128

Index

Petroleum, 280
Pfeifer, Gottfried, 141
Philip II, King of Spain, 216
Philippines, 217, 221, 222–223
Phillips, Walter S., 370
Physical geography: of Arizona, 19–20; on atmosphere versus physical forms of earth's surface, 170; in England and United States, 137; Hartshorne on, 12; historical nature of, 401; human agency in, 240; of New Guinea, 95–98; of Ozark Highlands, 90, 100, 325; and regional geography, 144–146; Sauer and Leighly's text on, 113–114, 119–135; Sauer on, 12, 106, 385; and soils, 20, 93, 106; of Upper Illinois Valley, 92, 114, 324–325
Physiography, 126–129, 324, 396. *See also* Geomorphology
Pigafetta, Antonio, 221, 222–223
Pineapple, 191, 207–208
Pinzón, Vicente Yáñez, 316, 319, 321
Pioneers and settlers, 4, 91, 107, 236, 282, 325, 388
Piperno, Dolores R., 190
Plant and Animal Exchanges (Sauer), 55
Plant ecology, 34, 170
Plantation system, 157, 179–181
Plants: in agricultural hearth north of equatorial area, 201; of Caribbean, 154; diversity of, 201; as form of landscape, 133–134; Humboldt on, 126; as medications, 209; and poisons, 209. *See also* Crops
Pliny the Elder, 220
Pohl, Frederick, 322n19
Political maps, 169
Political science, 163
Politische Geographie (Ratzel), 357
Polo, Marco, 123, 218, 275, 307, 309, 311, 312, 320
Population: of Caribbean, 157; changes in, 277–279; and decline in mineral prospects, 279–280; and diffusion of New World crops, 282–283; and empty lands, 288–289; of Indians, 276, 287–288, 327, 330–337; and irrigation and drainage engineering, 284; and lumbering, 285; pressures regarding, 246–247; Ratzel on, 165–166; redistribution of, 271–272, 277–291; and shrinking subsistence bases, 285–287; and spread of Old World crops, 280–282; and technologic advance, 289–290
Population genetics, 254
Portugal, 191, 194, 215, 217, 222–225, 319
Possibilisme, 137
Post, Lauren C., 412, 417
Postmodernism, 14, 31, 32, 34
Potatoes, 190, 199, 207, 208, 209, 282
Pound, Nathan Roscoe, 367
Powell, John Wesley, 112, 258
Prairies, 236, 239, 324–325
Pred, Alan R., 13–14
Prehistoric Settlements of Sonora, 54
President's Science Advisory Board, 7
Prester John, 224
Price, Edward T., xix, 24n6, 43, 345–351, 418, 435
Price, Marie D., 16–17
"Pristine Myth" (Denevan), 34–35
Proceedings of the International Congress of Americanists, xii
Processes and patterns, 17
Progress in Human Geography (Mitchell), 21–22
"Proposal of an Agricultural Survey on a Geographic Basis" (Sauer), 92–93
"The Prospect for Redistribution of Population" (Sauer), 271–272, 277–291
Protein, 204, 211
Pruitt, Evelyn L., 350
Ptolemy, Claudius, 122, 123
Public hygiene, 290
Pueblo people, 209, 257, 276, 331–334, 338
Puerto Rico, 155, 180, 313, 326
The Pulse of Progress (Huntington), 111
Pythagoreans, 121

"The Quality of Geography" (Sauer), 383, 385, 389–406
Quam, Louis O., 112
Quantification, 6, 13–14, 34
Quinoa, 190, 199
Quinton, Anthony, 232

The Races of Man (Peschel), 354
"Radical Critiques of Cultural Geography" (Blaut), 15
Ramusio, Giovanni, 191, 214, 223
Rand McNally, xiv
Ratzel, Friedrich: biography of, 345, 347, 356–357; book review by Sauer on, 89; citations to, by Sauer, 38; on diffusion of culture, 137, 153, 166, 356–357; on environmentalism, 137; influence of, on Semple, 360, 361; on population, 165–166; and traditions of geography generally, 388
—works: *Anthropogeographie*, 137, 165–166, 347, 356–357, 360; *Weltgeschichte*, 357
Raubbau, Raubwirtschaft (destructive exploitation), 238, 239, 242
Raup, Hallock F., 417
Rawlings, Marjorie, 39
Rebourg, Cècile, 196*n*1
"Recent Developments in Cultural Geography" (Sauer), xx, 114, 115, 116
Reclus, Elisée, 22, 388
Regional geography: emergence of, 112; in *Geographical Review*, 116, 144–146; and natural regions, 112–113, 358–359; of Ozark Highlands, 90–91, 99–102, 325; and physical geography, 144–146; Sauer on, 99–100, 112, 113, 114, 144–146, 153–154; Sauer's courses on, 113; of Upper Illinois Valley, 92, 114, 234, 324–325
"Regional Reality in Economy" (Sauer), xix–xx, 151–153, 162–172
Regionalism, 153–154, 252–253. *See also* Regional geography
Reid, Harry Fielding, 375
"The Reinvention of Cultural Geography" (Price and Lewis), 16–17, 24*n*6
"The Relation of Man to Nature in the Southwest" (Sauer), 237, 252–260
"Relevancy orientation," 33
Rice, 281–282
"Richard J. Russell, 1895–1971" (Sauer), 375–376
Richardson, Robert W., 372
Richardson, Miles, 16
Richthofen, Ferdinand von, 138, 146, 346
Riddall, Margaret, 328*n*1

Ritter, Carl, 36, 137, 345–346, 346, 355, 358–359
"The Road to Brown School" (Rostlund), 373
Rockefeller Foundation, 7, 18, 187
Roden, I. Y., 187
Romanticism, 232–233, 235, 239
Roorbach, George B., 115
Roosevelt, Franklin D., 20
Roosevelt, Theodore, 233
Rosenthal, Zeny, 24*n*9
Rossi, Pietro, 233
Rostlund, Erhard, 349, 350, 351, 371–374, 377, 401, 418
Rousseau, Jean-Jacques, 276, 341
Rousseau, Ronald, 49*n*7
Royal Geographical Society (London), 7
Ruel(lius), Jean, 214
"Ruliff S. Holway, 1857–1927" (Sauer), 352–353
Rural peoples, 14, 17, 37, 160–161, 373
Russell, J. M., 234
Russell, Richard J., 13, 112, 348, 350, 375–376
Rutgers University, 350

S.A.B. *See* Science Advisory Board (S.A.B.)
Salem Normal School, xiv–xv, 4, 91
Salisbury, Rollin D: citations by Sauer to, 38; as Sauer's mentor and dissertation advisor at University of Chicago, xiii–xiv, 4, 10, 19, 38, 90, 101, 234; and University of Chicago Department of Geography, 324; and University of Chicago teaching job for Sauer, xiv
—works: *The Elements of Geography*, 91, 113; *Geology*, 103, 234–235
Sapper, Karl, 36
Sauer, Carl: attitudes of, about geography and teaching, xi, xiii–xv; bibliographic citations to and by, 29–49; biography of, xviii, 3–8; book reviews by, 32, 89, 116; commentaries on, xii, xviii, xix, 8*n*1, 23*n*1, 53–86; critics of, 9–24; death of, xvii, 3, 22; dissertation by, xviii, 4, 22, 89, 90–91, 114, 236, 275; dissertations supervised by, 30, 417–419; early geography career of, xiii–xv, 89–94, 324–325, 328; education of, xiii, 4, 10, 19, 44, 231, 232–234; fieldwork by, 19–20, 89, 93, 101, 151, 154–155, 325–326;

foreign language skills of, 44, 49*n*6; German ancestry of, 3–4, 31, 345; honors and awards for, xvii, 7, 237; journalism/editing career considered by, xiii, xiv; parents of, xiii–xiv, 31; personality of, 7, 31; photograph of, *frontispiece*; poem on, 8*n*1; publications by, 8*n*1, 24*n*9, 30, 31–32, 54–56, 89–94, 421–431; research objective of, 387; retirement of, 30, 155; reviews of books and monographs by, 54–56, 91; romance and marriage of, xiv, xv, 4; and scholars in many disciplines generally, 5, 7, 13, 348; significance of, xviii, 3, 9, 22–23, 24*n*10, 29, 47–48; Spanish translation of works by, 24*n*9; students of, xii–xiii, xvii, xix, 4, 11, 13, 15, 16, 20, 24*n*9, 32, 38, 41–42, 44, 193, 350, 384, 387–388; teaching career of, xi, xiv–xv, 4–5, 30, 89, 91, 113, 115, 237–239; on university committee and administration work, xi; values of, xii; writers' and poets' interest in, 6, 7–8*n*1; writing style of, 6, 7–8*n*1, 31, 47. *See also* specific publications

Sauer, Jonathan D., 187
Sauer, Lorena Schowengerdt, xiv, xv, 4
"Sauer and 'Sauerology'" (Mikesell), xii
Des Sauvages (Champlain), 328
Savanna climates, 200
Sawatsky, Harry L., 419
Schiemann, Elizabeth, 186
Schlüter, Otto, 36–37, 46, 115, 138, 140
Schmieder, Oskar, 112
Schnädelbach, Herbert, 233
Schott, Carl, 38
Schowengerdt, Franklin, xiv
Schowengerdt, Lorena, xiv, xv, 4
Science Advisory Board (S.A.B.), 20, 94
Science Research Index (SRI), 41, 48*n*1
Sea resources. *See* Oceans
Seamon, David, 185
Seed agriculture, 189, 190, 198–199, 202, 210–212. *See also* Agriculture
Selected Essays (Sauer), xviii, 56
"The Seminar as Exploration" (Sauer), xix, 383, 385, 386, 387, 392–397
Seminars, 387, 392–397
Semple, Ellen Churchill, 38, 44, 114, 347–348, 360–361, 398–399

September 11 terrorist attack, 21
Settlers. *See* Pioneers and settlers
Seventeenth Century North America (Sauer), 5, 56, 275–276, 324–341
Sforza, Cardinal, 218, 219
Shantz, Homer L., 111, 348, 351
Sheep, 253, 257, 258, 263, 283, 348
Sidaway, James D., 23*n*1
Siedlungsgeographer (settlement geographer), 361
Siedlungskunde (settlement history), 140
Sigüenza y Góngora, Carlos, 338–339
Silver, 272, 279, 312
Simoons, Frederick J., 40, 418
Simpson, George Gaylord, 40
Sinaloa, 302
Sixteenth Century North America (Sauer), 56, 275, 327
Slavery, 157, 175, 179, 223–225, 310, 313, 314–315
Sluyter, Andrew, 24*n*10, 42
Smith, Bruce, 46
Smith, George Otis, 369
Smith, J. Russell, 150
Smith, Jonathan M., 46
Snowden, Joseph D., 220
Soares de Souza, João Teixeira, 223
Social science: as art, 168; critique of, 152–153, 162–172; geography as, 129–130; Hartshorne's disdain for positivistic social science, 12; overemphasis on theory and method in, 5–6; quantification in, 6; and regionalism, 252–253; research in, 6; Sauer on condition of, in mid-1930s, 152–153, 162–172
Social Science Research Council, 7
Social Science Research Index (SSRI), 41, 48*n*1
Social theory, 164. *See also* Social science
Sociology, 163
Soderini, Piero di Tommaso, 320
"Soil Conservation" (Sauer), xii, 236, 242–244
Soil Conservation Service, 7, 93, 151, 242, 243, 350, 378
Soil erosion: and agriculture, 243, 250–251; anthropogeographic approach to, 245–251;

Soil erosion (*continued*)
  and commercial exploitation of land, 242–243, 248–251; drainage pattern of, 249–250; human factors in, 139, 235–236, 242–244; importance of, 238; physical study of, 249; Sauer's support of research on, 20; and soil profile, 249; and truncation of soil column, 249. *See also* Erosion
Soil geography, 93, 105–108, 201, 238, 267
The Soil (Hall), 107–108
The Soil (King), 107
Soil profile, 249
Soil science, 93, 106, 107–108
Soil surveys, 89, 92–93, 151, 249
Soils (Hilgard), 107
Soils: Their Properties and Management (Lyon, Fippin, and Buckman), 108
Solot, Michael, 17
"Some Comments on Contemporary Geographic Methods" (Leighly), 11–12
Sonseca, Bishop, 315
Sopher, David E., 15, 418
Sorghum, 220–222
South America: agriculture in, 18–19, 187–189, 193–194, 199, 206–207, 212, 256; empty lands of, 288; influences of, 301–302; metallurgy in, 301; population of, 285; Sauer's trip to, 158. *See also* Latin America; and specific countries
South Carolina, 243
Southeast Asia, 18, 188
Southwest, 237, 252–260, 370. *See also* specific states
Space, 133, 194
Spain, 191, 193–194, 213, 215–217, 222, 246, 315
Spanish explorations and colonies: and agriculture, 207; Bolton on, 364–365; colonial policy of Spain, 246; and Cosa, 312, 315–318; and foods of natives, 216–217; by former sailors and soldiers of Columbus, 315–316; and gold and other precious metals, 279, 308, 309, 311–314; and Indians, 287, 330–333; and livestock, 181; and missionaries, 257, 315, 326, 363, 365; and New Spain, 276, 328, 330–331, 338–339; and slavery, 315–316; and Vespucci, 275, 306, 315–316, 318–321. *See also* Columbus, Christopher; and other explorers
Spatial positivism, 13–14
Spencer, Herbert, 15
Spencer, Joseph E., 411, 417
Spengler, Oswald A. G., 36, 37
Speth, William W., xix, xx, 10, 47, 53, 57, 231–241, 435
Squash, 202, 210–212, 256, 327
SRI. *See* Science Research Index (SRI)
SSRI. *See* Social Science Research Index (SSRI)
St. Cosme, Jean François Buisson de, 336
St. Kitt's, 155
St. Lawrence Valley, 334
Staeheli, 33
Stanford University, 364
Stanislawski, Dan, 24n9, 372, 418
Starch foods, 204, 205–207, 212
Starrs, Paul F., 40
Starved Rock State Park, Ill., 54, 89, 236, 325
Starved Rock State Park (Sauer), 54, 236
Statistics, 29, 151, 166–168, 378, 400, 411–412
Stea, David, 15
Stebbins, George Ledyard, Jr., 192
Stefansson, Vihjalmur, 327
Steinbeck, John, 153
Stevenson, Robert Louis, 39
Steward, Julian H., 187
Stewart, William Blair, 151
Stoddart, David R., 19–20
Stoner, Charles R., 192
Strabo, 121, 122–123, 124, 359
Street, John M., 155
Structuralist approaches to geography, 10, 14–15
Stuler, E., 224
Süddeutschland (Gradmann), 116, 144–145
Sugar and sugar cane, 179–182, 203, 280–281
Sumario de la naturale historia (Oviedo), 214
Superorganic, 15–16, 42
"The Superorganic in American Cultural Geography" (Duncan), 15
Survey method in geography, 112–113, 237. *See also* Land surveys; Soil surveys

"Survey Method in Geography and Its Objectives" (Sauer), 112, 117*n*1, 237
Sweden, 7, 141
Sweet potatoes, 200, 203, 207
Switzerland, 116, 144
Symanski, Richard, 16, 20–21, 24*n*7
Symbiosis, 132, 231, 236, 239, 244, 251*n*1, 257–259, 267

Taft, Lorado, 369
Talbot, Lee M., 419
Talon, Jean, 339
Taylor, Thomas Griffith, 112, 116, 144
Teleki, Pál, Count de Szék, 112
Tennessee Valley Authority, 153
"*Terra Firma: Orbis Novus*" (Sauer), xii, xviii, 158–159*n*3, 274–275, 306–323
Texas, 363
Theatrum orbis terrarum, 124
Thomas, William L., Jr., 5
Thornthwaite, C. Warren, 242, 350, 378, 417
Tierra firme, meaning of, 318
Time Must Have a Stop (Huxley), 415
Tobacco, 208, 209, 216, 251, 282–283, 327
Tönnies, Ferdinand, 43
Topography, 129
Torquemada, Juan de, 38, 49*n*6
Toscanelli, Paolo, 306, 307
Tower, Walter S., 91, 101, 113, 115
Transeau, Edgar N., 348
Trees. *See* Forests
Trewartha, Glenn T., 113
Trindade, Antonio, 227*n*8
Trinidad, 155, 173, 298, 299, 311
Turkeys, 199, 211
Turkish corn, 213–216
Turner, Frederick Jackson, 348, 362, 363
Turtle Island Foundation, 8*n*1
Turtles, 175, 204

UCLA, 378
United Fruit Company, 176
United States: agriculture in, 404–405; distressed regions of, 246–247; migration in, 246–247; past and present culture of, 390–391; shrinking of subsistence base in, 285–286; Southwest of, 237, 252–260. *See also* specific states; specific universities; and specific geographers
University of Arizona, 369, 370
University of California, Berkeley: Anthropology Department at, 372; doctoral dissertations at, 350, 372, 417–419; faculty of, 114, 348, 349, 350, 360, 364, 372–373, 379; Geography Department at, 348, 372; German geographers at, 11, 112; Hitchcock Lectures at, 345, 375, 376; libraries of, 192; Ph.D.s in geography from, 42, 53; Sauer at, xi, xvii, 4–5, 10, 19, 53, 114, 115, 240, 390; Sauer's honorary degree from, 7; and soil erosion research, 20, 236; students at, 349, 350, 371–372, 377–378. *See also* Berkeley school
University of California, Irvine, 20
University of California Publications in Geography, 348
University of Chicago: Department of Geography at, 324; geography faculty of, 10, 347; Salisbury as Sauer's mentor and dissertation advisor at, xiii–xiv, 4, 10, 19, 38, 90, 101, 234; Sauer's education at, 10, 231, 233–236; students at, 12, 44, 377
University of Michigan: anti-German wartime sentiment at, 49*n*5; and Bruhnes, 240; Kniffen as student at, 38; Sauer's field studies at, 92, 114, 151; Sauer's teaching at, 4, 30, 89, 236, 237, 383
University of Nebraska, 367–368, 410
University of North Carolina, Chapel Hill, 153
University of Pennsylvania, 363
University of Texas, 363
University of Wisconsin, 17, 158*n*3, 362
Unstead, John Frederick, 91
Der Untergang des Abedlandes (Spengler), 36
Unwin, Tim, 23*n*1
Upper Illinois Valley, 89, 92, 114, 234, 324–325
Upper Illinois Valley (Sauer), 54, 114, 234, 236, 324–325
Urlandschaft (original landscape), 36
Urquhart, Alvin W., 407

Uruguay, 283
"Urundi, Territory and People" (Shantz), 368
U.S. Department of Agriculture, 368
Uschmann, Georg, 233

Vaihinger, Hans, 36
Vallaux, Camille, 115, 139–140
Van Cleef, Eugene, 114, 144, 145
Van Hise, Charles R., 233–234
Van Valkenburg, Samuel, 37
Vance, Maurice M., 234
Varenius, Bernhardus (Bernhard Varen), 125, 127, 388
Vasco da Gama, 123
Vavilov, Nikolai, 40, 186, 187, 188
Veblen, Thomas T., 46, 185
Vegetables. *See* Crops; and specific crops
Vegetation. *See* Plants
The Vegetation and Soils of Africa (Shantz and Marbut), 111, 368
Vegetative planting, 189, 190, 198–199, 202–208
Venezuela, 155
Verrazzano, Giovanni, 196*n*1, 276, 327, 340
Vespucci, Amérigo, 275, 306, 315–316, 318–321, 322*n*19
Vicero, Ralph D., 415*n*
Vidal de la Blache, Paul, 37, 113, 115, 137, 388
Vikings, 327
Vinland, 327
Virginia, 243
Völkerkunde (Peschel), 354
Voyages (Ramusio), 214

Wagner, Hermann, 136
Wagner, Philip L., xix, xx, 29–30, 383–389, 418, 435
Wallach, Bret, xii, 24–25*n*10
Washington state, 247
Water, 253, 257, 258, 349
Weather. *See* Climate
Web of Science, 35, 41, 46, 48*n*1
Weeds, 39–40, 190, 211, 304
Wegener, Alfred, 46

Weltanschauung, 233
Weltgeschichte (Ratzel and Helmolt), 357
Weltverbesserung (betterment of the world), xi
West, Robert C., xviii, 24*n*9, 302, 418
West Indies, 207, 222, 298–299, 309–310
Wetlands, 199–200
Wheat, 257, 308
Whitaker, J. Russell, 240
Whitaker, Thomas W., 212
Whitehand, Jeremy W. R., 41, 45
Wildlife sanctuaries, 236
Wilgus, A. Curtis, 156
"Will Carl Sauer Make It across That Great Bridge to the Next Millennium?" (Sauer), xii
Willems-Braun, Bruce, 47
Williams, Michael, xviii, 53, 152, 435
Willis, Bailey, 90
Willis, John C., 189
Willmott, Curt J., 46
Wilson, Leonard S., 240
Wirtschaften, 246, 249
Wissler, Clark D., 37, 206
Wissmann, Hermann von, xii
Woeikof, Alexander I., 103
Worster, Donald E., 232–233, 239
Wright, Frank Lloyd, 153
Wright, John Kirtland, 111
Wright, Sewall G., 254
Wrigley, Gladys, 116

Yeung, Henry Wai-chung, 49*n*4
Yuca. *See* Manioc (yuca)
Yucatán, 154

Zaburro, 224–225, 227*n*8. *See also* Maize
Zajonc, Arthur, 185
Zea mays. *See* Maize
Zeder, Melinda A., 196*n*1
Zelinsky, Wilbur, 16, 32, 418
Zerega, Nyree J. C., 195
Zeuner, Frederick E., 40, 385
Zon, Raphael, 368
Zuñi, 332